Development and Optimization of Mathematical Models for Operations Research

Development and Optimization of Mathematical Models for Operations Research

Editors

Humberto Rocha
Ana Maria A. C. Rocha

MDPI • Basel • Beijing • Wuhan • Barcelona • Belgrade • Manchester • Tokyo • Cluj • Tianjin

Editors
Humberto Rocha
University of Coimbra
Portugal

Ana Maria A. C. Rocha
University of Minho
Portugal

Editorial Office
MDPI
St. Alban-Anlage 66
4052 Basel, Switzerland

This is a reprint of articles from the Special Issue published online in the open access journal *Mathematics* (ISSN 2227-7390) (available at: https://www.mdpi.com/si/mathematics/Optimization_Mathematical_Models_Operations_Research).

For citation purposes, cite each article independently as indicated on the article page online and as indicated below:

LastName, A.A.; LastName, B.B.; LastName, C.C. Article Title. *Journal Name* **Year**, *Volume Number*, Page Range.

ISBN 978-3-0365-6890-4 (Hbk)
ISBN 978-3-0365-6891-1 (PDF)

Contents

About the Editors

Humberto Rocha

Humberto Rocha is an Associate Professor at University of Coimbra and Researcher at CeBER and Inesc-Coimbra Research Centers. He received both his bachelor and master degrees in Applied Mathematics from the University of Coimbra in 1998 and 2001, respectively. In 2005, he completed a PhD in Computational and Applied Mathematics at Old Dominion University (ODU). His research centers on applications in physics, inverse engineering, aeronautics, medicine, and economics. This multidisciplinary approach lead to publications in a variety of journals of different areas, including Journal of Aircraft, Medical Physics, Physics in Medicine and Biology, Physica Medica, Physica A. Most of his publications are in top journals of his area (Applied Mathematics/Operations Research), including SIAM Journal on Optimization, European Journal of Operational Research, Annals of Operations Research, Journal of Global Optimization, Computational Optimization and Applications, Applied Mathematical Modeling, Applied Mathematics and Computation, Structural and Multidisciplinary Optimization, International Transactions of Operations Research.

Ana Maria A. C. Rocha

Ana Maria A. C. Rocha is an Associate Professor in the Department of Production and Systems at the University of Minho. She completed her PhD in Production and Systems Engineering in 2005, a Master's in Informatics Engineering in 1997 and a Degree in Systems and Informatics Engineering in 1993 from the University of Minho. She is a researcher at the ALGORITMI Research Centre, School of Engineering at the University of Minho and coordinator of the research group "Systems Engineering and Operational Research". Her scientific activity includes the publication of more than 100 scientific articles, with peer review, published in journals of recognized international merit and international conferences. She is a co-Editor of more than 20 Proceedings books. She has organized 2 international events as general chair. She received 2 awards from APDIO (Portuguese Association for Operational Research) for the quality of published articles. Her research interests are: Global Optimization; Non-Linear Optimization; Integer-mixed programming. Her main teaching interests are in the areas of Nonlinear Optimization, Numerical Methods and Statistics.

Preface to "Development and Optimization of Mathematical Models for Operations Research"

1. Introduction

The development of mathematical models and their optimization are fundamental for the effective resolution of many problems in operational research. In recent years, increased insights into real-world problems have led to the development of new mathematical models and new optimization algorithms, contributing to the development of a research area with increasing practical relevance. This Special Issue is dedicated to works at the interface of mathematical modeling, optimization, and operations research with a special focus on their real-world applications. The interest of the scientific community was significant, with submissions from authors from different countries from five continents, including Australia, China, Egypt, India, Israel, Portugal, Russia, Saudi Arabia, and the United States of America. Ten papers were accepted for publication after thorough peer-review by dedicated reviewers with expertise in the fields of the papers.

2. Description of Published Papers

This section presents a brief overview of the published papers.

S. Singh et al. [1] proposed a multi-time generalized Nash equilibrium problem and proved its equivalence with a multi-time quasi-variational inequality problem. Moreover, the authors proved an existence theorem for equilibria. Next, they applied the considered model to a traffic network problem. Finally, they studied the multi-time generalized Nash equilibrium problem as a projected dynamical system and showed a numerical example with its approximated solution.

H. Zhou et al. [2] presented a genetic algorithm for optimizing the formation and schedule of heavy-haul trains, which is a special case of a more general Train Formation Problem. The authors established two integer programming models in stages involving the train service plan problem model and the train time-tabling problem model. They then implemented a number of experiments to illustrate the feasibility and effectiveness of the proposed approaches.

R. Mondal et al. [3] formulated an inventory model that combines two important components of inventory management: the demand of a product and the uncertainty of customers' behavior. They derived a mathematical formulation that maximized the profit of the inventory system. They considered three numerical examples to validate the model and solved them using different variants of quantum-behaved particle swarm optimization techniques in order to determine the duration of stock-in time and the preservation technology cost.

L. Alkhalifa and H. Mittelmann [4] introduced Piecewise Linear Approximation (PLA), one of most popular methods used to transform nonlinear problems into linear ones. Following a brief background and literature review, authors proposed two piecewise linear approximation helpers in mixed-integer nonlinear programming: one related to domain partitioning and another with a partial application of PLA to MINLP. They used quadratically constrained quadratic programming (QCQP) and MIQCQP to demonstrate that problems under PLA with nonuniform partition resulted in more accurate solutions and required less time compared to PLA with uniform partition.

N. Krivulin et al. [5] presented an application of pairwise comparisons method on decision making: the determination of consumer preferences in hotel selection. They demonstrated several known methods on a sample of 202 university students, evaluating their preferred criteria when selecting a hotel for accommodation during a professional development program in a foreign country. The comparison of the solutions produced showed a high degree of similarity in results.

P. Ruan et al. [6] considered a two-echelon supply chain with an up-stream supplier and down-stream retailer during the COVID-19 period. They constructed an inventory model considering the following four elements: ordering cost, holding cost, deterioration cost, and purchasing cost. A computer program provided a numerical solution indicating the minimum total cost per unit time.

R. Etgar and Y. Cohen [7] presented a novel approach to solve a current problem that R&D companies face: multi-annual project portfolio selection. The suggested approach was to expand and improve common meta-heuristic methods and, thus, solve this NP-hard problem of determining the roadmap of multi-annual portfolio planning. This study developed an efficient tool that can provide both practical and academic benefits.

A. Antunes et al. [8] focused on single-stage scheduling problems occurring in parallel machine environments. They applied a genetic algorithm (GA) to the scheduling problem of unrelated parallel machines in order to minimize the makespan of a set of tasks that was subject to varying setup times. A comparative statistical analysis of small and large instances of scheduling problems showed the advantage of the GA proposed.

K. Alnowibet et al. [9] presented modified algorithms based on conjugate gradient (CG) principles to solve local and global minimization problems. First, they improved the existing CG algorithm to enhance its global and local optimization capacities. Then, they obtained a hybrid stochastic conjugate gradient algorithm using the improved CG method. The performance profiles used to compare the proposed hybrid approach and four other hybrid stochastic conjugate gradient algorithms showed the competitiveness of the former. The authors tested both convex and non-convex problems.

B. Emambocus et al. [10] proposed an optimized discrete adapted Dragonfly Algorithm (DA) using the Steepest Ascent Hill-Climbing algorithm as a local search. They applied the proposed DA to a traveling salesman problem, modeling a package delivery system in Kuala Lumpur. The improved DA showed better performance than the discrete adapted DA and other studied swarm intelligence algorithms.

3. Conclusions

As guest editors of the Special Issue 'Development and Optimization of Mathematical Models for Operations Research', we express our gratitude to all the authors who sent their articles for publication in this issue. We also cordially thank all anonymous referees and staff of MDPI for contributing to the creation of this Special Issue. Special thanks are due to the Managing Editor of the Special Issue, Ms. Linn Li, for her excellent collaboration and valuable assistance. We are confident that the papers selected for this Special Issue will attract a significant audience in the scientific community and further stimulate research involving the development of mathematical models and their optimization.

References

1. Singh, S.; Gibali, A.; Reich, S. Multi-Time Generalized Nash Equilibria with Dynamic Flow Applications. Mathematics **2021**, *9*, 1658. https://doi.org/10.3390/math9141658.

2. Zhou, H.; Zhou, L.; Guo, B.; Bai, Z.; Wang, Z.; Yang, L. A Scheduling Approach for the Combination Scheme and Train Timetable of a Heavy-Haul Railway. Mathematics **2021**, *9*, 3068. https://doi.org/10.3390/math9233068.

3. Mondal, R.; Shaikh, A.; Bhunia, A.; Hezam, I.; Chakrabortty, R. Impact of Trapezoidal Demand and Deteriorating Preventing Technology in an Inventory Model in Interval Uncertainty under Backlogging Situation. *Mathematics* **2022**, *10*, 78. https://doi.org/10.3390/math10010078.

4. Alkhalifa, L.; Mittelmann, H. New Algorithm to Solve Mixed Integer Quadratically Constrained Quadratic Programming Problems Using Piecewise Linear Approximation. *Mathematics* **2022**, *10*, 198. https://doi.org/10.3390/math10020198.

5. Krivulin, N.; Prinkov, A.; Gladkikh, I. Using Pairwise Comparisons to Determine Consumer Preferences in Hotel Selection. *Mathematics* **2022**, *10*, 730. https://doi.org/10.3390/math10050730.

6. Ruan, P.; Huang, Y.; Weng, M. Impact of COVID-19 on Supply Chains: A Hybrid Trade Credit Policy. *Mathematics* **2022**, *10*, 1209. https://doi.org/10.3390/math10081209.

7. Etgar, R.; Cohen, Y. Roadmap Optimization: Multi-Annual Project Portfolio Selection Method. *Mathematics* **2022**, *10*, 1601. https://doi.org/10.3390/math10091601.

8. Antunes, A.; Matos, M.; Rocha, A.; Costa, L.; Varela, L. A Statistical Comparison of Metaheuristics for Unrelated Parallel Machine Scheduling Problems with Setup Times. *Mathematics* **2022**, *10*, 2431. https://doi.org/10.3390/math10142431.

9. Alnowibet, K.; Mahdi, S.; Alshamrani, A.; Sallam, K.; Mohamed, A. A Family of Hybrid Stochastic Conjugate Gradient Algorithms for Local and Global Minimization Problems. *Mathematics* **2022**, *10*, 3595. https://doi.org/10.3390/math10193595.

10. Emambocus, B.; Jasser, M.; Amphawan, A.; Mohamed, A. An Optimized Discrete Dragonfly Algorithm Tackling the Low Exploitation Problem for Solving TSP. *Mathematics* **2022**, *10*, 3647. https://doi.org/10.3390/math10193647.

Humberto Rocha and Ana Maria A. C. Rocha
Editors

mathematics

Article

Multi-Time Generalized Nash Equilibria with Dynamic Flow Applications

Shipra Singh [1,†], Aviv Gibali [2,3,*,†] and Simeon Reich [1,†]

1 Department of Mathematics, The Technion—Israel Institute of Technology, Haifa 3200003, Israel; shiprasingh384@gmail.com (S.S.); sreich@technion.ac.il (S.R.)
2 Department of Mathematics, ORT Braude College, Karmiel 2161002, Israel
3 The Center for Mathematics and Scientific Computation, University of Haifa, Haifa 3498838, Israel
* Correspondence: avivg@braude.ac.il
† These authors contributed equally to this work.

Abstract: We propose a multi-time generalized Nash equilibrium problem and prove its equivalence with a multi-time quasi-variational inequality problem. Then, we establish the existence of equilibria. Furthermore, we demonstrate that our multi-time generalized Nash equilibrium problem can be applied to solving traffic network problems, the aim of which is to minimize the traffic cost of each route and to solving a river basin pollution problem. Moreover, we also study the proposed multi-time generalized Nash equilibrium problem as a projected dynamical system and numerically illustrate our theoretical results.

Keywords: multi-time generalized Nash equilibrium problem; projected dynamical system; river basin pollution problem; traffic network equilibrium problem; variational inequality problem

Citation: Singh, S.; Gibali, A.; Reich, S. Multi-Time Generalized Nash Equilibria with Dynamic Flow Applications. *Mathematics* **2021**, *9*, 1658. https://doi.org/10.3390/math9141658

Academic Editors: Humberto Rocha and Ana Maria Rocha

Received: 26 May 2021
Accepted: 12 July 2021
Published: 14 July 2021

1. Introduction

The concept of an equilibrium problem originated with Cournot [1] in the context of an oligopolistic economy, but formally it was introduced by Nash [2,3]. Therefore, it is called a Nash equilibrium problem. Nash equilibrium problems have been extensively studied and employed as powerful and flexible tools. However, the Nash equilibrium only deals with the dependency of the payoff function of each player on the strategies of the other (rival) players. Later, this notion was extended to the generalized Nash equilibrium by Arrow and Debreu [4], where each player's strategy set also depends on the strategies of the other players. Generalized Nash equilibrium problems are important in mathematical modeling because of their usefulness in the modeling of economic systems [5], routing problems in communication networks [6], and in engineering applications [7]. The survey papers [8,9] give a complete overview of the state of the art regarding theoretical results and numerical methods for solving generalized Nash equilibrium problems. One of the popular strategies for solving the generalized Nash equilibrium problem is to first bridge the gap between this problem and well-known variational tools in the literature, and then to use these well-developed tools to solve it. Instances of such methods are reformulations of generalized Nash equilibrium problems as suitable variational inequalities and quasi-variational inequalities.

A variational inequality problem comprises an inequality which must be satisfied for all the elements of a given (convex) set. The study of variational inequality problems in traffic analysis was initiated by Smith [10] and Dafermos [11]. They set up the traffic assignment problem in terms of a finite-dimensional variational inequality problem. Presently variational inequalities constitute an important modeling tool in economics [12,13], optimization [14] and game theory [15–17]. A quasi-variational inequality problem is an extension of the concept of a variational inequality problem, where the feasible set is also allowed to vary. Such problems have been used to model more complex phenomena. It was Bensoussan [18] who first recognized the connection between the generalized Nash equilibrium and quasi-variational inequality problems, and studied them in Hilbert space. Thereafter, Harker [17] investigated these problems in Euclidean spaces. Aussel et al. [19] studied the time-dependent generalized Nash equilibrium problem and reformulated it as an evolutionary quasi-variational inequality problem. More relevant papers are well documented in [20–22].

In recent decades the notion of multi-time has frequently been used in optimization theory and in multi-time control problems. Indeed, several science and engineering problems can be converted into optimization problems that are defined as m-flow type PDEs (multi-time evolution systems) and the associated cost functionals are expressed as path-independent curvilinear integrals or multiple integrals. Udrişte and Ţevy [23] introduced the basic optimization problems involving path-independent curvilinear integrals and multiple integrals. Mathematically speaking, these integrals are equivalent, but their meanings are completely different in real life problems. Thereafter, a systematic study of multi-time problems was initiated by the research group of Udrişte [24–27]. In this way, a multi-time parameter of evolution approach in optimization theory started to be used and this concept has extensively been explored; see [28–30]. Apart from optimization, the concept of a multi-time parameter of evolution is also used in space theory. A space coordinate is merely an index numbering degrees of freedom and the time coordinate is usually the physical time in which the system evolves. In some physical problems, two-time $t = (t_1; t_2)$ is used, where t_1 means the intrinsic time and t_2 the observer time. These prominent roles of multi-time parameters in different areas of science show the necessity of some new formulations of this concept. To pursue further explorations and present novel results involving multi-time parameters, particularly in optimization and noncooperative games, we formulate the multi-time generalized Nash equilibrium problem (MGNEP), which is a generalized form of the time-dependent generalized Nash equilibrium problem studied by Aussel et al. [19], and reformulate it as a multi-time quasi-variational inequality problem. We also establish the existence of equilibria. To provide an application of the formulated multi-time generalized Nash equilibrium problems in traffic analysis, we interpret a traffic network model in terms of such an equilibrium problem for a courier service company with the intention of minimizing the traffic cost of each route. Moreover, we also provide an application to solving river basin pollution problems. We formulate a river basin pollution problem in terms of our multi-time generalized Nash equilibrium problem, and demonstrate how the industrial factories (agents) situated along a river can maximize their profit by following the particular norms and restrictions of reducing the river water pollution imposed by the basin authorities. Finally, we propose a method for solving the multi-time generalized Nash equilibrium problem via projected dynamical system theory. To exhibit the utility of our proposed method, we solve the well-known Nguyen traffic network problem [31] using this method and numerically illustrate our results.

Our paper is organized as follows: preliminaries and the formulations of the problem are presented in Section 2. The equivalence of the multi-time generalized Nash equilibrium problem with the multi-time quasi-variational inequality problem is established in Section 3. The existence of equilibria is obtained in Section 4. The applications of multi-time generalized Nash equilibrium problems in traffic network analysis and to solving river basin pollution problems are demonstrated in Section 5. Explorations of the multi-time generalized Nash equilibrium problem as a projected dynamical system, as well as numerical illustrations regarding the Nguyen traffic network, are presented in Section 6. Section 7 concludes our paper.

2. Preliminaries and Problem Formulations

We start with the formulation of a multi-time generalized Nash equilibrium problem. To this end, we first introduce important notation and mathematical tools. We consider a hyperparallelepiped $\Omega_{l_\circ,l_1}$ in $\mathbb{R}^m$ with the opposite diagonal points $l_\circ = (l_\circ^1, l_\circ^2, \ldots, l_\circ^m)$

and $l_1 = (l_1^1, l_1^2, \dots, l_1^m)$. Using the product order relation on $\mathbb{R}^m$, we can view this hyperparallelepiped as an interval $[l_\circ, l_1]^m$. This interval is sometimes called the planning horizon of players for preparing their strategies. We denote by t the multi-time parameter of evolution. It is defined as $t = (t^\alpha) \in \Omega_{l_\circ, l_1}$, where $\alpha = (1, 2, \dots, m)$. The multi-time generalized Nash equilibrium problem comprises N players and each player μ controls the variable $x^\mu(t) \in L^2(\Omega_{l_\circ, l_1}, \mathbb{R}^{n_\mu})$. This variable is the multi-time-dependent vector of strategies of the player μ. Let $x^{-\mu}(t) \in L^2(\Omega_{l_\circ, l_1}, \mathbb{R}^{n-n_\mu})$ be the vector of strategies set up by the decision variables of all the players except the player μ at a given multi-time $t \in \Omega_{l_\circ, l_1}$ and let $x(t) \in L^2(\Omega_{l_\circ, l_1}, \mathbb{R}^n)$ be the multi-time-dependent vector of strategies of all players. Here $n = \sum\limits_{\mu=1}^{N} n_\mu$. When we wish to put the strategy vector of the player μ under a spotlight, we write the strategy vector $x(t)$ of all the players as

$$x(t) = (x^\mu(t), x^{-\mu}(t)).$$

This is still the vector $x(t) = (x^1(t), x^2(t), \dots, x^{\mu-1}(t), x^\mu(t), x^{\mu+1}(t), \dots, x^N(t))$ which belongs to $L^2(\Omega_{l_\circ, l_1}, \mathbb{R}^n)$. Please note that the notation $(x^\mu(t), x^{-\mu}(t))$ does not mean that the block components of $x(t)$ are reordered in such a way so that $x^\mu(t)$ becomes the first block. We hark back to $L^2(\Omega_{l_\circ, l_1}, \mathbb{R}^n) = L^2(\Omega_{l_\circ, l_1}, \mathbb{R}^{n_\mu}) \times L^2(\Omega_{l_\circ, l_1}, \mathbb{R}^{n-n_\mu})$. Now let K be a nonempty, closed and convex subset of $L^2(\Omega_{l_\circ, l_1}, \mathbb{R}^n)$. For any given strategy vector $x^{-\mu}(t)$ of the rival players, we denote the nonempty, closed and convex feasible set (or strategy set) of the player μ by $K_\mu(x^{-\mu}(t))$. This is a subset of $L^2(\Omega_{l_\circ, l_1}, \mathbb{R}^{n_\mu})$. Each player has an objective function which is called the cost (or loss, or payoff) function. It depends on the player's own variables $x^\mu(t)$, as well as on those of the rival players $x^{-\mu}(t)$. In our formulation of the multi-time generalized Nash equilibrium problem, we write the total cost function $F^\mu : L^2(\Omega_{l_\circ, l_1}, \mathbb{R}^n) \to \mathbb{R}$ of the player μ as a multiple integral. More precisely,

$$F^\mu(x(t)) = \int_{\Omega_{l_\circ, l_1}} f^\mu(x^\mu(s), x^{-\mu}(s)) ds,$$

where $f^\mu(x^\mu(s), x^{-\mu}(s))$ is a real-valued continuously differentiable function which denotes the running cost (loss) function the player μ bears when the rival players have chosen the strategy $x^{-\mu}(s)$ at a given time $s \in \Omega_{l_\circ, l_1}$, and $ds = ds^1 \dots ds^m$ denotes the volume element of $\Omega_{l_\circ, l_1}$. We use the following notation to denote the value of the function represented by $p(t)$ at the point $q(t)$:

$$\langle\langle p(t), q(t) \rangle\rangle = \int_{\Omega_{l_\circ, l_1}} \langle p(s), q(s) \rangle ds \ \forall\ p(t),\ q(t) \in L^2(\Omega_{l_\circ, l_1}, \mathbb{R}^n),$$

where $\langle ., . \rangle$ represents the Euclidean inner product. Throughout the paper, the abbreviation "a.e." stands for "almost everywhere" and $\mathbb{R}_+^p$ represents the set of non-negative vectors in the Euclidean space $\mathbb{R}^p$.

Remark 1. *In our paper, we do not impose any special structure on the feasible set $K_\mu(x^{-\mu}(t))$ of each player. For studies of generalized Nash equilibrium problems with special structures on the feasible set of each player, we refer the interested readers to [8,21].*

Now we are able to introduce our multi-time generalized Nash equilibrium problem (MGNEP). For a given strategy vector $x^{-\mu}(t)$ of rival players, the aim of a player μ is to choose a strategy vector $x^\mu(t)$ such that it solves the following multi-time minimization problem:

$$\textbf{(MGNEP)} \qquad \min_{x^\mu(t)} \int_{\Omega_{l_\circ, l_1}} f^\mu(x^\mu(t), x^{-\mu}(t)) dt,$$

$$\text{subject to } x^\mu(t) \in K_\mu(x^{-\mu}(t)).$$

Problem (MGNEP) can also be interpreted in the following way.

Definition 1. *A strategy vector $y(t) \in K$ is said to be a multi-time generalized Nash equilibrium if and only if for each player μ, we have $y^{\mu}(t) \in K_{\mu}(y^{-\mu}(t))$ and*

$$\int_{\Omega_{l_\circ,l_1}} f^{\mu}(y^{\mu}(t), y^{-\mu}(t))dt \le \int_{\Omega_{l_\circ,l_1}} f^{\mu}(x^{\mu}(t), y^{-\mu}(t))dt \ \forall\, x^{\mu}(t) \in K_{\mu}(y^{-\mu}(t)).$$

Special Case. If $m = 1$, then $\Omega_{l_\circ,l_1}$ is simply the closed real interval $[l_\circ, l_1]$. Furthermore, for more convenience, we put $l_\circ = 0$ and $l_1 = T$ (which denotes an arbitrary time) and then $\Omega_{l_\circ,l_1} = [0, T]$ (a fixed time interval). Consequently, in this case (MGNEP) reduces to the time-dependent generalized Nash equilibrium problem, studied by Aussel et al. [19].

To formulate our multi-time quasi-variational inequality problem, we define the set-valued map $\Gamma : K \to 2^K$ by

$$\Gamma(x(t)) := \prod_{\mu=1}^{N} K_{\mu}(x^{-\mu}(t))$$

for each $x(t) \in K$. We also let $J : L^2(\Omega_{l_\circ,l_1}, \mathbb{R}^n) \to L^2(\Omega_{l_\circ,l_1}, \mathbb{R}^n)$ be a single-valued map. Our multi-time quasi-variational inequality problem is defined as follows:

(MQVIP) to find a vector $y(t) \in K$ such that $y(t) \in \Gamma(y(t))$ and

$$\int_{\Omega_{l_\circ,l_1}} \langle J(y(t)), x(t) - y(t)\rangle dt \ge 0 \ \forall\, x(t) \in \Gamma(y(t)).$$

Taking into account the definitions of generalized convexities formulated in [32,33], we define the following notion of convexity for a multi-time functional $H : K \to \mathbb{R}$ of the form $H(x(t)) = \int_{\Omega_{l_\circ,l_1}} h(x(s))ds$, where h is a real-valued continuously differentiable function.

Definition 2. *The multi-time functional H is said to be convex on the set K if for all $x(s)$, $y(s) \in K$, the following inequality holds:*

$$\int_{\Omega_{l_\circ,l_1}} h(x(s))ds - \int_{\Omega_{l_\circ,l_1}} h(y(s))ds \ge \int_{\Omega_{l_\circ,l_1}} \left\langle \frac{\partial h}{\partial x}(y(s)), x(s) - y(s) \right\rangle ds.$$

Next, we recall the following definitions and theorem (compare [34,35]).

Definition 3. *The nearest point projection of a point $x(t) \in L^2(\Omega_{l_\circ,l_1}, \mathbb{R}^n)$ onto the set K is defined by*

$$proj_K(x(t)) := arg \min_{y(t)\in K} \|x(t) - y(t)\|.$$

Remark 2. *For each $x(t) \in L^2(\Omega_{l_\circ,l_1}, \mathbb{R}^n)$, $proj_K(x(t))$ enjoys the following property:*

$$\langle\langle x(t) - proj_K(x(t)), y(t) - proj_K(x(t))\rangle\rangle \le 0 \ \forall\, y(t) \in K.$$

Definition 4. *The polar set K° associated with K is defined by*

$$K^{\circ} := \{y(t) \in L^2(\Omega_{l_\circ,l_1}, \mathbb{R}^n) : \langle\langle y(t), x(t)\rangle\rangle \le 0 \ \forall\, x(t) \in K\}.$$

Definition 5. *The tangent cone to the set K at a point $x(t) \in K$ is defined by*

$$T_K(x(t)) := \mathrm{cl}\left(\bigcup_{\lambda>0} \frac{K - x(t)}{\lambda} \right),$$

where cl *denotes the closure operation.*

Definition 6. *The normal cone of K at a point $x(t)$ is defined by*

$$N_K(x(t)) := \begin{cases} \{y(t) \in L^2(\Omega_{l_\circ,l_1}, \mathbb{R}^n) : \langle\langle y(t), z(t) - x(t)\rangle\rangle \leq 0 \,\forall\, z(t) \in K\}, & x(t) \in K, \\ \emptyset, & x(t) \notin K. \end{cases}$$

Please note that $T_K(x(t)) = [N_K(x(t))]^\circ$.

Definition 7. *A subset D of K is said to be compactly open (respectively, compactly closed) in K if for any nonempty compact subset L of K, the intersection $D \cap L$ is open (respectively, closed) in L.*

Remark 3. (a) *It is evident from the above definition that every open (respectively, closed) set is compactly open (respectively, compactly closed).*
(b) *The union or intersection of a finite number of compactly open (respectively, compactly closed) sets is compactly open (respectively, compactly closed).*
(c) *If $A \subset K_1$ and $B \subset K_2$ are compactly open (respectively, compactly closed) in K_1 and K_2, respectively, then $A \times B \subset K_1 \times K_2$ is compactly open (respectively, compactly closed) in $K_1 \times K_2$.*

Definition 8. *A family $\{g^\mu\}_{\mu=1}^N$ of maps $g^\mu : K \to L^2(\Omega_{l_\circ,l_1}, \mathbb{R}^n)$ is called hemicontinuous if for all $x(t), y(t) \in K$ and $\lambda \in [0,1]$, the mapping $\lambda \mapsto \sum_{\mu=1}^{N} \langle\langle g^\mu(x(t) + \lambda z(t)), z^\mu(t)\rangle\rangle$ with $z^\mu(t) = y^\mu(t) - x^\mu(t)$ is continuous, where $z^\mu(t)$ is the μth component of $z(t)$.*

Theorem 1 ([34]). *Assume that $S, T : K \to 2^K$ are set-valued maps and that the following hypotheses are satisfied:*

1. $\forall\, x(t) \in K$, $S(x(t)) \subset T(x(t))$,
2. $\forall\, x(t) \in K$, $S(x(t)) \neq \emptyset$,
3. $\forall\, x(t) \in K$, $T(x(t))$ *is convex,*
4. $\forall\, y(t) \in K$, $S^{-1}(\{y(t)\}) = \{x(t) \in K : y(t) \in S(x(t))\}$ *is compactly open,*
5. *there exists a nonempty, closed and compact subset D of K and $\overline{y}(t) \in D$ such that $K \setminus D \subset S^{-1}(\{\overline{y}(t)\})$.*

Then there exists $\overline{x}(t) \in K$ such that $\overline{x}(t) \in T(\overline{x}(t))$.

3. An Equivalent Form of the Multi-Time Generalized Nash Equilibrium Problem

We begin our analysis by presenting an equivalent form of our multi-time generalized Nash equilibrium problem in terms of a multi-time quasi-variational inequality problem. This equivalent formulation turns out to be useful for proving further results in the sequel. From now onwards, the symbol $\frac{\partial f^\mu}{\partial x^\mu}(y(t))$ stands for the partial derivative of the running cost function f^μ of the player μ with respect to the argument $x^\mu(t)$ at the strategy vector $y(t) \in K$.

Theorem 2. *Assume that $J(x(t)) = \left(\frac{\partial f^\mu}{\partial x^\mu}(x(t))\right)_{\mu=1}^N$ for each $x(t) \in K$, and that for each $\mu \in \{1, 2, \ldots, N\}$ and each $x^{-\mu}(t)$, the multi-time cost functional F^μ is convex on K in the argument $x^\mu(t)$. Then $y(t) \in K$ is a multi-time generalized Nash equilibrium if and only if it is a solution to (MQVIP).*

Proof. Let $y(t) \in K$ be a solution of (MQVIP). We shall first prove that for each $\mu \in \{1, 2, \ldots, N\}$, $y^{\mu}(t) \in K_{\mu}(y^{-\mu}(t))$ satisfies the following inequality:

$$\left\langle \frac{\partial f^{\mu}}{\partial x^{\mu}}(y(t)), x^{\mu}(t) - y^{\mu}(t) \right\rangle \geq 0 \;\; \forall\, x^{\mu}(t) \in K_{\mu}(y^{-\mu}(t)) \text{ and a.e. on } \Omega_{l_\circ, l_1}. \tag{1}$$

To this end, suppose on the contrary that this inequality does not hold. Then it follows that there exists a $\nu \in \{1, 2, \ldots, N\}$, and a strategy vector $z^{\nu}(t) \in K_{\nu}(y^{-\nu}(t))$ together with a set $G \subset \Omega_{l_\circ, l_1}$ of positive measure such that for $y^{\nu}(t) \in K_{\nu}(y^{-\nu}(t))$ the following inequality holds:

$$\left\langle \frac{\partial f^{\nu}}{\partial x^{\nu}}(y(t)), z^{\nu}(t) - y^{\nu}(t) \right\rangle < 0 \text{ a.e. on } G. \tag{2}$$

Next, we consider the strategy vector $h(t) \in L^2(\Omega_{l_\circ, l_1}, \mathbb{R}^n)$ defined by

$$h(t) := \begin{cases} h^{\mu}(t) = y^{\mu}(t), \; t \in \Omega_{l_\circ, l_1} \text{ and } \mu \neq \nu, \\ h^{\mu}(t) = z^{\nu}(t), \; t \in G \text{ and } \mu = \nu, \\ h^{\mu}(t) = y^{\nu}(t), \; t \in \Omega_{l_\circ, l_1} \setminus G \text{ and } \mu = \nu. \end{cases}$$

We have $h(t) \in \Gamma(y(t))$ and

$$\begin{aligned} \int_{\Omega_{l_\circ, l_1}} \langle J(y(t)), h(t) - y(t) \rangle dt &= \sum_{\mu=1}^{N} \int_{\Omega_{l_\circ, l_1}} \left\langle \frac{\partial f^{\mu}}{\partial x^{\mu}}(y(t)), h^{\mu}(t) - y^{\mu}(t) \right\rangle dt \\ &= \sum_{\mu=1\; (\mu \neq \nu)}^{N} \int_{\Omega_{l_\circ, l_1}} \left\langle \frac{\partial f^{\mu}}{\partial x^{\mu}}(y(t)), h^{\mu}(t) - y^{\mu}(t) \right\rangle dt \\ &\quad + \int_{\Omega_{l_\circ, l_1}} \left\langle \frac{\partial f^{\nu}}{\partial x^{\nu}}(y(t)), h^{\nu}(t) - y^{\nu}(t) \right\rangle dt \\ &= \int_{G} \left\langle \frac{\partial f^{\nu}}{\partial x^{\nu}}(y(t)), z^{\nu}(t) - y^{\nu}(t) \right\rangle dt. \end{aligned} \tag{3}$$

It now follows from Inequalities (2) and (3) that

$$\int_{\Omega_{l_\circ, l_1}} \langle J(y(t)), h(t) - y(t) \rangle dt < 0,$$

which contradicts the fact that $y(t)$ is a solution to (MQVIP). Thus, inequality (1) holds, as claimed. Furthermore, for each μ, the convexity of the multi-time cost functional F^{μ} on the set K in the argument $x^{\mu}(t)$ implies that

$$\begin{aligned} &\int_{\Omega_{l_\circ, l_1}} f^{\mu}(x^{\mu}(t), y^{-\mu}(t)) dt - \int_{\Omega_{l_\circ, l_1}} f^{\mu}(y^{\mu}(t), y^{-\mu}(t)) dt \\ &\quad \geq \int_{\Omega_{l_\circ, l_1}} \left\langle \frac{\partial f^{\mu}}{\partial x^{\mu}}(y(t)), x^{\mu}(t) - y^{\mu}(t) \right\rangle dt \;\forall\, x^{\mu}(t) \in K_{\mu}(y^{-\mu}(t)). \end{aligned} \tag{4}$$

By combining inequalities (1) and (4), we obtain

$$\int_{\Omega_{l_\circ, l_1}} f^{\mu}(x^{\mu}(t), y^{-\mu}(t)) dt - \int_{\Omega_{l_\circ, l_1}} f^{\mu}(y^{\mu}(t), y^{-\mu}(t)) dt \geq 0 \;\forall\, x^{\mu}(t) \in K_{\mu}(y^{-\mu}(t)), \tag{5}$$

which implies that $y(t)$ is a multi-time generalized Nash equilibrium, as asserted.

Conversely, let $y(t) \in K$ be a multi-time generalized Nash equilibrium. Then we have inequality (5). Since $K_\mu(y^{-\mu}(t))$ is a convex set, for all $x^\mu(t) \in K_\mu(y^{-\mu}(t))$ and $\lambda \in [0,1]$, inequality (5) can be rewritten as

$$\int_{\Omega_{l_o,l_1}} [f^\mu(y^\mu(t) + \lambda(x^\mu(t) - y^\mu(t)), y^{-\mu}(t)) - f^\mu(y^\mu(t), y^{-\mu}(t))]dt \geq 0.$$

Dividing the above inequality by λ, taking the limit as $\lambda \to 0$, and using Taylor's series, we arrive at

$$\int_{\Omega_{l_o,l_1}} \left\langle \frac{\partial f^\mu}{\partial x^\mu}(y(t)), x^\mu(t) - y^\mu(t) \right\rangle dt \geq 0 \; \forall \; x^\mu(t) \in K_\mu(y^{-\mu}(t)).$$

Since by hypothesis, $J(y(t)) = \left(\frac{\partial f^\mu}{\partial x^\mu}(y(t))\right)_{\mu=1}^N$, we have for any $x(t) \in \Gamma(y(t))$,

$$\int_{\Omega_{l_o,l_1}} \langle J(y(t)), x(t) - y(t)\rangle dt = \sum_{\mu=1}^N \int_{\Omega_{l_o,l_1}} \left\langle \frac{\partial f^\mu}{\partial x^\mu}(y(t)), x^\mu(t) - y^\mu(t) \right\rangle dt \geq 0.$$

Since we already have $y(t) \in \Gamma(y(t))$, it follows that $y(t)$ is a solution to (MQVIP). □

Remark 4. *The set $G_{l_o,\frac{l_o+l_1}{2}}$ of positive measure in the converse part of the proof of Theorem 3.1 of [29] should also be simply denoted by G.*

4. Existence of Equilibria

Our aim in this section is to establish the existence of multi-time generalized Nash equilibria. In view of the equivalence of the multi-time generalized Nash equilibrium problem with the multi-time quasi-variational inequality problem, we can take advantage of techniques for proving existence results regarding quasi-variational inequality problems, which were investigated in [34,35]. Throughout this section, for better understanding of the strategy vector $x^\mu(t)$ of each player $\mu \in \{1,2,\ldots,N\}$ and of the strategy vectors $x^{-\mu}(t)$ of the rival players excluding the player μ, the subset $K \subset L^2(\Omega_{l_o,l_1}, \mathbb{R}^n))$ is given as $K = \prod_{\mu=1}^N X_\mu$ and $X_{-\mu} = \prod_{\nu=1,\,(\nu\neq\mu)}^N X_\nu$, where $\{X_\mu\}_{\mu=1}^N$ is a family of nonempty, closed and convex subsets with each $X_\mu \subset L^2(\Omega_{l_o,l_1}, \mathbb{R}^{n_\mu})$. With this notation, the entire strategy vector $x(t) \in K$ can be written as $x(t) = (x^\mu(t), x^{-\mu}(t)) \in X_\mu \times X_{-\mu}$. For all $x^{-\mu}(t) \in X_{-\mu}$, the strategy set of each player μ is a subset of X_μ, i.e., $K_\mu(x^{-\mu}(t)) \subset X_\mu$ and for each $y^\mu(t) \in X_\mu$, $K_\mu^{-1}(\{y^\mu(t)\}) \subset X_{-\mu}$.

Using the above mathematical framework, it is not difficult to see that

$$\Gamma^{-1}(\{y(t)\}) = \bigcap_{\mu=1}^N [X_\mu \times K_\mu^{-1}(\{y^\mu(t)\})] \quad \forall \, y(t) \in K.$$

We assume that for each $\mu \in \{1,2,\ldots,N\}$, X_μ is compactly open and for all $y^\mu(t) \in X_\mu$, the set $K_\mu^{-1}(\{y^\mu(t)\})$ is compactly open in $X_{-\mu}$. Therefore, Remark 3(b) and (c) yield that $\Gamma^{-1}(\{y(t)\})$ is compactly open for all $y(t) \in K$. Moreover, we also assume that the set $A = \{x(t) \in K : x(t) \in \Gamma(x(t))\}$ is compactly closed.

Theorem 3. *Let $y(t) \in K$ be an arbitrary strategy vector, $J(y(t)) = \left(\frac{\partial f^\mu}{\partial x^\mu}(y(t))\right)_{\mu=1}^N$, and for each $\mu \in \{1,2,\ldots,N\}$ and a given $y^{-\mu}(t)$, let the multi-time cost functional F^μ be convex on the set K in the argument $y^\mu(t)$. Assume that there exist a nonempty, closed and compact subset $D \subset K$ and $\hat{y}(t) \in D$ such that*

$$\sum_{\mu=1}^{N}\int_{\Omega_{t_0,t_1}}\left\langle \frac{\partial f^{\mu}}{\partial x^{\mu}}(\hat{y}(t)), \hat{y}^{\mu}(t) - x^{\mu}(t)\right\rangle dt < 0 \ \forall\, x(t) \in K \setminus D \text{ with } \hat{y}(t) \in \Gamma(x(t)). \tag{6}$$

Then (MQVIP) has a solution.

Proof. First, we define two set-valued maps $\Gamma_1, \Gamma_2 : K \to 2^K$ as follows: for each $x(t) \in K$, let

$$\Gamma_1(x(t)) := \left\{ y(t) \in K : \sum_{\mu=1}^{N}\int_{\Omega_{t_0,t_1}}\left\langle \frac{\partial f^{\mu}}{\partial x^{\mu}}(y(t)), y^{\mu}(t) - x^{\mu}(t)\right\rangle dt < 0 \right\},$$

$$\Gamma_2(x(t)) := \left\{ y(t) \in K : \sum_{\mu=1}^{N}\int_{\Omega_{t_0,t_1}}\left\langle \frac{\partial f^{\mu}}{\partial x^{\mu}}(x(t)), y^{\mu}(t) - x^{\mu}(t)\right\rangle dt < 0 \right\}.$$

Since each multi-time cost functional F^{μ} is convex on the set K in the arguments of X_{μ}, we have, for all $x_1(t)$ and $x_2(t) \in K$,

$$\int_{\Omega_{t_0,t_1}} f^{\mu}(x_1(t))dt - \int_{\Omega_{t_0,t_1}} f^{\mu}(x_2(t))dt \geq \int_{\Omega_{t_0,t_1}}\left\langle \frac{\partial f^{\mu}}{\partial x^{\mu}}(x_2(t)), x_1^{\mu}(t) - x_2^{\mu}(t)\right\rangle dt. \tag{7}$$

By interchanging the variables $x_1(t)$ and $x_2(t)$ in inequality (7), we get

$$\int_{\Omega_{t_0,t_1}} f^{\mu}(x_2(t))dt - \int_{\Omega_{t_0,t_1}} f^{\mu}(x_1(t))dt \geq \int_{\Omega_{t_0,t_1}}\left\langle \frac{\partial f^{\mu}}{\partial x^{\mu}}(x_1(t)), x_2^{\mu}(t) - x_1^{\mu}(t)\right\rangle dt. \tag{8}$$

Adding inequalities (7) and (8), we obtain the inequality

$$\int_{\Omega_{t_0,t_1}}\left\langle \frac{\partial f^{\mu}}{\partial x^{\mu}}(x_1(t)), x_2^{\mu}(t) - x_1^{\mu}(t)\right\rangle dt \leq \int_{\Omega_{t_0,t_1}}\left\langle \frac{\partial f^{\mu}}{\partial x^{\mu}}(x_2(t)), x_2^{\mu}(t) - x_1^{\mu}(t)\right\rangle dt,$$

which yields the following inequality:

$$\sum_{\mu=1}^{N}\int_{\Omega_{t_0,t_1}}\left\langle \frac{\partial f^{\mu}}{\partial x^{\mu}}(x_1(t)), x_2^{\mu}(t) - x_1^{\mu}(t)\right\rangle dt \leq \sum_{\mu=1}^{N}\int_{\Omega_{t_0,t_1}}\left\langle \frac{\partial f^{\mu}}{\partial x^{\mu}}(x_2(t)), x_2^{\mu}(t) - x_1^{\mu}(t)\right\rangle dt. \tag{9}$$

Inequality (9) implies that $\Gamma_1(x(t)) \subset \Gamma_2(x(t))$ for all $x(t) \in K$. Next, we define two more set-valued maps $S, T : K \to 2^K$ as follows:

$$S(x(t)) := \begin{cases} \Gamma(x(t)) \cap \Gamma_1(x(t)), & \text{if } x(t) \in A \\ \Gamma(x(t)), & \text{if } x(t) \in K \setminus A \end{cases}.$$

$$T(x(t)) := \begin{cases} \Gamma(x(t)) \cap \Gamma_2(x(t)), & \text{if } x(t) \in A \\ \Gamma(x(t)), & \text{if } x(t) \in K \setminus A \end{cases}.$$

Clearly, the point images of the set-valued maps Γ and Γ_2, i.e., $\Gamma(x(t))$ and $\Gamma_2(x(t))$, are convex for all $x(t) \in K$. Therefore, the point images of the set-valued map T, i.e., $T(x(t))$, are also convex for all $x(t) \in K$. Moreover, $S(x(t)) \subset T(x(t))$ for all $x(t) \in K$. Now, for all $y(t) \in K$, we have

$$
\begin{aligned}
S^{-1}(\{y(t)\}) &= \{x(t) \in K : y(t) \in S(x(t))\} \\
&= \{x(t) \in A : y(t) \in \Gamma(x(t)) \cap \Gamma_1(x(t))\} \cup \{x(t) \in K \setminus A : y(t) \in \Gamma(x(t))\} \\
&= [A \cap (\Gamma^{-1}(\{y(t)\}) \cap \Gamma_1^{-1}(\{y(t)\}))] \cup [K \setminus A \cap \Gamma^{-1}(\{y(t)\})] \\
&= [(A \cap (\Gamma^{-1}(\{y(t)\}) \cap \Gamma_1^{-1}(\{y(t)\}))) \cup K \setminus A] \\
&\cap [(A \cap (\Gamma^{-1}(\{y(t)\}) \cap \Gamma_1^{-1}(\{y(t)\}))) \cup \Gamma^{-1}(\{y(t)\})] \\
&= [K \cap ((\Gamma^{-1}(\{y(t)\}) \cap \Gamma_1^{-1}(\{y(t)\})) \cup K \setminus A)] \\
&\cap [(A \cup \Gamma^{-1}(\{y(t)\})) \cap (\Gamma^{-1}(\{y(t)\}) \cap \Gamma_1^{-1}(\{y(t)\}))] \\
&= [(\Gamma^{-1}(\{y(t)\}) \cap \Gamma_1^{-1}(\{y(t)\})) \cup K \setminus A] \cap \Gamma^{-1}(\{y(t)\}) \\
&= (\Gamma^{-1}(\{y(t)\}) \cap \Gamma_1^{-1}(\{y(t)\})) \cup (K \setminus A \cap \Gamma^{-1}(\{y(t)\})).
\end{aligned}
$$

Furthermore, for each $y(t) \in K$, the complement of $\Gamma_1^{-1}(\{y(t)\})$ in K can be written as

$$
[\Gamma_1^{-1}(\{y(t)\})]^c = \left\{ x(t) \in K : \sum_{\mu=1}^{N} \int_{\Omega_{l_0,l_1}} \left\langle \frac{\partial f^\mu}{\partial x^\mu}(y(t)), y^\mu(t) - x^\mu(t) \right\rangle dt \geq 0 \right\},
$$

which is closed in K. Consequently, the set $\Gamma_1^{-1}(\{y(t)\})$ is open in K. Remark 3(a) implies that $\Gamma_1^{-1}(\{y(t)\})$ is compactly open for all $y(t) \in K$. We also note that for all $y(t) \in K$, the sets $\Gamma^{-1}(\{y(t)\})$ and $K \setminus A$ are compactly open. Hence the set $S^{-1}(\{y(t)\})$ is now seen to also be compactly open for each $y(t) \in K$. We now claim that there is a point $\hat{x}(t) \in A$ such that $\Gamma(\hat{x}(t)) \cap \Gamma_1(\hat{x}(t)) = \emptyset$. Suppose on the contrary that for each $x(t) \in A$, the set $\Gamma(x(t)) \cap \Gamma_1(x(t)) \neq \emptyset$. Since we already know that the set $\Gamma(x(t))$ is nonempty and convex for each $x(t) \in K$, it follows that $S(x(t)) \neq \emptyset$ for each $x(t) \in K$. Our hypothesis yields that there exist a nonempty, closed and compact subset $D \subset K$ and a point $\hat{y}(t) \in D$ such that $K \setminus D \subset S^{-1}(\{\hat{y}(t)\})$. Thus, all the conditions of Theorem 1 are satisfied and so we conclude that there exists a point $z(t) \in K$ such that $z(t) \in T(z(t))$. The definitions of the set A and the set-valued map T imply that $\{x(t) \in K : x(t) \in T(x(t))\} \subset A$. Hence $z(t) \in A$, $z(t) \in \Gamma(z(t)) \cap \Gamma_2(z(t))$ and consequently, $z(t) \in \Gamma_2(z(t))$. It follows that

$$
\sum_{\mu=1}^{N} \int_{\Omega_{l_0,l_1}} \left\langle \frac{\partial f^\mu}{\partial x^\mu}(z(t)), z^\mu(t) - z^\mu(t) \right\rangle dt < 0,
$$

which is impossible. The contradiction we have reached shows that there indeed exists a point $\hat{x}(t) \in A$ such that $\Gamma(\hat{x}(t)) \cap \Gamma_1(\hat{x}(t)) = \emptyset$, as claimed. That is, $\hat{x}(t) \in \Gamma(\hat{x}(t))$ and

$$
\sum_{\mu=1}^{N} \int_{\Omega_{l_0,l_1}} \left\langle \frac{\partial f^\mu}{\partial x^\mu}(y(t)), y^\mu(t) - \hat{x}^\mu(t) \right\rangle dt \geq 0 \ \ \forall\, y(t) \in \Gamma(\hat{x}(t)).
$$

Using the convexity of both $K_\mu(\hat{x}^{-\mu}(t))$ and $\Gamma(\hat{x}(t))$, we can rewrite the above inequality as follows:

$$
\sum_{\mu=1}^{N} \int_{\Omega_{l_0,l_1}} \left\langle \frac{\partial f^\mu}{\partial x^\mu}(\hat{x}(t) + \lambda(y(t) - \hat{x}(t))), y^\mu(t) - \hat{x}^\mu(t) \right\rangle dt \geq 0 \ \forall\, y(t) \in \Gamma(\hat{x}(t)) \text{ and } \lambda \in [0,1].
$$

Since $\frac{\partial f^\mu}{\partial x^\mu}(\cdot)$ is hemicontinuous, by taking the limit as $\lambda \to 0^+$ in the above inequality, we obtain

$$
\sum_{\mu=1}^{N} \int_{\Omega_{l_0,l_1}} \left\langle \frac{\partial f^\mu}{\partial x^\mu}(\hat{x}(t)), y^\mu(t) - \hat{x}^\mu(t) \right\rangle dt \geq 0 \ \ \forall\, y(t) \in \Gamma(\hat{x}(t)).
$$

Using our hypothesis, we can rewrite the above inequality as follows:

$$\langle\langle J(\hat{x}(t)), y(t) - \hat{x}(t)\rangle\rangle \geq 0 \ \ \forall\, y(t) \in \Gamma(\hat{x}(t)).$$

In other words, $\hat{x}(t)$ is a solution to (MQVIP). □

5. Applications

5.1. Traffic Network Problem

Motivated by the network equilibrium problems studied by Nagurney et al. [36], and Daniele [37], in this section we study in more detail our multi-time generalized Nash equilibrium problem and demonstrate how this concept can apply to traffic network problems. To this end, we assume that the traffic network is made of M nodes, which represent the airports, railway stations, crossings, etc., and that the nodes are connected by the set of directed links L. Furthermore, W represents the set of origin–destination pairs while V represents the entire set of routes. Let r be a path consisting of a sequence of links which connect an origin–destination pair of nodes. Let p be the number of paths in the network. We assume that each route $r \in V$ links exactly one origin–destination pair. The set of all $r \in V$ which link a given $w \in W$ is denoted by $V(w)$, where $|W| = l$ and $p > l$. Let $x(t)$ be the entire traffic flow vector over the multi-time $t \in \Omega_{l_0,l_1}$ and be an element of $L^2(\Omega_{l_0,l_1}, \mathbb{R}^p_+)$. To emphasize the rth route flow vector in $x(t)$, sometimes we write $(x^r(t), x^{-r}(t))$ in place of $x(t)$. Bear in mind that this is still the vector $x(t) = (x^1(t), x^2(t), \ldots, x^{r-1}(t), x^r(t), x^{r+1}(t), \ldots, x^p(t))$. Indeed, $x^r(t) \in L^2(\Omega_{l_0,l_1}, \mathbb{R}_+)$ is the flow vector over the route r over the multi-time $t \in \Omega_{l_0,l_1}$ and $x^{-r}(t) \in L^2(\Omega_{l_0,l_1}, \mathbb{R}^{p-1}_+)$ denotes the flow vector of all the routes excluding the route r. We adhere to the restrictions that every traffic flow vector $x(t)$ satisfies the multi-time-dependent capacity constraints

$$\eta(t) \leq x(t) \leq \theta(t), \text{ a.e. on } \Omega_{l_0,l_1}$$

and the traffic conservation law/demand requirements

$$\phi x(t) = \rho(t), \text{ a.e. on } \Omega_{l_0,l_1},$$

where $\eta(t),\ \theta(t) \in L^2(\Omega_{l_0,l_1}, \mathbb{R}^p_+)$ are given bounds with $\eta(t) \leq \theta(t)$ and the function $\rho(t) \in L^2(\Omega_{l_0,l_1}, \mathbb{R}^l_+)$ is the given demand. Here $\rho(t) \geq 0$ and $\phi = \phi_{r,w}$ is the pair-route incidence matrix, the entries of which are 1 if route r links the pair w and 0 otherwise. We also have

$$\phi\eta(t) \leq \rho(t) \leq \phi\theta(t), \text{ a.e. on } \Omega_{l_0,l_1},$$

which implies that the set of all feasible flows

$$K := \{x(t) \in L^2(\Omega_{l_0,l_1}, \mathbb{R}^p_+) : \eta(t) \leq x(t) \leq \theta(t) \text{ and } \phi x(t) = \rho(t), \text{ a.e. on } \Omega_{l_0,l_1}\}$$

is not empty. For any given $x^{-r}(t)$, the nonempty, closed and convex feasible traffic flow set of each route r is denoted by $K_r(x^{-r}(t))$. This is a subset of $L^2(\Omega_{l_0,l_1}, \mathbb{R}_+)$.

The multi-time cost functional of each route r, $H^r : K \to \mathbb{R}$, is defined by the multiple integral

$$H^r(x(t)) = \int_{\Omega_{l_0,l_1}} C^r(x^r(s), x^{-r}(s))ds,$$

where $C^r(x^r(s), x^{-r}(s))$ denotes the running cost function of the route r that depends on both its own variable $x^r(s)$ (the flow in the route r) and $x^{-r}(s)$ (the flow in the other routes except route r). It is assumed to be a real-valued continuously differentiable function. Our aim is to compute the entire traffic flow vector $x(t) \in K$ so as to minimize the cost of each route r in the time period Ω_{l_0,l_1} when the flow vectors of the other routes except that of the route r, $x^{-r}(t)$, are given, i.e., to solve the following multi-time optimization problem:

$$\begin{aligned}&\min_{x^r(t)} \int_{\Omega_{l_0,l_1}} C^r(x^r(t), x^{-r}(t))dt \\ &\text{subject to } x^r(t) \in K_r(x^{-r}(t)).\end{aligned} \tag{10}$$

Evidently, an equilibrium of the multi-time optimization problem (10) is a multi-time generalized Nash equilibrium in the sense of our (MGNEP). To present a realistic demonstration of the traffic network model that can be reformulated as a multi-time generalized Nash equilibrium problem (10), we consider a general traffic network structure, displayed in Figure 1, for a courier service company which wishes to minimize the traffic cost of routes for delivering packages at their destinations. The traffic network pattern of Figure 1 comprises 13 nodes and 16 links. We assume that a branch of the courier service company is situated at a node, say O, which must deliver the packages at the nodes P_1 and P_2. Thus, we have two origin–destination pairs $w_1 = (O, P_1)$ and $w_2 = (O, P_2)$, which are respectively connected by the following routes:

$$w_1 : \begin{cases} r_1 = & (O, a_1) \cup (a_1, P_1) \\ r_2 = & (O, a_9) \cup (a_9, a_{10}) \cup (a_{10}, P_1) \\ r_3 = & (O, a_4) \cup (a_4, a_3) \cup (a_3, a_2) \cup (a_2, a_1) \cup (a_1, P_1), \end{cases}$$

$$w_2 : \begin{cases} r_4 = & (O, a_9) \cup (a_9, a_8) \cup (a_8, a_7) \cup (a_7, a_6) \cup (a_6, a_5) \cup (a_5, P_2) \\ r_5 = & (O, a_6) \cup (a_6, a_5) \cup (a_5, P_2) \\ r_6 = & (O, a_4) \cup (a_4, P_2). \end{cases}$$

We have explicitly $p = 6$ paths in the given traffic network. The courier company must find the entire traffic flow vector to minimize the cost of each route $\{r_1, r_2, r_3, r_4, r_5, r_6\}$. This can be calculated by solving the multi-time optimization problem (10).

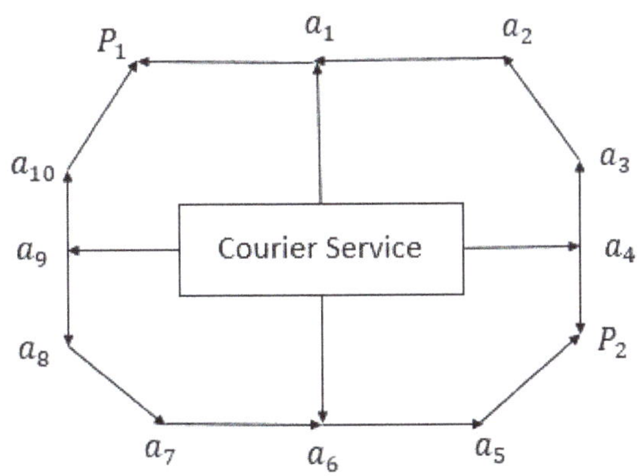

Figure 1. Traffic network pattern with 6 routes, i.e., $p = 6$.

5.2. River Basin Pollution Problem

In this subsection we show how our multi-time generalized Nash equilibrium problem can be applied to solving the river basin pollution problem [38]. For studies of the river basin pollution problem, we refer the interested reader to [39–41]. We use the term time $t \in \Omega_{l_0,l_1}$ as defined in Section 2. We assume that n industrial factories (paper mills, chemical factories, pharmaceutical manufacturing companies, etc.) are located along a river. In the sequel we call them industrial agents. Presently, it is a very common scenario that industrial factories often dump waste garbage, such as dirty water, used chemicals and oils, sewage, and cafeteria waste, directly into a community water source (river, lake or stream). Waste dumped contains several contaminants which mix and create pollution concentration along the river. We assume that m basin authorities (monitoring stations) are located along the river. Each basin authority is empowered to set a maximum pollutant concentration level at time t which we denote by $\chi_s(t) \in L^2(\Omega_{l_o,l_1}, \mathbb{R}_+)$ where $s \in \{1,2,\ldots,m\}$. Furthermore, we let $e^f(t) \in L^2(\Omega_{l_o,l_1}, \mathbb{R}_+)$ be the pollutant emission coefficient for the industrial agent $f \in \{1,2,\ldots,n\}$. Let $x^f(t) \in L^2(\Omega_{l_o,l_1}, \mathbb{R}_+)$ be the chosen emitted pollutant concentration level at time t by the industrial agent f and let $x^{-f}(t) \in L^2(\Omega_{l_o,l_1}, \mathbb{R}^{n-1}_+)$ be the chosen emitted pollutant concentration level at time t by all the industrial agents except the agent f. Furthermore, let $x(t) \in L^2(\Omega_{l_o,l_1}, \mathbb{R}^n_+)$ be the chosen emitted pollutant concentration level at time t by all the industrial agents. To put the chosen emitted pollutant concentration level $x^f(t)$ of the industrial agent f in focus, we write $x(t)$ as $x(t) = (x^f(t), x^{-f}(t))$. Please note that $x(t)$ is still the vector $x(t) = (x^1(t), x^2(t), \ldots, x^{f-1}(t), x^f(t), x^{f+1}(t), \ldots, x^n(t))$. Waste materials, dumped by the industrial agents into the river, spread, decay and then finally reach the basin authorities. Thus, the amount of pollution received by the basin authority $s \in \{1,2,\ldots,m\}$ is $\sum_{f=1}^{n} \delta_s^f(t)e^f(t)x^f(t)$, where $\delta_s^f(t)$ is the decay-and-transportation coefficient from the agent f to the monitoring station s. The basin authorities impose constraints on the pollution, so that industrial agents control their pollutant emission into the river. Thus, the pollution constraint set imposed by the authority s is given by

$$\sum_{f=1}^{n} \delta_s^f(t)e^f(t)x^f(t) \leq \chi_s(t) \quad \text{for } s \in \{1,2,\ldots,m\} \text{ and a.e. on } \Omega_{l_o,l_1}.$$

The nonempty set of entire feasible pollution concentration levels is given by

$$K = \left\{ x(t) \in L^2(\Omega_{l_o,l_1}, \mathbb{R}^n_+) : \sum_{f=1}^{n} \delta_s^f(t)e^f(t)x^f(t) \leq \chi_s(t) \quad \text{for } s \in \{1,2,\ldots,m\} \right.$$
$$\left. \text{and a.e. on } \Omega_{l_o,l_1} \right\}.$$

We bear in the mind that industrial agents are dependent on each other, at least because of the finiteness of the amount of dumping pollutants into the river. Therefore, for any given $x^{-f}(t)$, the nonempty, closed and convex feasible pollution concentration level set of each industrial agent f is denoted by $K_f(x^{-f}(t))$. This is a subset of $L^2(\Omega_{l_o,l_1}, \mathbb{R}_+)$. Each agent wishes to maximize its profit in the time period t, where the profit is defined by the difference between the revenue $[p_1 - p_2 \sum_{f=1}^{n} x^f(t)]x^f(t)$ and the cost $[a^f(t) + b^f(t)x^f(t)]x^f(t)$. Here p_1 and p_2 are economic constants which follow the inverse demand law and $a^f(t), b^f(t) \in L^2(\Omega_{l_o,l_1}, \mathbb{R}_+)$ are the cost coefficient functions. Now, for a given $x^{-f}(t)$, the aim of the industrial agent f is to choose an emitted pollutant concentration level $x^f(t)$ such that it solves the following multi-time maximization problem:

$$\max_{x^f(t)} \int_{\Omega_{l_o,l_1}} \left[\left\{ \left(p_1 - p_2 \sum_{f=1}^{n} x^f(t) \right) x^f(t) \right\} - \{(a^f(t) + b^f(t)x^f(t))x^f(t)\} \right] dt,$$
$$\text{subject to } x^f(t) \in K_f(x^{-f}(t)).$$

An equilibrium of the above defined multi-time maximization problem is a multi-time generalized Nash equilibrium in the sense of our (MGNEP), where

$$f^\mu(x^\mu(t), x^{-\mu}(t)) = \left[\{(a^\mu(t) + b^\mu(t)x^\mu(t))x^\mu(t)\} - \left\{ \left(p_1 - p_2 \sum_{\mu=1}^{n} x^\mu(t) \right) x^\mu(t) \right\} \right].$$

6. The Multi-Time Generalized Nash Equilibrium Problem as a Projected Dynamical System

In this section, we investigate our multi-time generalized Nash equilibrium problem via a projected dynamical system. We also propose a method for finding the equilibria of our multi-time generalized Nash equilibrium problem. Our work is motivated by [12,20,42]. We also refer the interested reader to the more recent work [29], which deals with a projected dynamical system pertaining to a single-valued map in the context of a multi-time variational inequality problem, while the projected dynamical system of this section pertains to a set-valued map in the context of a multi-time quasi-variational inequality problem. More precisely, in the present section we consider the following projected dynamical system (PDS) pertaining to the set-valued map Γ, where $x(\cdot) \in \Gamma(x(\cdot))$:

$$\begin{aligned} \frac{dx(\cdot,\tau)}{d\tau} &= \Pi_{\Gamma(x(\cdot,\tau))}(x(\cdot,\tau), -J(x(\cdot,\tau))), \\ x(\cdot,0) &= x_\circ(\cdot) \in \Gamma(x(\cdot)), \end{aligned} \tag{11}$$

where $J : L^2(\Omega_{l_\circ,l_1}, \mathbb{R}^n) \to L^2(\Omega_{l_\circ,l_1}, \mathbb{R}^n)$ is a Lipschitz continuous vector field and $\Pi_{\Gamma(x(\cdot))} : L^2(\Omega_{l_\circ,l_1}, \mathbb{R}^n) \times L^2(\Omega_{l_\circ,l_1}, \mathbb{R}^n) \to L^2(\Omega_{l_\circ,l_1}, \mathbb{R}^n)$ is the operator defined by

$$\Pi_{\Gamma(x(\cdot))}(x(\cdot), v(\cdot)) := \lim_{\delta \to 0^+} \frac{\mathrm{proj}_{\Gamma(x(\cdot))}(x(\cdot) + \delta v(\cdot)) - x(\cdot)}{\delta},$$

where $x(\cdot) \in K$ and $v(\cdot) \in L^2(\Omega_{l_\circ,l_1}, \mathbb{R}^n)$. The characteristics of the times t and τ in (PDS) are different. Indeed, for all $t \in \Omega_{l_\circ,l_1}$, a solution of (MQVIP) specifies the static states of the underlying system and the static states defined by one or more equilibrium curves when t varies over the set $\Omega_{l_\circ,l_1}$. On the other hand, τ represents the dynamics of the system over the interval $[0, \infty)$ until it reaches one of the equilibria on the curves. It is evident that the solutions of (PDS) lie in the class of absolutely continuous functions with respect to τ, which take $[0, \infty)$ into $\Gamma(x(\cdot))$. Moreover, to avoid any possible confusion between t and τ, we represent elements of the sets $L^2(\Omega_{l_\circ,l_1}, \mathbb{R}^n)$ at fixed moments $t \in \Omega_{l_\circ,l_1}$ by $x(\cdot)$. To describe the connection of our (MGNEP) with (PDS), we need the following definition, which is inspired by [42].

Definition 9. *$y(\cdot) \in K$ is called a critical point of (PDS) if $y(\cdot) \in \Gamma(y(\cdot))$ and*

$$\Pi_{\Gamma(y(\cdot))}(y(\cdot), -J(y(\cdot))) = 0.$$

Lemma 1 ([43], Lemma 1.2.8). *For each $i = \{1, 2, \ldots, p\}$, let H_i be a Hilbert space and let $C_i \subset H_i$ be closed and convex, Set $C = C_1 \times C_2 \times \ldots \times C_p \subset H_1 \times H_2 \times \ldots \times H_p = H$ and let $x = (x_1, x_2, \ldots, x_p) \in H$. Let $proj_{C_i}$ denote the metric projection onto C_i. Then the metric projection $proj_C(x)$ is given by*

$$proj_C(x) = (proj_{C_1}(x_1), proj_{C_2}(x_2), \ldots, proj_{C_p}(x_p)).$$

The following propositions concerning each player $\mu \in \{1, 2, \ldots, N\}$ and the strategies $y^{-\mu}(\cdot) \in L^2(\Omega_{l_\circ,l_1}, \mathbb{R}^{n-n_\mu})$ of the rival players except player μ are direct consequences of Proposition 2.1 and 2.2 in [42].

Proposition 1. *For all $y^\mu(\cdot) \in K_\mu(y^{-\mu}(\cdot))$ and $v^\mu(\cdot) \in L^2(\Omega_{l_\circ,l_1}, \mathbb{R}^{n_\mu})$, $\Pi_{K_\mu(y^{-\mu}(\cdot))}(y^\mu(\cdot), v^\mu(\cdot))$ exists and $\Pi_{K_\mu(y^{-\mu}(\cdot))}(y^\mu(\cdot), v^\mu(\cdot)) = proj_{T_{K_\mu(y^{-\mu}(\cdot))}(y^\mu(\cdot))}(v^\mu(\cdot))$.*

Proposition 2. *For each $y^\mu(\cdot) \in K_\mu(y^{-\mu}(\cdot))$, there exists $n^\mu(\cdot) \in N_{K_\mu(y^{-\mu}(\cdot))}(y^\mu(\cdot))$ such that $\Pi_{K_\mu(y^{-\mu}(\cdot))}(y^\mu(\cdot), v^\mu(\cdot)) = v^\mu(\cdot) - n^\mu(\cdot)$ for all $v^\mu(\cdot) \in L^2(\Omega_{l_\circ,l_1}, \mathbb{R}^{n_\mu})$.*

The following theorem establishes a strong connection between (MGNEP) and (PDS).

Theorem 4. *Assume that* $J(x(\cdot)) = \left(\frac{\partial f^\mu}{\partial x^\mu}(x(\cdot))\right)_{\mu=1}^N$ *for any* $x(\cdot) \in K$*, and that for each* $\mu \in \{1,2,\dots,N\}$ *and* $x^{-\mu}(\cdot)$*, the multi-time cost functional* F^μ *is convex on* K *in the argument* $x^\mu(\cdot)$*. Then* $y(\cdot) \in K$ *is a multi-time generalized Nash equilibrium if and only if it is a critical point of (PDS).*

Proof. Suppose that $y(\cdot) \in K$ is a multi-time generalized Nash equilibrium. Then it follows from Theorem 2 that for each $\mu \in \{1,2,\dots,N\}$, we have

$$\int_{\Omega_{t_0,t_1}} \left\langle \frac{\partial f^\mu}{\partial x^\mu}(y(\cdot)), x^\mu(\cdot) - y^\mu(\cdot) \right\rangle dt \geq 0 \,\forall\, x^\mu(\cdot) \in K_\mu(y^{-\mu}(\cdot)).$$

The above inequality can be rewritten as

$$\left\langle\left\langle \frac{\partial f^\mu}{\partial x^\mu}(y(\cdot)), x^\mu(\cdot) - y^\mu(\cdot) \right\rangle\right\rangle \geq 0 \,\forall\, x^\mu(\cdot) \in K_\mu(y^{-\mu}(\cdot)),$$

which leads to the following inclusion:

$$-\frac{\partial f^\mu}{\partial x^\mu}(y(\cdot)) \in N_{K_\mu(y^{-\mu}(\cdot))}(y^\mu(\cdot)).$$

Proposition 2 now yields

$$\Pi_{K_\mu(y^{-\mu}(\cdot))}\left(y^\mu(\cdot), -\frac{\partial f^\mu}{\partial x^\mu}(y(\cdot))\right) = 0. \tag{12}$$

Since $K_\mu(y^{-\mu}(\cdot))$ is convex for each $\mu \in \{1,2,\dots,N\}$, Lemma 1 implies that

$$\Pi_{\Gamma(y(\cdot))}(y(\cdot), -J(y(\cdot))) = 0. \tag{13}$$

Therefore $y(\cdot)$ is indeed a critical point of (PDS), as asserted.

Conversely, assume that $y(\cdot) \in K$ is a critical point of (PDS). Then $y(\cdot) \in \Gamma(y(\cdot))$ and inequality (13) holds. Consequently, (12) holds too. Proposition 1 implies that

$$\mathrm{proj}_{T_{K_\mu(y^{-\mu}(\cdot))}(y^\mu(\cdot))}\left(-\frac{\partial f^\mu}{\partial x^\mu}(y(\cdot))\right) = 0.$$

In view of Remark 2, the above expression leads to

$$\left\langle\left\langle -\frac{\partial f^\mu}{\partial x^\mu}(y(\cdot)), z^\mu(\cdot) \right\rangle\right\rangle \leq 0 \;\;\forall\, z^\mu(\cdot) \in T_{K_\mu(y^{-\mu}(\cdot))}(y^\mu(\cdot)),$$

which in turn implies that

$$-\frac{\partial f^\mu}{\partial x^\mu}(y(\cdot)) \in N_{K_\mu(y^{-\mu}(\cdot))}(y^\mu(\cdot)),$$

which yields that $y(\cdot)$ is a solution of (MQVIP) with $J(y(\cdot)) = \left(\frac{\partial f^\mu}{\partial x^\mu}(y(\cdot))\right)_{\mu=1}^N$. Thus, the first part of the proof of Theorem 2 implies that $y(\cdot)$ is a multi-time generalized Nash equilibrium. □

Remark 5. *In view of the proof techniques of Theorem 4 and the fact that the normal cone of a product set is equal to the product of the normal cones ([44], Proposition 6.41), we can write*

(PDS) (11) as a concatenation of the following N dynamical systems $(PDS)_\mu$ for $J(x(\cdot,\tau)) = \left(\frac{\partial f^\mu}{\partial x^\mu}(x(\cdot,\tau))\right)_{\mu=1}^{N}$:

$$\frac{dx^\mu(\cdot,\tau)}{d\tau} = \Pi_{K_\mu(x^{-\mu}(\cdot,\tau))}\left(x^\mu(\cdot,\tau), -\frac{\partial f^\mu}{\partial x^\mu}(x(\cdot,\tau))\right),$$
$$x^\mu(\cdot,0) = x^\mu_\circ(\cdot) \in K_\mu(x^{-\mu}(\cdot)).$$

Numerical Illustrations

In this subsection we turn our attention to demonstrating a method for solving (MGNEP) by taking advantage of (PDS) (11). Theorem 4 tells us that any point on a curve of equilibria in the set K is a critical point of (PDS) and vice versa. Please note that the existence of equilibria has already been established. Next, we take the following discretization of $\Omega_{l_\circ,l_1}$: $(l^1_\circ, l^2_\circ, \ldots, l^m_\circ) = (t^1_\circ, t^2_\circ, \ldots, t^m_\circ) < (t^1_1, t^2_1, \ldots, t^m_1) < \ldots < (t^1_i, t^2_i, \ldots, t^m_i) < \ldots < (t^1_n, t^2_n, \ldots, t^m_n) = (l^1_1, l^2_1, \ldots, l^m_1)$. Then for each $t_i = (t^1_i, t^2_i, \ldots, t^m_i)$, $i = \{0, 1, \ldots, n\}$, we obtain a sequence of (PDS) on the distinct, nonempty, finite-dimensional and convex sets K_{t_i}. After computing all the critical points of each (PDS), we obtain a sequence of critical points which by interpolation yields the curves of equilibria. To apply this procedure in practice, we consider a Nguyen traffic network [31] which comprises 13 nodes, 19 links, four origin–destination pairs and 24 paths. See Figure 2. Here we use the terminology of Section 5. Every origin–destination pair of our Nguyen traffic network is connected by six paths. Let $m = 2$ and $\Omega_{l_\circ,l_1} = \Omega_{0,4} = [0,4]^2$. The set of feasible flows is given by

$$K = \left\{ x(t) \in L^2(\Omega_{0,4}, \mathbb{R}^{24}_+) : (0,\ldots,0) \le (x^r(t))_{r=1}^{24} \le (t^1 + t^2 + r)_{r=1}^{24} \right.$$
$$\text{and } \sum_{r=1}^{6} x^r(t) = 6t^1 + 6t^2 + 21, \sum_{r=7}^{12} x^r(t) = 6t^1 + 6t^2 + 57,$$
$$\left. \sum_{r=13}^{18} x^r(t) = 6t^1 + 6t^2 + 93, \sum_{r=19}^{24} x^r(t) = 6t^1 + 6t^2 + 129 \text{ a.e. on } \Omega_{0,4} \right\}.$$

To simplify our calculations, we assume the following special structure of the feasible traffic flow set of each route $r \in \{1, 2, \ldots, 24\}$. This is motivated by Rosen [5].

$$K_r(x^{-r}(t)) = \{x(t) \in L^2(\Omega_{0,4}, \mathbb{R}_+) : (x^r(t), x^{-r}(t)) \in K\}$$

and the multi-time cost functional of each route $r \in \{1, 2, \ldots, 24\}$ is defined by

$$H^r(x(t)) = \int_{\Omega_{0,4}} (x^r(s) + (x^r(s))^2) ds,$$

where $x(t) = (x^1(t), x^2(t), \ldots, x^{r-1}(t), x^r(t), x^{r+1}(t), \ldots, x^{24}(t))$. It can easily be proven that for each $r \in \{1, 2, \ldots, 24\}$, the multi-time cost functional $H^r(x(t))$ is convex in the argument $x^r(t)$ and that there exists a nonempty, closed and compact subset $D \subset K$ which is given by

$$D = \left\{ x(t) \in L^2(\Omega_{0,4}, \mathbb{R}^{24}_+) : (0,\ldots,0) \le (x^r(t))_{r=1}^{24} \le \left(t^1 + t^2 + \frac{r}{2}\right)_{r=1}^{24} \right.$$
$$\text{and } \sum_{r=1}^{6} x^r(t) = 6t^1 + 6t^2 + 21, \sum_{r=7}^{12} x^r(t) = 6t^1 + 6t^2 + 57,$$
$$\left. \sum_{r=13}^{18} x^r(t) = 6t^1 + 6t^2 + 93, \sum_{r=19}^{24} x^r(t) = 6t^1 + 6t^2 + 129 \text{ a.e. on } \Omega_{0,4} \right\}$$

such that for $\hat{y}(t) \in D$, inequality (6) is satisfied. Thus, all the hypotheses of Theorem 3 are fulfilled. Therefore, for $J(y(t)) = (1 + 2y^r(t))_{r=1}^{24}$, (MDVIP) has a solution. Consequently, by Theorem 2, (MGNEP) has a solution too.

Selecting points $t_i \in \left\{\left(\frac{k}{8}, \frac{k}{8}\right) : k \in \{0, 1, 2, \ldots, 32\}\right\}$, we obtain a sequence of (PDS) defined on the sets

$$K_{t_i} = \left\{ x(t_i) \in L^2(\Omega_{0,4}, \mathbb{R}^{24}_+) : (0, \ldots, 0) \leq (x^r(t_i))_{r=1}^{24} \leq (t_i^1 + t_i^2 + r)_{r=1}^{24} \right.$$

$$\text{and} \sum_{r=1}^{6} x^r(t_i) = 6t_i^1 + 6t_i^2 + 21, \sum_{r=7}^{12} x^r(t_i) = 6t_i^1 + 6t_i^2 + 57,$$

$$\left. \sum_{r=13}^{18} x^r(t_i) = 6t_i^1 + 6t_i^2 + 93, \sum_{r=19}^{24} x^r(t_i) = 6t_i^1 + 6t_i^2 + 129 \text{ a.e. on } \Omega_{0,4} \right\}.$$

We have

$$K_r(x^{-r}(t_i)) = \{x(t_i) \in L^2(\Omega_{0,4}, \mathbb{R}_+) : (x^r(t_i), x^{-r}(t_i)) \in K_{t_i}\},$$

$$\text{and } \Gamma(y(t_i)) = \prod_{r=1}^{24} K_r(y^{-r}(t_i)).$$

For calculating the equilibrium, we consider the following system at the points t_i: to find the point $y(t_i) = (y^1(t_i), y^2(t_i), \ldots, y^{r-1}(t_i), y^r(t_i), y^{r+1}(t_i), \ldots, y^{24}(t_i)) \in K_{t_i}$ such that

$$-(1 + 2y^r(t_i)) \in N_{K_r(y^{-r}(t_i))}(y^r(t_i)).$$

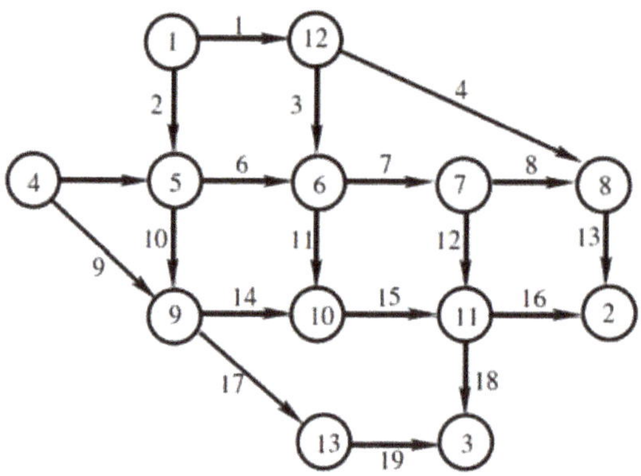

Figure 2. The Nguyen traffic network (13 nodes, 19 links, 4 origin–destination pairs).

After a simple calculation, we find the equilibrium points which are given in Tables 1–4. We then interpolate the points of these tables and finally obtain the curves of equilibria. They are displayed in Figure 3.

Table 1. Numerical Results.

$t_i = \{t_i^1, t_i^2\}$	$y^1(t_i)$	$y^2(t_i)$	$y^3(t_i)$	$y^4(t_i)$	$y^5(t_i)$	$y^6(t_i)$
$\{0,0\}$	1	2	3	4	5	6
$\{\frac{1}{8},\frac{1}{8}\}$	1.25	2.25	3.25	4.25	5.25	6.25
$\{\frac{1}{4},\frac{1}{4}\}$	1.5	2.5	3.5	4.5	5.5	6.5
$\{\frac{3}{8},\frac{3}{8}\}$	1.75	2.75	3.75	4.75	5.75	6.75
$\{\frac{1}{2},\frac{1}{2}\}$	2	3	4	5	6	7
$\{\frac{5}{8},\frac{5}{8}\}$	2.25	3.25	4.25	5.25	6.25	7.25
$\{\frac{3}{4},\frac{3}{4}\}$	2.5	3.5	4.5	5.5	6.5	7.5
$\{\frac{7}{8},\frac{7}{8}\}$	2.75	3.75	4.75	5.75	6.75	7.75
$\{1,1\}$	3	4	5	6	7	8
$\{\frac{9}{8},\frac{9}{8}\}$	3.25	4.25	5.25	6.25	7.25	8.25
$\{\frac{5}{4},\frac{5}{4}\}$	3.5	4.5	5.5	6.5	7.5	8.5
$\{\frac{11}{8},\frac{11}{8}\}$	3.75	4.75	5.75	6.75	7.75	8.75
$\{\frac{3}{2},\frac{3}{2}\}$	4	5	6	7	8	9
$\{\frac{13}{8},\frac{13}{8}\}$	4.25	5.25	6.25	7.25	8.25	9.25
$\{\frac{7}{4},\frac{7}{4}\}$	4.5	5.5	6.5	7.5	8.5	9.5
$\{\frac{15}{8},\frac{15}{8}\}$	4.75	5.75	6.75	7.75	8.75	9.75
$\{2,2\}$	5	6	7	8	9	10
$\{\frac{17}{8},\frac{17}{8}\}$	5.25	6.25	7.25	8.25	9.25	10.25
$\{\frac{9}{4},\frac{9}{4}\}$	5.5	6.5	7.5	8.5	9.5	10.5
$\{\frac{19}{8},\frac{19}{8}\}$	5.75	6.75	7.75	8.75	9.75	10.75
$\{\frac{5}{2},\frac{5}{2}\}$	6	7	8	9	10	11
$\{\frac{21}{8},\frac{21}{8}\}$	6.25	7.25	8.25	9.25	10.25	11.25
$\{\frac{11}{4},\frac{11}{4}\}$	6.5	7.5	8.5	9.5	10.5	11.5
$\{\frac{23}{8},\frac{23}{8}\}$	6.75	7.75	8.75	9.75	10.75	11.75
$\{3,3\}$	7	8	9	10	11	12
$\{\frac{25}{8},\frac{25}{8}\}$	7.25	8.25	9.25	10.25	11.25	12.25
$\{\frac{13}{4},\frac{13}{4}\}$	7.5	8.5	9.5	10.5	11.5	12.5
$\{\frac{27}{8},\frac{27}{8}\}$	7.75	8.75	9.75	10.75	11.75	12.75
$\{\frac{7}{2},\frac{7}{2}\}$	8	9	10	11	12	13
$\{\frac{29}{8},\frac{29}{8}\}$	8.25	9.25	10.25	11.25	12.25	13.25
$\{\frac{15}{4},\frac{15}{4}\}$	8.5	9.5	10.5	11.5	12.5	13.5
$\{\frac{31}{8},\frac{31}{8}\}$	8.75	9.75	10.75	11.75	12.75	13.75
$\{4,4\}$	9	10	11	12	13	14

Table 2. Numerical Results.

$t_i = \{t_i^1, t_i^2\}$	$y^7(t_i)$	$y^8(t_i)$	$y^9(t_i)$	$y^{10}(t_i)$	$y^{11}(t_i)$	$y^{12}(t_i)$
$\{0,0\}$	7	8	9	10	11	12
$\{\frac{1}{8},\frac{1}{8}\}$	7.25	8.25	9.25	10.25	11.25	12.25
$\{\frac{1}{4},\frac{1}{4}\}$	7.5	8.5	9.5	10.5	11.5	12.5
$\{\frac{3}{8},\frac{3}{8}\}$	7.75	8.75	9.75	10.75	11.75	12.75
$\{\frac{1}{2},\frac{1}{2}\}$	8	9	10	11	12	13
$\{\frac{5}{8},\frac{5}{8}\}$	8.25	9.25	10.25	11.25	12.25	13.25
$\{\frac{3}{4},\frac{3}{4}\}$	8.5	9.5	10.5	11.5	12.5	13.5
$\{\frac{7}{8},\frac{7}{8}\}$	8.75	9.75	10.75	11.75	12.75	13.75
$\{1,1\}$	9	10	11	12	13	14
$\{\frac{9}{8},\frac{9}{8}\}$	9.25	10.25	11.25	12.25	13.25	14.25
$\{\frac{5}{4},\frac{5}{4}\}$	9.5	10.5	11.5	12.5	13.5	14.5
$\{\frac{11}{8},\frac{11}{8}\}$	9.75	10.75	11.75	12.75	13.75	14.75
$\{\frac{3}{2},\frac{3}{2}\}$	10	11	12	13	14	15
$\{\frac{13}{8},\frac{13}{8}\}$	10.25	11.25	12.25	13.25	14.25	15.25
$\{\frac{7}{4},\frac{7}{4}\}$	10.5	11.5	12.5	13.5	14.5	15.5
$\{\frac{15}{8},\frac{15}{8}\}$	10.75	11.75	12.75	13.75	14.75	15.75
$\{2,2\}$	11	12	13	14	15	16
$\{\frac{17}{8},\frac{17}{8}\}$	11.25	12.25	13.25	14.25	15.25	16.25
$\{\frac{9}{4},\frac{9}{4}\}$	11.5	12.5	13.5	14.5	15.5	16.5
$\{\frac{19}{8},\frac{19}{8}\}$	11.75	12.75	13.75	14.75	15.75	16.75
$\{\frac{5}{2},\frac{5}{2}\}$	12	13	14	15	16	17
$\{\frac{21}{8},\frac{21}{8}\}$	12.25	13.25	14.25	15.25	16.25	17.25
$\{\frac{11}{4},\frac{11}{4}\}$	12.5	13.5	14.5	15.5	16.5	17.5
$\{\frac{23}{8},\frac{23}{8}\}$	12.75	13.75	14.75	15.75	16.75	17.75
$\{3,3\}$	13	14	15	16	17	18
$\{\frac{25}{8},\frac{25}{8}\}$	13.25	14.25	15.25	16.25	17.25	18.25
$\{\frac{13}{4},\frac{13}{4}\}$	13.5	14.5	15.5	16.5	17.5	18.5
$\{\frac{27}{8},\frac{27}{8}\}$	13.75	14.75	15.75	16.75	17.75	18.75
$\{\frac{7}{2},\frac{7}{2}\}$	14	15	16	17	18	19
$\{\frac{29}{8},\frac{29}{8}\}$	14.25	15.25	16.25	17.25	18.25	19.25
$\{\frac{15}{4},\frac{15}{4}\}$	14.5	15.5	16.5	17.5	18.5	19.5
$\{\frac{31}{8},\frac{31}{8}\}$	14.75	15.75	16.75	17.75	18.75	19.75
$\{4,4\}$	15	16	17	18	19	20

Table 3. Numerical Results.

$t_i = \{t_i^1, t_i^2\}$	$y^{13}(t_i)$	$y^{14}(t_i)$	$y^{15}(t_i)$	$y^{16}(t_i)$	$y^{17}(t_i)$	$y^{18}(t_i)$
$\{0,0\}$	13	14	15	16	17	18
$\{\frac{1}{8},\frac{1}{8}\}$	13.25	14.25	15.25	16.25	17.25	18.25
$\{\frac{1}{4},\frac{1}{4}\}$	13.5	14.5	15.5	16.5	17.5	18.5
$\{\frac{3}{8},\frac{3}{8}\}$	13.75	14.75	15.75	16.75	17.75	18.75
$\{\frac{1}{2},\frac{1}{2}\}$	14	15	16	17	18	19
$\{\frac{5}{8},\frac{5}{8}\}$	14.25	15.25	16.25	17.25	18.25	19.25
$\{\frac{3}{4},\frac{3}{4}\}$	14.5	15.5	16.5	17.5	18.5	19.5
$\{\frac{7}{8},\frac{7}{8}\}$	14.75	15.75	16.75	17.75	18.75	19.75
$\{1,1\}$	15	16	17	18	19	20
$\{\frac{9}{8},\frac{9}{8}\}$	15.25	16.25	17.25	18.25	19.25	20.25
$\{\frac{5}{4},\frac{5}{4}\}$	15.5	16.5	17.5	18.5	19.5	20.5
$\{\frac{11}{8},\frac{11}{8}\}$	15.75	16.75	17.75	18.75	19.75	20.75
$\{\frac{3}{2},\frac{3}{2}\}$	16	17	18	19	20	21
$\{\frac{13}{8},\frac{13}{8}\}$	16.25	17.25	18.25	19.25	20.25	21.25
$\{\frac{7}{4},\frac{7}{4}\}$	16.5	17.5	18.5	19.5	20.5	21.5
$\{\frac{15}{8},\frac{15}{8}\}$	16.75	17.75	18.75	19.75	20.75	21.75
$\{2,2\}$	17	18	19	20	21	22
$\{\frac{17}{8},\frac{17}{8}\}$	17.25	18.25	19.25	20.25	21.25	22.25
$\{\frac{9}{4},\frac{9}{4}\}$	17.5	18.5	19.5	20.5	21.5	22.5
$\{\frac{19}{8},\frac{19}{8}\}$	17.75	18.75	19.75	20.75	21.75	22.75
$\{\frac{5}{2},\frac{5}{2}\}$	18	19	20	21	22	23
$\{\frac{21}{8},\frac{21}{8}\}$	18.25	19.25	20.25	21.25	22.25	23.25
$\{\frac{11}{4},\frac{11}{4}\}$	18.5	19.5	20.5	21.5	22.5	23.5
$\{\frac{23}{8},\frac{23}{8}\}$	18.75	19.75	20.75	21.75	22.75	23.75
$\{3,3\}$	19	20	21	22	23	24
$\{\frac{25}{8},\frac{25}{8}\}$	19.25	20.25	21.25	22.25	23.25	24.25
$\{\frac{13}{4},\frac{13}{4}\}$	19.5	20.5	21.5	22.5	23.5	24.5
$\{\frac{27}{8},\frac{27}{8}\}$	19.75	20.75	21.75	22.75	23.75	24.75
$\{\frac{7}{2},\frac{7}{2}\}$	20	21	22	23	24	25
$\{\frac{29}{8},\frac{29}{8}\}$	20.25	21.25	22.25	23.25	24.25	25.25
$\{\frac{15}{4},\frac{15}{4}\}$	20.5	21.5	22.5	23.5	24.5	25.5
$\{\frac{31}{8},\frac{31}{8}\}$	20.75	21.75	22.75	23.75	24.75	25.75
$\{4,4\}$	21	22	23	24	25	26

Table 4. Numerical Results.

$t_i = \{t_i^1, t_i^2\}$	$y^{19}(t_i)$	$y^{20}(t_i)$	$y^{21}(t_i)$	$y^{22}(t_i)$	$y^{23}(t_i)$	$y^{24}(t_i)$
$\{0, 0\}$	19	20	21	22	23	24
$\{\frac{1}{8}, \frac{1}{8}\}$	19.25	20.25	21.25	22.25	23.25	24.25
$\{\frac{1}{4}, \frac{1}{4}\}$	19.5	20.5	21.5	22.5	23.5	24.5
$\{\frac{3}{8}, \frac{3}{8}\}$	19.75	20.75	21.75	22.75	23.75	24.75
$\{\frac{1}{2}, \frac{1}{2}\}$	20	21	22	23	24	25
$\{\frac{5}{8}, \frac{5}{8}\}$	20.25	21.25	22.25	23.25	24.25	25.25
$\{\frac{3}{4}, \frac{3}{4}\}$	20.5	21.5	22.5	23.5	24.5	25.5
$\{\frac{7}{8}, \frac{7}{8}\}$	20.75	21.75	22.75	23.75	24.75	25.75
$\{1, 1\}$	21	22	23	24	25	26
$\{\frac{9}{8}, \frac{9}{8}\}$	21.25	22.25	23.25	24.25	25.25	26.25
$\{\frac{5}{4}, \frac{5}{4}\}$	21.5	22.5	23.5	24.5	25.5	26.5
$\{\frac{11}{8}, \frac{11}{8}\}$	21.75	22.75	23.75	24.75	25.75	26.75
$\{\frac{3}{2}, \frac{3}{2}\}$	22	23	24	25	26	27
$\{\frac{13}{8}, \frac{13}{8}\}$	22.25	23.25	24.25	25.25	26.25	27.25
$\{\frac{7}{4}, \frac{7}{4}\}$	22.5	23.5	24.5	25.5	26.5	27.5
$\{\frac{15}{8}, \frac{15}{8}\}$	22.75	23.75	24.75	25.75	26.75	27.75
$\{2, 2\}$	23	24	25	26	27	28
$\{\frac{17}{8}, \frac{17}{8}\}$	23.25	24.25	25.25	26.25	27.25	28.25
$\{\frac{9}{4}, \frac{9}{4}\}$	23.5	24.5	25.5	26.5	27.5	28.5
$\{\frac{19}{8}, \frac{19}{8}\}$	23.75	24.75	25.75	26.75	27.75	28.75
$\{\frac{5}{2}, \frac{5}{2}\}$	24	25	26	27	28	29
$\{\frac{21}{8}, \frac{21}{8}\}$	24.25	25.25	26.25	27.25	28.25	29.25
$\{\frac{11}{4}, \frac{11}{4}\}$	24.5	25.5	26.5	27.5	28.5	29.5
$\{\frac{23}{8}, \frac{23}{8}\}$	24.75	25.75	26.75	27.75	28.75	29.75
$\{3, 3\}$	25	26	27	28	29	30
$\{\frac{25}{8}, \frac{25}{8}\}$	25.25	26.25	27.25	28.25	29.25	30.25
$\{\frac{13}{4}, \frac{13}{4}\}$	25.5	26.5	27.5	28.5	29.5	30.5
$\{\frac{27}{8}, \frac{27}{8}\}$	25.75	26.75	27.75	28.75	29.75	30.75
$\{\frac{7}{2}, \frac{7}{2}\}$	26	27	28	29	30	31
$\{\frac{29}{8}, \frac{29}{8}\}$	26.25	27.25	28.25	29.25	30.25	31.25
$\{\frac{15}{4}, \frac{15}{4}\}$	26.5	27.5	28.5	29.5	30.5	31.5
$\{\frac{31}{8}, \frac{31}{8}\}$	26.75	27.75	28.75	29.75	30.75	31.75
$\{4, 4\}$	27	28	29	30	31	32

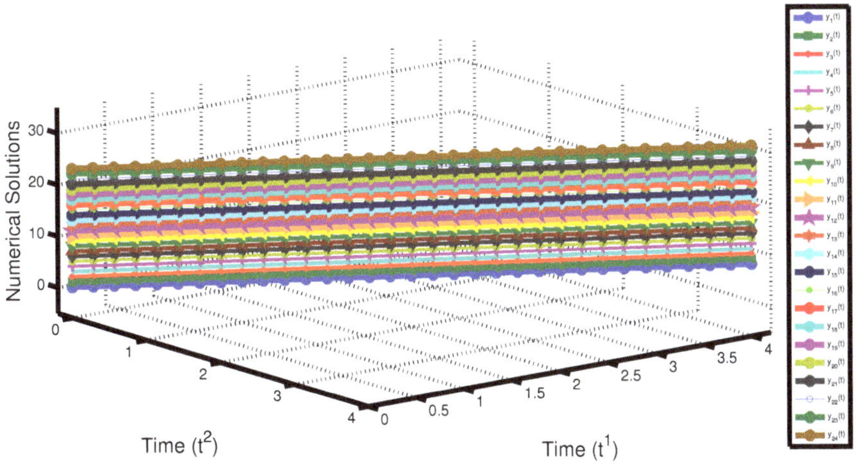

Figure 3. Nguyen traffic network pattern with 24 routes, i.e., $p = 24$.

7. Conclusions and Further Developments

This paper is a contribution to the field of noncooperative games. Using a multi-dimensional approach to time, we have first formulated a multi-time generalized Nash equilibrium problem and a multi-time quasi-variational inequality problem, and then we have established an equivalence between these two problems. Next, we have proved the existence of an equilibrium. As applications of our multi-time generalized Nash equilibrium problem, we have formulated a traffic network model for a courier service company with the aim of minimizing the traffic cost of routes and a river basin pollution problem in the terms of such problems. We have also provided a method for finding equilibria using projected dynamical system theory and have solved the well-known Nguyen traffic network problem by applying our method to it. Indeed, since the decision maker (the company) in this problem aims to minimize the total delivery cost, an optimization reformulation is perhaps more natural than a generalized Nash equilibrium problem (10). Nevertheless, it can be noted that both the feasible traffic flow set and the cost function of each route also depend on the traffic flow of other routes in the generalized Nash equilibrium model (10), but this scenario is not present in an optimization model. In essence, the outcomes of our generalized Nash equilibrium model (10) provide a new approach to solving traffic network problems. We also intend to develop more practical theories and experiments to ascertain that generalized Nash equilibrium models are more efficient than optimization models when applied to traffic network problems. Moreover, we also intend to further explore certain aspects of the river basin pollution problem.

Author Contributions: Conceptualization, S.S., A.G. and S.R.; methodology, S.S.; validation, S.S., A.G. and S.R.; writing—original draft preparation, S.S.; writing—review and editing, A.G. and S.R. All authors have read and agreed to the published version of the manuscript.

Funding: Simeon Reich was partially supported by the Israel Science Foundation (Grant No. 820/17), by the Fund for the Promotion of Research at the Technion and by the Technion General Research Fund.

Institutional Review Board Statement: Not applicable.

Informed Consent Statement: Not applicable.

Data Availability Statement: Not applicable.

Acknowledgments: All the authors are very grateful to two anonymous referees for their valuable suggestions and helpful comments, which have greatly improved the results and the presentation of the paper. In particular, our study has been enhanced by their drawing our attention to the river basin pollution problem.

Conflicts of Interest: The authors declare no conflict of interest.

References

1. Cournot, A.A. *Recherches sur les Principes Mathématiques de la théorie des Richesses*; Hachette: Paris, France, 1838.
2. Nash, J.F. Equilibrium points in n-person games. *Proc. Natl. Acad. Sci. USA* **1950**, *36*, 48–49. [CrossRef] [PubMed]
3. Nash, J.F. Non-cooperative games. *Ann. Math.* **1951**, *54*, 286–295. [CrossRef]
4. Arrow, K.J.; Debreu, G. Existence of an equilibrium for a competitive economy. *Econometrica* **1954**, *22*, 265–290. [CrossRef]
5. Rosen, J.B. Existence and uniqueness of equilibrium points for concave n-person games. *Econometrica* **1965**, *33*, 520–534. [CrossRef]
6. Kesselman, A.; Leonardi, S.; Bonifaci, V. Game-theoretic analysis of internet switching with selfish users. In *Internet and Network Economics*; Lecture Notes in Computer Science; Deng, X., Ye, Y., Eds.; Springer: Berlin/Heidelberg, Germany, 2005; Volume 3828.
7. Pang, J.-S.; Scutari, G.; Facchinei, F.; Wang, C. Distributed power allocation with rate constraints in Gaussian parallel interference channels. *IEEE Trans. Inf. Theory* **2008**, *54*, 3471–3489. [CrossRef]
8. Facchinei, F.; Kanzow, C. Generalized Nash equilibrium problems. *Ann. Oper. Res.* **2010**, *175*, 177–211. [CrossRef]
9. Fischer, A.; Herrich, M.; Schönefeld, K. Generalized Nash equilibrium problems-recent advances and challenges. *Pesqui. Oper.* **2014**, *34*, 521–558. [CrossRef]
10. Smith, M.J. The existence, uniqueness and stability of traffic equilibrium. *Transp. Res.* **1979**, *13*, 295–304. [CrossRef]
11. Dafermos, S. Traffic equilibrium and variational inequalities. *Transp. Sci.* **1980**, *14*, 42–54. [CrossRef]
12. Cojocaru, M.G.; Daniele, P.; Nagurney, A. Projected dynamical systems and evolutionary variational inequalities via Hilbert spaces with applications. *J. Optim. Theory Appl.* **2005**, *127*, 549–563. [CrossRef]
13. Nagurney, A. *Network Economics: A Variational Inequality Approach*; Springer Science and Business Media: Berlin/Heidelberg, Germany, 2013.
14. Facchinei, F.; Pang, J.S. *Finite-Dimensional Variational Inequalities and Complementarity Problems*; Springer Science and Business Media: Berlin/Heidelberg, Germany, 2007.
15. Barbagallo, A.; Cojocaru, M.G. Dynamic equilibrium formulation of the oligopolistic market problem. *Math. Comput. Model.* **2009**, *49*, 966–976. [CrossRef]
16. Cojocaru, M.G.; Bauch, C.T.; Johnston M.D. Dynamics of vaccination strategies via projected dynamical systems. *Bull. Math. Biol.* **2007**, *69*, 453–476. [CrossRef]
17. Harker, P.T. Generalized Nash games and quasi-variational inequalities. *Eur. J. Oper. Res.* **1991**, *54*, 81–94. [CrossRef]
18. Bensoussan, A. Points de Nash dans le cas de fontionnelles quadratiques et jeux differentiels linéaires a N personnes. *SIAM J. Control* **1974**, *12*, 460–499. [CrossRef]
19. Aussel, D.; Gupta, R.; Mehra, A. Evolutionary variational inequality formulation of the generalized Nash equilibrium problem. *J. Optim. Theory Appl.* **2016**, *169*, 74–90. [CrossRef]
20. Migot, T.; Cojocaru, M.-G. Nonsmooth dynamics of generalized Nash games. *J. Nonlinear Var. Anal.* **2020**, *4*, 27–44.
21. Pang, J.-S.; Fukushima, M. Quasi-variational inequalities, generalized Nash equilibria, and multi-leader-follower games. *Comput. Manag. Sci.* **2005**, *2*, 21–56. [CrossRef]
22. Shehu, Y.; Gibali, A.; Sagratella, S. Inertial projection-type methods for solving quasi-variational inequalities in real Hilbert spaces. *J. Optim. Theory Appl.* **2020**, *184*, 877–894. [CrossRef]
23. Udrişte, C.; Ţevy, I. Multi-time Euler-Lagrange-Hamilton theory. *WSEAS Trans. Math.* **2007**, *6*, 701–709.
24. Udrişte, C. Multitime controllability, observability and bang-bang principle. *J. Optim. Theory Appl.* **2008**, *139*, 141–157. [CrossRef]
25. Udrişte, C. Simplified multitime maximum principle. *Balkan J. Geom. Appl.* **2009**, *14*, 102–119.
26. Udrişte, C.; Dogaru, O.; Ţevy, I. Null Lagrangian forms and Euler-Lagrange PDEs. *J. Math. Study* **2008**, *1*, 143–156.
27. Udrişte, C.; Ferrara, M. Multitime models of optimal growth. *WSEAS Trans. Math.* **2008**, *7*, 51–55.
28. Singh, S.; Pitea, A.; Qin, X. An iterative method and weak sharp solutions for multitime-type variational inequalities. *Appl. Anal.* **2019**. [CrossRef]
29. Singh, S.; Reich, S. A multidimensional approach to traffic analysis. *Pure Appl. Funct. Anal.* **2021**, *6*, 383–397.
30. Treanţă, S.; Singh, S. Weak sharp solutions associated with a multidimensional variational-type inequality. *Positivity* **2020**. [CrossRef]
31. Nguyen, S. *Estimating Origin Destination Matrices from Observed Flows*; Elsevier Science Publishers BV: Amsterdam, The Netherlands, 1984.
32. Jayswal, A.; Singh, S.; Kurdi, A. Multitime multiobjective variational problems and vector variational-like inequalities. *Eur. J. Oper. Res.* **2016**, *254*, 739–745. [CrossRef]
33. Mititelu, Ş.; Treanţă, S. Efficiency conditions in vector control problems governed by multiple integrals. *J. Appl. Math. Comput.* **2018**, *57*, 647–665. [CrossRef]
34. Ansari, Q.H.; Schaible, S.; Yao, J.C. Generalized vector quasi-variational inequality problems over product sets. *J. Glob. Optim.* **2005**, *32*, 437–449. [CrossRef]

35. Ding, X.P. Existence of solutions for quasi-equilibrium problems in noncompact topological spaces. *Comput. Math. Appl.* **2000**, *39*, 13–21. [CrossRef]
36. Nagurney, A.; Liu, Z.; Cojocaru, M.-G.; Daniele, P. Dynamic electric power supply chains and transportation networks: An evolutionary variational inequality formulation. *Transp. Res. E Logist. Transp. Rev.* **2007**, *43*, 624–646. [CrossRef]
37. Daniele, P. Evolutionary variational inequalities and applications to complex dynamic multi-level models. *Transp. Res. E Logist. Transp. Rev.* **2010**, *46*, 855–880. [CrossRef]
38. Haurie, A.; Krawczyk, J.B. Optimal charges on river effluent from lumped and distributed sources. *Environ. Model. Assess.* **1997**, *2*, 177–189. [CrossRef]
39. Krawczyk, J.B. Coupled constraint Nash equilibria in environmental games. *Resour. Energy Econ.* **2005**, *27*, 157–181. [CrossRef]
40. Krawczyk, J.B.; Uryasev, S. Relaxation algorithms to find Nash equilibria with economic applications. *Environ. Model. Assess.* **2000**, *5*, 63–73. [CrossRef]
41. Kubota, K.; Fukushima, M. Gap function approach to the generalized Nash equilibrium problem. *J. Optim. Theory Appl.* **2010**, *144*, 511–531. [CrossRef]
42. Cojocaru, M.G.; Jonker, L.B. Existence of solutions to projected differential equations in Hilbert spaces. *Proc. Am. Math. Soc.* **2003**, *132*, 183–193. [CrossRef]
43. Cegielski, A. *Iterative Methods for Fixed Point Problems in Hilbert Spaces*; Springer: Heidelberg, Germany, 2012.
44. Rockafellar, R.T.; Wets, R.J.-B. *Variational Analysis*; Springer Science and Business Media: Berlin/Heidelberg, Germany, 2009.

mathematics

MDPI

Article

A Scheduling Approach for the Combination Scheme and Train Timetable of a Heavy-Haul Railway

Hanxiao Zhou [1], Leishan Zhou [1], Bin Guo [2,*], Zixi Bai [3], Zeyu Wang [1] and Lu Yang [1]

1 School of Traffic and Transportation, Beijing Jiaotong University, Beijing 100044, China; 16114181@bjtu.edu.cn (H.Z.); lshzhou@bjtu.edu.cn (L.Z.); 18114045@bjtu.edu.cn (Z.W.); 15114207@bjtu.edu.cn (L.Y.)
2 State Research Center of Rail Transit Technology Education and Service, Beijing Jiaotong University, Beijing 100044, China
3 Beijing Key Laboratory of Traffic Engineering, Faculty of Architecture, Civil and Transportation Engineering, Beijing University of Technology, Beijing 100124, China; zixibai@bjut.edu.cn
* Correspondence: guobin@bjtu.edu.cn; Tel.: +86-138-1029-3178

Abstract: Heavy-haul railway transport is a critical mode of regional bulk cargo transport. It dramatically improves the freight transport capacity of railway lines by combining several unit trains into one combined train. In order to improve the efficiency of the heavy-haul transport system and reduce the transportation cost, a critical problem involves arranging the combination scheme in the combination station (CBS) and scheduling the train timetable along the trains' journey. With this consideration, this paper establishes two integer programming models in stages involving the train service plan problem (TSPP) model and train timetabling problem (TTP) model. The TSPP model aims to obtain a train service plan according to the freight demands by minimizing the operation cost. Based on the train service plan, the TTP model is to simultaneously schedule the combination scheme and train timetable, considering the utilization optimal for the CBS. Then, an effective hybrid genetic algorithm (HGA) is designed to solve the model and obtain the combination scheme and train timetable. Finally, some experiments are implemented to illustrate the feasibility of the proposed approaches and demonstrate the effectiveness of the HGA.

Keywords: freight transportation; heavy-haul railway; combination scheme; train timetable; genetic algorithm

Citation: Zhou, H.; Zhou, L.; Guo, B.; Bai, Z.; Wang, Z.; Yang, L. A Scheduling Approach for the Combination Scheme and Train Timetable of a Heavy-Haul Railway. *Mathematics* **2021**, *9*, 3068. https://doi.org/10.3390/math9233068

Academic Editors: Humberto Rocha and Ana Maria Rocha

Received: 7 November 2021
Accepted: 26 November 2021
Published: 29 November 2021

1. Introduction

1.1. Background

Heavy-haul railways have been the backbone of the coal transportation system because of their high capacity [1] and high efficiency for a long time. With the continuous growth of freight volume, the number of trains in the heavy-haul railway system keeps increasing, which brings great difficulties to the transportation plans' scheduling and operation management in both stations and trunk lines. Therefore, one of the emergency issues that railway operators are concerned about is how to schedule transportation plans to reduce operating costs while ensuring transportation demand.

After North American railways took the lead in adopting heavy-haul railway transportation in the 1950s, this transportation mode quickly adapted to the needs of bulk cargo transport such as for coal, ore, etc., and developed rapidly in the world. The United States, Canada, Russia, Brazil, China, South Africa, Australia, Sweden, and more than ten other countries have carried out heavy-haul railway transportation. The heavy-haul railway transportation mode dramatically improves the freight transport capacity of railway lines by combining several unit trains into one combined train. Taking China as an example, the supply and demand for coal are extremely uneven geographically. The eastern region of China has been economically developed and its industrial production has a great need for coal resources. However, coal resources are mainly distributed in the western region [2]. Under this particular distribution pattern of natural resources, developing a heavy-haul railway can effectively guarantee coal supply and transportation.

With the rapid growth of the demand for heavy-haul transportation, railway operators have proposed a series of measures to guarantee the transport capacity of the heavy-haul railway, including increasing the load of combined trains, running different types of combined trains, and reducing the headway of adjacent trains. Under the condition that railway infrastructure cannot be revolutionarily upgraded, the flexible organization of several types of combined trains is one of the most effective ways for railway operators to satisfy the freight demand at a lower cost. From a transportation system-wide perspective, a reasonable heavy-haul railway transportation plan should meet the transportation demand, reduce the station operation workload, and reduce the operation cost.

A complete heavy-haul railway transportation plan consists of several sub-plans. It includes the train service plan, train combination scheme, train timetable, locomotive circulation scheme, empty train return scheme, decomposition scheme, etc. The heavy-haul railway operators are involved in various steps of the decision-making process to obtain a complete heavy-haul railway transportation plan. Six main stages in the decision-making process of the heavy-haul railway transportation plan are listed below.

(1) Collect the transport demand. The railway operators collect the buyers' transport requirements, including the number and the destination of the required bulk goods.

(2) Schedule the train service plan. Determine the number and type of heavy-haul trains running between each station pair under the given unloading station requirements and loading station capacity conditions.

(3) Schedule the train combination scheme. In a heavy-haul railway system, the unit trains must be combined at a combination station (CBS) to run towards the unloading stations. Therefore, a detailed combination scheme should be arranged for the unit trains in the CBS.

(4) Schedule the train timetable. Based on the train combination scheme, determine all trains' arrival and departure times at each station.

(5) Schedule the locomotive circulation scheme and train maintenance scheme. Heavy-haul trains generally require multi-locomotive traction. Set up the trains' locomotive circulation and train maintenance schemes according to the preset locomotive routing and maintenance procedures.

(6) Schedule the empty train return scheme and the decomposition scheme. After the combined trains' unloading at the unloading stations, the empty trains must return to the CBS for decomposition operations and return to the loading stations.

This paper focuses on developing a joint scheduling approach to schedule the (2)–(4) stages listed above. The decision-making process of the (2)–(4) stages is a complex process guided by transport demand. Figure 1 illustrates the relations between the stages in the decision-making process. It starts by collecting the loading capacity of each loading station and the requests for each unloading station's demand. Then, heavy-haul railway operators need to generate executable combination schemes and train timetables based on the collected information according to the actual situation of the heavy-haul railway. The decision-making process at these stages needs to comprehensively consider all kinds of information of the heavy-haul railway transportation system and reflect it into the transportation plan.

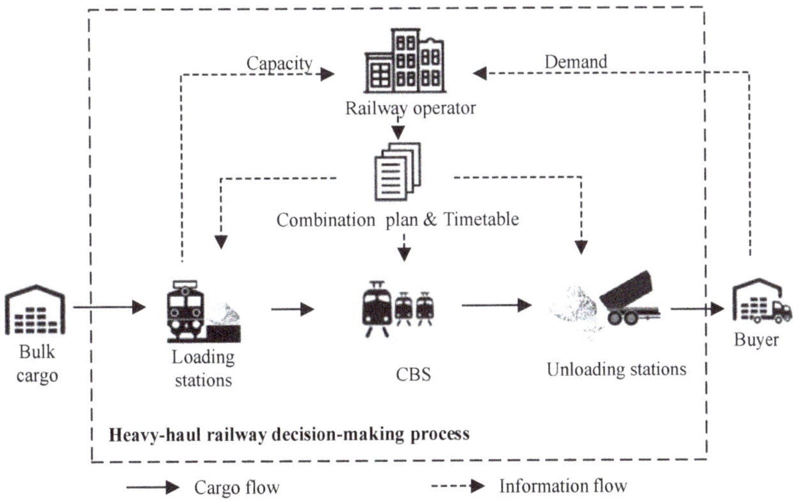

Figure 1. The relations between stages in the decision-making process of the heavy-haul railway transportation plan.

1.2. Literature Review

In the early stage, heavy-haul transportation was mainly carried out to relieve the tight transportation capacity on busy trunk lines. Thus, the research focused mainly on freight transportation organization from a macro perspective. Fan [3] took the selectivity of the traffic flow path, the line capacity, and the unloading area of the heavy-haul railway into the optimization process and constructed an optional optimization model for the loading area. Yang [4] considered the logical matching relationship for the weight, speed, and density of heavy-haul railways to be a critical factor in freight railways and the matching relations of these three elements should be assured according to the country economy, society needs, and routes' essential condition. Ma [5] discussed the external and internal factors that affect heavy-haul railways. The external factors include economic development, transportation price, and production factor price. In contrast, the internal factors include loading capacity, station capacity, unloading capacity, and maintenance capacity. Sun [6] analyzed the traffic capacity of coal transportation in the Baotou-Shenmu Railway and suggested strengthening the carrying capacity according to the present and further freight volume. Zhou [7] put forward strategies to improve the heavy-haul railway carrying capacity of the Baotou-Shenmu Railway in stations, improve train transportation efficiency and traction quality, and optimize the heavy-haul transportation organization plan, among other strategies as well.

With the increase of the freight volume of the heavy-haul railway, scholars have begun discussing the organization of the railcar flow in heavy-haul transport systems, focusing mainly on the relationship between the demand of cargo and the handling capacity of the railway system. Xue [8] proposed a method calculating the coupling degree between the station stages plan and the given dynamic railcar flow. Wang [9] studied the heavy-haul train operation plan problem with a multi-objective programming model, taking the prescriptive transportation volume index, the capacity of the railway line, and the available locomotives as the constraints. Zhao [10] established an optimization model of railcar flow organization in the loading area of the heavy-haul railway to minimize the combination time consumption and to maximize the flow of the heavy-haul railway. Then, he solved the model with the minimum cost and maximum flow algorithm. Tang [11] also built a minimum cost maximum flow model, of which the optimization goal is to maximize the operation time saved by direct transportation. Wang [12] focused on the cost of cargo flow in transit and established a 0–1 non-linear programming model with the objective of achieving the least cost. Jing [13] established the optimization model of direct flow in the loading area with the least operation time and lowest running consumption.

The increasing freight volume also increased the workload of stations in the heavy-haul railway systems. However, the manual approaches are still widely used in the actual railway operation, which are time-consuming and highly rely on the experience of the design managers [14]. Thus, scholars also paid attention to the organizational optimization of heavy-haul railway stations. The station technical operation is an essential part of the heavy-haul railway transportation organization. The stations of heavy-haul railways can be divided into four types: loading stations, CBS, unloading stations, and intermediate stations [15], which mainly handle the loading, unloading, combined, and decomposed operations. Among these types, the CBS is primarily responsible for the combined operations of heavy trains based on the combination scheme [16]. Hence, it is an important node of the heavy-haul railway at the macro level. After realizing the critical role of station technical schemes in the operation and the management of heavy-haul railway systems, scholars began to focus on the research about the CBS.

Along this line, Wei [17] studied the influence of turnout selection on the passing capacity of heavy-haul railway stations, the additional start or stop time of trains, the period of the train timetable, and the headway of tracks. Ye [18] analyzed the reasons for the station's passing capacity based on the operation of 20,000 tons of combined trains in Hudong Station. He put forward an optimized organization plan to ensure heavy-haul trains' safe transportation in Hudong Station. Liang [19] calculated the time that the trains occupied the station track by analyzing the operation process of the combined trains in the CBS and checked the number of the station arrival–departure tracks setting in Hudong Station. Tang [11] studied the operation organization of Hudong Station on the Daqin Railway and made an in-depth analysis of the organization mode of train queuing. In the cargo flow organization model established by him, the optimization goal is to maximize the adaptation time saved by direct transportation. Under constraints of the combination regulations, train's weight, the latest permissible time for the combination of the departure trains, etc., the non-linear 0–1 programming model was established by Han [20] with the objective of the minimum station dwell time and decomposition time of the heavy-haul train at CBS. Ma [21] constructed the linear 0–1 programming model and quadratic 0–1 programming model, setting the utilization equilibrium and the track's selection tendency as the objective functions.

The heavy-haul railway transport is a particular pattern of railway freight transport, which is different from the general passenger and freight railway transport. From the whole network scale perspective, the heavy-haul railway is a radial tree-shaped network with strong system independence rather than a large-scale network structure. From the view of station operation in the technical station, the heavy-haul trains need to be combined or decomposed in the CBS with unit trains as the minimum unit, which is similar to the freight train's marshaling. The scheduling of the combined scheme in the CBS of the heavy-haul railway can be regarded as a special case of the train formation problem (TFP). The TFP determines the routing and frequency of trains and assigns the demands to trains [22].

The TFP has received research attention from a mathematical point of view. Therefore, the transportation organization of heavy-haul railway stations can also learn from some research studies based on marshaling yards. For example, a neural network is examined for obtaining good solutions in short time periods for the TFP by Martinelli [23]. Xiao [14] established the comprehensive optimization model of the train formulation plan using both the single-block trains and two-block trains, aimed at the minimization of the total car–hour consumption at all yards. As for a discrete deterministic-controlled system that simulates the operation of the flat yard, Kozachenko [24] obtained a mathematical statement of the problem of choosing the optimal order of multi-group train formation. Lin [25] built a bi-level linear integer model to solve the train service network problem of the Chinese railway system. Later, he [26] presented the formulation of a train formation problem in rail loading stations from a systematic perspective. Yaghini [27] proposed a hybrid algorithm of the simplex method and simulated annealing for the train formation problem. Murali [28] presented a decision tool to aid train planners to obtain good quality routes and schedules quickly for short-time horizons. Lazarev [29] provided the integer linear programming statement to form trains and define both their routes and schedules by minimizing the total weighted delivery time of all orders.

There are mainly two differences between heavy-haul railway transport and general freight railway transport. Firstly, at the beginning of the transport, the loading stations need to load a particular length of unit trains without specifying the destination of each railcar. In contrast, in the general freight railway system, different railcars have specific destinations, thus railway operators need to combine the railcars with similar destinations into a train. Secondly, at the end of the transport, considering that the unloading stations in the heavy-haul railway have a large unloading capacity, the railway operators transport the combined trains into specific unloading stations directly.

To our knowledge, however, there are few demand-oriented scheduling approaches of heavy-haul railway transportation plans. This paper is particularly interested in proposing a new approach to simultaneously schedule the combination scheme and heavy-haul train timetable on a heavy-haul railway considering the demand of the unloading area.

The rest of this paper is organized as follows. Section 2 introduces the general form of a heavy-haul railway network and the operation process of trains in the heavy-haul transport system. We also explain several terms of the heavy-haul transport system. Section 3 describes the problem and proposes two optimization models to solve the heavy-haul railway's combination scheme and train timetable. Section 4 designs a hybrid algorithm based on the genetic algorithm framework to solve the proposed model. In Section 5, a small case and a real-world case are solved by the proposed approach to prove the method's effectiveness. Section 6 presents some conclusions.

2. Conceptual Illustration

This section will introduce the composition of a heavy-haul railway and the operation process of trains in the railway system.

2.1. Composition of the Heavy-Haul Railway

The heavy-haul railway usually connects the railway to the production side and the demand side of bulk cargo like a corridor. Thus, it is also called the heavy-haul railway corridor. It has the characteristics of high independence and large carrying capacity. The heavy-haul railway corridor is usually a special freight railway line used to transport a particular type of bulk cargo. This kind of heavy-haul railway generally transports a single type of cargo and operates massive railcars' flow every day.

Figure 2 is a simplified schematic diagram of a heavy-haul railway. As shown in the figure, a typical heavy-haul railway can be divided into three parts: loading area, transport area, and unloading area.

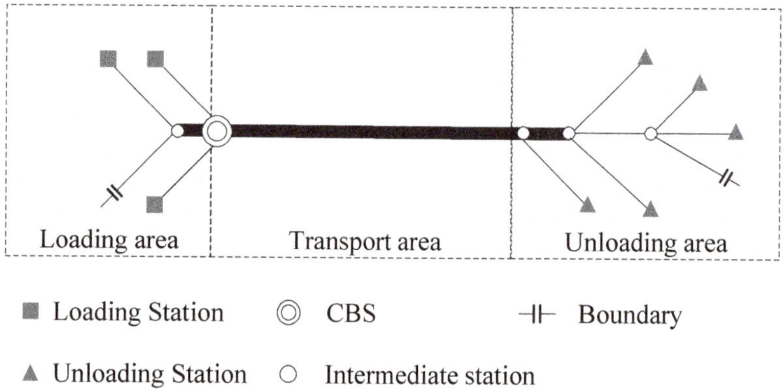

Figure 2. Simplified schematic diagram of a heavy-haul railway.

1. Loading area

The loading area contains two types of technical stations. One is the loading station, which is the place where the unit trains are loaded. The loading stations in the loading area mainly connect to the trunk line through branch railway lines. Empty railcars are loaded at loading stations and become heavy unit trains (in the following text, we call it unit trains). Part of the heavy unit trains also directly turn into the heavy-haul railway through the boundary (the boundary connects the heavy-haul railway to other railway lines). Another technical station is the CBS, the most critical technical station in the heavy-haul railway system. The unit trains would be sent here for combined operation. The unit trains are assembled here and combined into combined trains for the unloading area.

2. Transport area

The transport area is the trunk line between the CBS and the unloading area. According to the train timetable, the combined trains run on the trunk line to transport the bulk cargo to the unloading area.

3. Unloading area

The unloading area is the end of a heavy-haul railway. Here, the buyer submits the demand and hands over the bulk cargo. Part of the branch line in the unloading area is not directly connected to the unloading station but is a boundary connecting other railway lines. The combined trains will arrive at the unloading stations to unload the bulk cargo or to be transferred from the boundary to other places. The combined trains become empty trains after unloading at the unloading station.

For the sake of simplification, this paper considers that all unit trains will be combined in the CBS. However, there were also a number of direct trains on the heavy-haul railways that only need a simple inspection operation at the CBS. As running trains directly connect to the destination, it is helpful to reduce the workload of the CBS [30], and some loading stations load and send specific types of combined trains. These trains do not need combined operation when through the CBS. For this scenario, when these direct trains run between the loading area and the CBS, we regard them as unit trains. When they run between the CBS and unloading area, we regard them as combined trains. We also treat the inspection operation of direct trains in CBS as a combined operation.

2.2. Operating Process of Trains

In order to intuitively introduce the operational process of unit trains and the combined trains in a heavy-haul railway system, an illustration of a small heavy-haul railway network is given below.

As shown in Figure 3, A is the CBS in a small-scale heavy-haul railway network. The unloading station d requires 10 kt of coal. Therefore, two unit trains with 5 kt loads are sent to the unloading station d from both loading station a and loading station b. Unit train u_1 (blue dotted arcs) departs from loading station a, and unit train u_2 (red dotted arcs) departs from loading station b. At the CBS A, u_1 and u_2 are combined into a combined train c_1 (purple arcs) with a 10 kt total load.

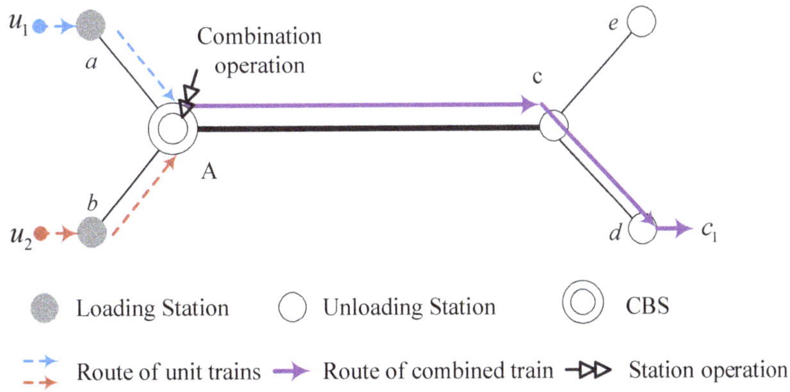

Figure 3. The operating process of trains in a heavy-haul railway network.

When we extend this physical railway network on the time axis, we can find more temporal details from the time–geography perspective. As shown in Figure 4, the vertical axis corresponds to time and we discretize consecutive periods into small increments. The benefit of adopting a space–time network framework is to precisely capture the temporal and spatial interaction of the transportation system [31]. In this way, we can intuitively understand the moving trajectory of the trains in the heavy-haul railway. The process of train operation in the heavy-haul railway can be divided into two types: travel in the section and station technical operation.

4. Travel in the section

The straight arcs in Figure 4 are the space–time trajectories of the trains. They show the spatial and temporal displacement of the train in the sections. The start of the straight arcs means the section's origin station and the train's departure time. The end of the straight arcs means the section's destination station and the train's arrival time. Thus, the duration of the straight arcs reflects the train's running time in the section. The running time of the trains in a section is generally constant because the trains do not need to perform additional operations in the section.

5. Station technical operation

Train station operations in the heavy-haul railway include loading operations at the loading station, unloading operations at the unloading station, and technical operations at the CBS. After arriving at the CBS, the unit train needs to connect other pre-designated unit trains as a combined train. The whole combined operation includes the process of arriving, changing locomotives, connecting trains, and train inspection. A combined operation usually takes more than 2 h. However, the direct trains only need to carry out the train inspection at the CBS rather than participate in the combined operation. Here, we treat train inspection as a special combination of operations. All heavy-haul trains need to finish the technical operation at the CBS before they can be dispatched to the trunk line.

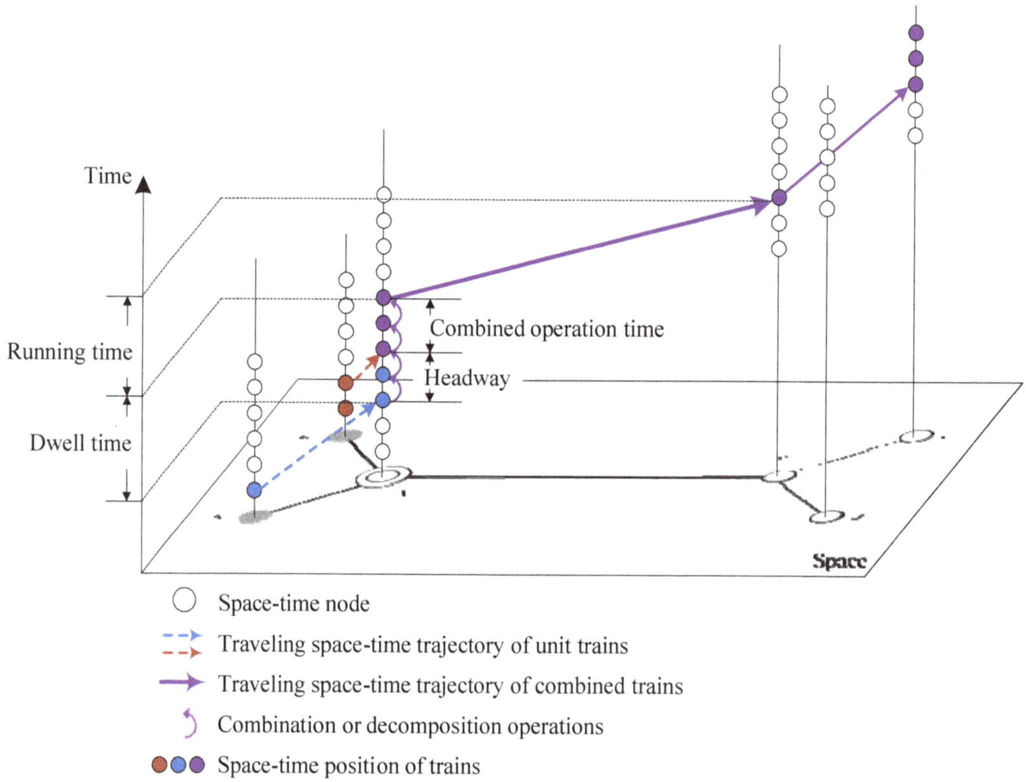

Figure 4. The operating process of heavy-haul trains in a space–time network.

As shown in Figure 4, the duration of a unit train's staying in the CBS is the dwell time from the unit train's arrival to its departure after combining into a combined train, including the headway and all combined operation times. Obviously, the longer unit trains stay in the CBS, the longer cargo will be in transit, increasing delivery time. At the same time, the unit trains occupy the resources such as tracks in the CBS, which will reduce the station's utilization. Therefore, in actual operation, we hope to reduce the total dwell time of unit trains in the CBS as much as possible so as to improve the efficiency of CBS operations and to reduce the delivery time of cargo.

The technical operations of the CBS are based on a combination scheme. The combination scheme will specify when the unit trains arrive at the CBS, will assign the specific constituent unit trains for the combined train, and will specify when the combined train will be dispatched from the CBS. The formulation of the combination scheme determines the total dwell time of the unit trains in the CBS. Therefore, optimizing the combination scheme is the key point to improve the efficiency of the heavy-haul transport system.

2.3. Framework

This paper proposes a new methodology using a hybrid genetic algorithm to simultaneously account for cargo transportation demand and station technical operation to optimize both train combination scheme and timetable. Thus, the heavy-haul transport system can meet the transportation service at a lower cost. The framework of our proposed methodology is illustrated in Figure 5.

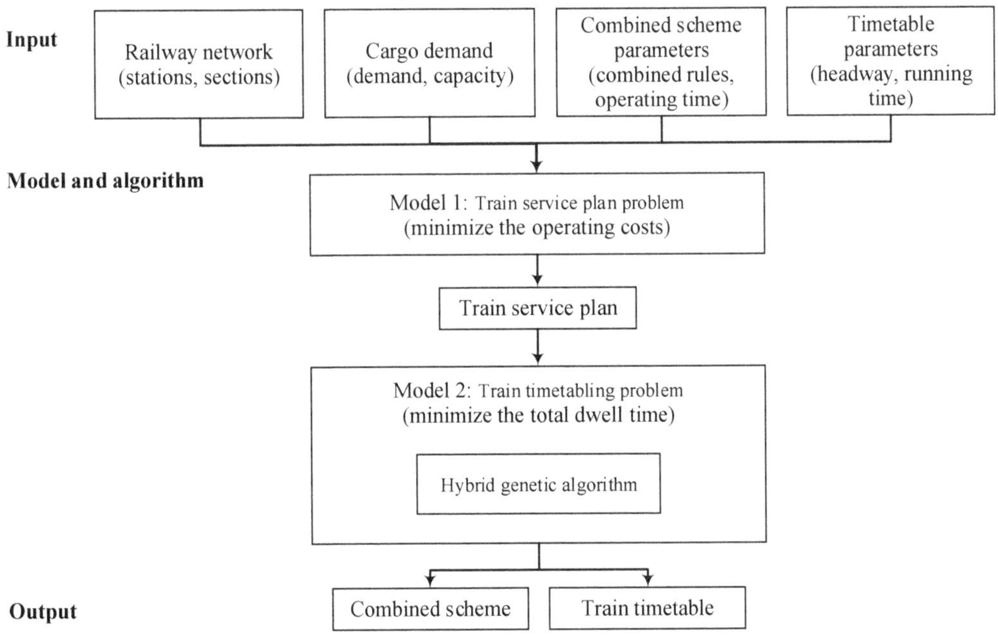

Figure 5. The framework of the heavy-haul train timetabling methodology.

To simplify the problem, we divide the heavy-haul railway transportation plan problem into two stages. In the first stage, the heavy-haul train service plan is designed. The cargo demand and heavy-haul railway's operating conditions determine the number of each type of heavily loaded train between different stations. The second stage is to schedule an appropriate combination scheme and train timetable for the heavy-haul train service plan generated in the first stage.

3. Optimization Model

To provide high-performance transportation services, railway operators need to optimize the heavy-haul railway transportation plan based on the required demand. In this section, we describe the train service plan problem (TSPP) and the train timetabling problem (TTP). Table 1 lists the sets, indices, and parameters used in this paper.

Table 1. Sets, indices, and parameters.

Symbol	Definition
T	The set of combined trains and unit trains
T^{com}	The set of combined trains, $T^{com} \in T$
T^{uni}	The set of unit trains, $T^{uni} \in T$
S	The set of stations
V	The set of sections
K	The set of combined train types
U	The set of unit train types
p	The index of combined trains, $p \in T^{com}$
q	The index of unit trains, $q \in T^{uni}$
i, j	The index of stations, $i, j \in S$
s_z	The label of the CBS, $s_z \in S$

Table 1. *Cont.*

Symbol	Definition
S_{loa}	The set of loading stations, $S_{loa} \in S$
S_{unl}	The set of unloading stations, $S_{unl} \in S$
(i,j)	The label of the section that connects station i and station j, $(i,j) \in V$
k	The index of combined train types, $k \in K$
u	The index of unit train types, $u \in U$
$p(i,j)$	$p(i,j) \in \{0,1\}$; if combined train p, choose the section (i,j), $p(i,j) = 1$, otherwise $p(i,j)= 0$
$q(i,j)$	$q(i,j) \in \{0,1\}$; if unit train q, choose the section (i,j), $q(i,j) = 1$, otherwise $q(i,j)= 0$
p_k	$p_k \in \{0,1\}$; if the type of combined train is k, $p_k = 1$, otherwise $p_k= 0$
I_k	The minimum headway of the combined train when its type is k
I_u	The minimum headway of the unit train when its type is u
q_u	$q_u \in \{0,1\}$; if the type of unit train q is u, $q_u = 1$, otherwise $q_u= 0$
$\varphi_{k,u}$	The number of u-type unit trains required to combine a k-type combined train, $\varphi_{k,u} \in \mathbb{Z}$
t_k^{tec}	The total dwell time of the k-type combined train in the CBS
$r(i,j)$	The running time of trains running on the section (i,j)
w_u	The number of railcars in a u-type unit train
l_i^{cap}	The unloading capacity of the unloading station i
l_i^{dem}	The demand for unloading station i
μ_k	The cycle length of train turn-around
M	A big enough positive number
c_k	The cost of running a k-type combined train
λ	The section capacity utilization
t_λ	Longest service time of the railway line
D^{start}, D^{end}	The allowable starting and ending time for combined trains
A^{start}, A^{end}	The allowable starting and ending time for unit trains

3.1. Train Service Plan Problem (TSPP) Model

The heavy-haul railway train service plan should include the following parts: the type and quantity of the unit trains departing from each loading station, and the type and quantity of the combined trains arriving at each unloading station.

In a heavy-haul railway system, the combined trains take on the transport between the CBS and unloading stations. The combined trains require different operation times and human resources, and cause different track wear due to their different total loads. Thus, the operating cost of the different types of the combined trains is different. However, the heavy-haul railway train service plan involves a key criterion, which is railway operators' profitability [32]. To this end, we introduce an optimization model that intends to minimize the total cost requirements. The problem also aims to determine the number of u-type unit trains departing from loading station i to the CBS, which can be defined by variable y_i^u, and to determine the number of k-type combined trains departing from the CBS to unloading station j, which can be defined by variable x_j^k. Table 2 lists the two types of variables used in this model.

Table 2. Decision variables for the TSPP model.

Symbol	Definition
y_i^u	The number of u-type unit trains departing from loading station i to the CBS, $i \in S_{loa}$
x_j^k	The number of k-type combined trains departing from the CBS to unloading station j, $j \in S_{unl}$

1. Problem statement

A heavy-haul railway system has several loading and unloading stations. Given the loading stations' capacity, the unloading stations' demand, and the unloading capacity. Find a train service plan to meet the cargo demand to make the total operation cost minimal.

2. Objective function

The objective function is to minimize the operating costs of the heavy-haul railway. c_k denotes the cost of running a k-type combined train. The parameter c_k can be calibrated by the experience of the railway operator.

$$\min z = \sum_{j \in S_{unl}} \sum_{k} c_k x_j^k \tag{1}$$

3. Demand constraints

Constraint (2) ensures that the cargo that can be transported by the combined trains, arriving at the unloading station within one day, shall meet the demand for the cargo at the unloading station.

$$\sum_{k} \sum_{u} x_j^k \varphi_{k,u} w_u \geq l_j^{dem}, \forall j \in S_{unl} \tag{2}$$

4. Unloading capacity constraints

Due to the limited unloading capacity of each unloading station, the quantity of cargo to be carried by the combined trains arriving at each unloading station shall be less than the daily unloading capacity of the unloading station.

$$\sum_{k} \sum_{u} x_j^k \varphi_{k,u} w_u \leq l_j^{cap}, \forall j \in S_{unl} \tag{3}$$

5. Loading capacity constraints

Due to the limited loading capacity of each loading station, the number of unit trains departing from each loading station shall be less than the daily loading capacity of the loading station.

$$\sum_{u} y_i^u w_u \leq l_i^{cap}, \forall i \in S_{loa} \tag{4}$$

6. Section carrying capacity constraints

In a heavy-haul railway's trunk line and branch lines, the headway between any two adjacent trains should be greater than the minimum headway. As a result, the number of trains going through the heavy-haul railway every day is limited. Therefore, the number of unit trains or combined trains passing through any section should be less than the section carrying capacity.

$$\sum_{j} \sum_{k} I_k y_j^k \leq t_\lambda \tag{5}$$

7. Flow balance constraint for the CBS

The unit trains arriving at the CBS are combined into the combined trains and sent into the trunk line. The CBS cannot be used as a storage place for trains, thus the number of railcars arriving at the CBS is equal to the railcars departing from it.

$$\sum_{j \in S_{unl}} \sum_{k} x_j^k \varphi_{k,u} = \sum_{i \in S_{loa}} y_i^u, \forall u \in U \tag{6}$$

3.2. Train Timetabling Problem (TTP) Model

In our approach, the result of the TSPP model provides input parameters related to the train timetable. Since we already know the number of different types of unit trains and combined trains that need to be operated in the heavy-haul railway, we need to further develop appropriate schedules for these trains.

The goal of the TTP model is to schedule the trains to minimize the total dwell time in the CBS. The model also aims to determine the combination relationship between the unit trains and combined trains, which can be defined by variable $\theta_{p,q}$. Here, we use two groups of decision variables. The unit train schedule variable $d_q(i,j)$ and $a_q(i,j)$ denote the departure time and arrival time of unit train q in station i and station j. The combined train schedule variable $d_p(i,j)$ and $a_p(i,j)$ denote the departure time and the arrival time of combined train p in station i and station j. Table 3 lists the variables used in this model.

Table 3. Decision variables for the TTP model.

Symbol	Definition
$a_p(i,j)$	The arrival time of combined train p arriving at station j from section (i,j)
$a_q(i,j)$	The arrival time of unit train q arriving at station j from section (i,j)
$d_p(i,j)$	The departure time of combined train p departing from station i to section (i,j)
$d_q(i,j)$	The departure time of unit train q departing from station i to section (i,j)
$\theta_{p,q}$	Unit train assignment variable (if the combined train p is combined by unit train q, $\theta_{p,q}=1$; otherwise, $\theta_{p,q}=0$)

1. Problem statement

Given the number and type of unit trains and combined trains that need to be operated in the heavy-haul railway system, determine the trains' combination scheme in the CBS and the train timetable of each station to make the total dwell time minimal.

2. Objective function

The objective function is to minimize the total dwell time of the trains in the CBS. The function $\sum_p d_p(i,j)\theta_{p,q}$ denotes the departure time of combined train p. Additionally, suppose the combined train p is composed of unit train q. The dwell time of unit train q in the CBS can be calculated by the function $\sum_p d_p(i,j)\theta_{p,q} - a_q(i,j)$. Thus, the total dwell time of unit trains in the CBS can be expressed by (7).

$$\min z_2 = \sum_q \left[\sum_p d_p(s_z,j)\theta_{p,q} - a_q(i,s_z) \right], \forall p \in T^{com}, \forall q \in T^{uni} \tag{7}$$

3. Minimum station operating time constraints

The dwell time of trains must be longer than its minimum station operating time to ensure the necessary combined operation before departure. Constraint (8) guarantees the minimum station operating time of combined train q in the CBS. M is a big enough positive number. If combined train p is not composed of unit train q, the right-hand side of the inequality is equal to a big enough negative number, thus the inequality is always true. If combined train p is composed of unit train q, the right-hand side of the inequality is equal to the minimum station operating time.

$$d_p(s_z,j) - a_q(i,s_z) \geq \theta_{p,q} t^{tec} + (\theta_{p,q} - 1)M, \forall p \in T^{com}, \forall q \in T^{uni} \tag{8}$$

4. Marshalling constraints

Each type of combined train has a prescribed rule of marshaling which specifies the type and number of the component unit trains. For example, a 15 kt combined train is composed of three 5 kt unit trains and a 10 kt combined train is composed of two 5 kt unit trains. It is worth noting that a 10 kt unit train can be sent into the trunk line directly without marshaling. For the sake of description in the model, we describe it as a 10 kt combined train composed of a 10 kt unit train in this scenario.

$$\sum_q \theta_{p,q} q_u = \sum_k p_k \varphi_{k,u}, \forall p \in P, \forall u \in U \tag{9}$$

5. Running time constraints

$r(i,j)$ denotes the running time of trains in section (i,j). Equation constraints (10) and (11) enforce the rule that the running time of combined trains and unit trains should equal to $r(i,j)$ if (i,j) is in its route.

$$a_p(i,j) - d_p(i,j) = r(i,j)p(i,j), \forall (i,j) \in V, \forall p \in T^{com} \quad (10)$$

$$a_q(i,j) - d_q(i,j) = r(i,j)q(i,j), \forall (i,j) \in V, \forall q \in T^{uni} \quad (11)$$

6. Headway constraints

The headways of different types of trains are different. The function $\sum_k I_k p_k$ denotes the minimum headway of combined train p and constraint (13) ensures that the interval between two adjacent combined trains in section (i,j) must not be less than the minimum headway.

$$\left| a_{q'}(i,j) - a_q(i,j) \right| \geq \sum_u q_u I_u, \forall q, q' \in T^{uni} \quad (12)$$

$$\left| a_{p'}(i,j) - a_p(i,j) \right| \geq \sum_k p_k I_k, \forall p, p' \in T^{com} \quad (13)$$

7. Operating period constraint

Both combined trains and unit trains must be operated within the permitted period.

$$a_q(i,s_z) \in \left[A^{start}, A^{end}\right], \forall q \in T^{uni} \quad (14)$$

$$d_p(s_z,j) \in \left[A^{start}, A^{end}\right], \forall p \in T^{com} \quad (15)$$

4. Hybrid Genetic Algorithm

To solve the TSPP model and TTP model for a heavy-haul railway system, we propose the following solution approach by solving the TSPP model using commercial solver GUROBI and solving the TTP model using a hybrid genetic algorithm (HGA). In this section, we introduce the detailed solving process of the HGA.

The TTP model is the optimal integration of a heavy-haul railway combined scheme and train timetable. It is a difficult problem since both the train formation problem and the train timetabling problem are NP-hard problems [22,33,34]. Therefore, to solve the TTP model, we developed a genetic algorithm-based framework.

Genetic algorithms (GA), first proposed by Holland [35], are inspired by natural genetic and evolutionary mechanisms to find high-quality solutions for complex problems. The genetic algorithm constructs a fitness function according to the objective function of the problem. The fitness value may directly or indirectly represent a solution to the original problem. The algorithm performs evaluation, genetic calculation, and selection on a population composed of multiple solutions (each solution corresponds to a chromosome), and after multiple generations of reproduction, the individual with the best fitness is the optimal solution to the problem. A population contains multiple individuals and for each their chromosome includes one or several gene fragments. The GA includes the following steps [36]: (1) chromosome representations; (2) initial populations generate; (3) fitness function; (4) genetic operations (including crossover and mutation); (5) selection mechanisms; and (6) termination condition.

The GA is widely used in industrial engineering, artificial intelligence, automatic control, and other fields because of its great potential in solving complex optimization problems [37].

Specifically, the HGA to solve the TTP model is set up as explained below.

4.1. Representation Scheme

In this solution approach, we represent solutions indirectly by parameters that are later used to obtain a solution by a special decoding procedure. Each solution chromosome is made of two gene fragments.

Gene fragment I: The first gene fragment represents the departure sequence of the combined trains departing from the CBS. It is made of N_c genes, where N_c is the number of the combined trains. After obtaining the type and number of trains that need to run between the CBS and each loading station or unloading station, we create a unique index for each train. The value of the jth gene of gene fragment I represents the departure sequence of train j.

Gene fragment II: The second gene fragment is a matrix to represent the connection relationship between the unit trains and combined trains. When the number of combined trains is N_c and the number of unit trains is N_u and a $N_u \times N_c$, a matrix can be built and we call it the combination matrix. The combination matrix is a 0–1 matrix. If the value of row i and column j is 1, that means combined train j is composed of unit train i.

Figure 6 depicts a small example. As shown in the figure, seven unit trains are assigned into four combined trains in the CBS. Four combined trains, namely c_1, c_2, c_3, and c_4, depart the CBS in the sequence as 2, 1, 3, and 4, respectively. Thus, the departure sequence can be represented by the vector $(2,1,3,4)$. For combined train c_1, it consists of unit trans u_1 and u_2. Thus, its corresponding row in the combination matrix should be $(1,0,0,1,0,0,0)$.

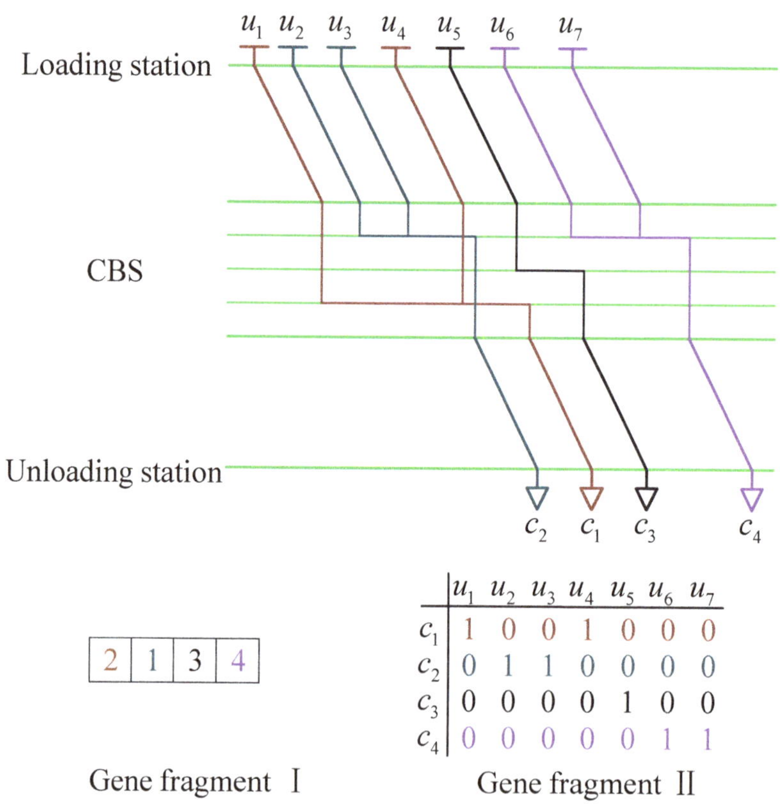

Figure 6. Gene fragments of the combined operation.

4.2. Initial Populations and Infeasible Solution Adjustment

The combination matrix is a gene fragment of a chromosome generated randomly by initialization or through crossover and mutation. It may not satisfy the constraints of train marshaling. Therefore, infeasible combination matrix adjustment strategies are designed for chromosomes to obtain an available connection relationship. The infeasible combination matrix adjustment process is illustrated in Algorithm 1.

Algorithm 1 Infeasible combination matrix adjustment

Step 1. Calculate the initial number and type of unit trains that are combined into a combined train.

Create set C_{com}, calculate the current number of unit trains connected to the combined train p, and let c_p denote it. Then, add c_p to C_{com}.

Step 2. Adjust chromosomes so that each combined train consists of a specified number of unit trains.

For each combined train $p \in T^{com}$
- If $c_p > \varphi_{k,u}$
  - Create index set R_{p1}
  - Find out all indices r where $\theta_{p,r} = 1$ and add r to R_{p1}
  - Create subset R_{p2}
  - Randomly choose $\varphi_{k,u} - c_p$ elements r' in R_{p1} and add r' to R_{p2}
  - Let $\theta_{p,r'} = 0$ for all r' in R_{p2}
- If $c_p < \varphi_{k,u}$
  - Create index set R_{p1}
  - Find out all indices r where $\theta_{p,r} = 0$ and add r to R_{p1}
  - Create subset R_{p2}
  - Randomly choose $c_p - \varphi_{k,u}$ elements r' in R_{p1} and add r' to R_{p2}
  - Let $\theta_{p,r'} = 1$ for all r' in R_{p2}

Step 3. Calculate the current number and type of unit trains that are combined into a combined train.

Create set C_{unit}, calculate the current number of combined trains connected to unit train q, and let c_q denote it. Then, add c_q to C_{unit}.

Step 4. Adjust chromosomes so that each unit train can be combined into, at most, one combination train.

Create index set R_{q0}

Find out all indices s where $c_q = 0$ and add s to R_{q0}

For each unit train $q \in T^{uni}$
- If $c_q > 1$
  - Create index set R_{q1}
  - Find out all indices r where $\theta_{r,q} = 1$ and add r to R_{q1}
  - Create subset R_{q2}, R_{q3}
  - Randomly choose $c_q - 1$ elements r' in R_{q1} and add r' to R_{q2}
  - Randomly choose $c_q - 1$ elements s' in $R_{q0} - \sum_{q=1}^{q-1} R_{q3}$ and add s' to R_{q3}
  - Let $\theta_{r',q} = 0$ for all r' in R_{q2} and let $\theta_{r',s} = 0$ for all s' in R_{q3}

4.3. Decoding

The chromosomes are made of two gene fragments that represent the combined trains' departure sequence and the connection relationship; thus, the chromosomes have to be decoded to derive the combination scheme and train timetable. The chromosome decoding process is divided into three main steps. First, we transform the two gene fragments of the chromosome into the operation sequence of the trains. Then, we use a linear program to determine the combination scheme of the unit trains and the combined trains in the CBS. Finally, we calculate the train timetable according to the combination scheme.

First of all, we determine the order of any two combined trains departing from the CBS according to gene fragment I of the chromosome. Next, determine the order of any two unit trains arriving at the CBS according to gene fragment II. Here, we introduce two parameters, namely $O^{com}(p, p')$ and $O^{uni}(q, q')$. $O^{com}(p, p')$ denotes the departure order of combined trains p and p'. When combined train p departs from the CBS before combined train p', then $O^{com}(p, p') = 1$; otherwise, $O^{com}(p, p') = 0$. $O^{uni}(q, q')$ denotes the arrival order of unit trains q and q'. When unit train q arrives at the CBS before combined train q', then $O^{com}(q, q') = 1$; otherwise, $O^{com}(q, q') = 0$.

The process of determining the value of the parameters $O^{com}(p, p')$ and $O^{uni}(q, q')$ is shown in Algorithm 2. The calculation process of determining the operation sequence of trains can be divided into two steps. In step 1, we roughly calculate all trains' arrival and departure times in the COS according to the departure sequence in gene fragment I and the connection relationship in gene fragment II. Then, in step 2, we determine the departure order of any combined train pair by sequencing in gene fragment I. Similarly, we compare the arrival times in step 1 to determine the arrival order of any unit train pair. As a result, we derive an assignment to O^{com} and O^{uni}.

Algorithm 2 Determine the operation sequence of trains in the CBS

Input: Gene fragment I $\mathbf{g1}$, gene fragment II$\mathbf{G2}$
Output: O^{com}, O^{uni}
Step 1: Initialize the departure time of the combined trains and the arrival time of the unit trains according to gene fragments I and II.
  Create set D^{com} to denote the initialized departure time of the combined trains
  Create set A^{uni} to denote the initialized arrival time of the unit trains
  For each combined train $p \in T^{com}$
    If $\mathbf{g1}(p) = 1$
      Let $d_p = 0$ and add d_p to D^{com}
    Else
      If $\mathbf{g1}(p) = \mathbf{g1}(p') + 1$
      Let $d_p = d_{p'} + \sum_k I_k p'_k$ and add d_p to D^{com}
  For each unit train $q \in T^{uni}$
    $u = 0$
    For each combined train $p \in T^{com}$
      If $\mathbf{G2}(p, q) = 1$
      Let $a_q = d_p - \sum_k p_k t_k^{tec} - I_u u$ and add a_q to A^{uni}
      $u = u + 1$
Step 2: Determine departure order O^{com} for any combination train pair and arrival order O^{uni} for any unit train pair.
  Create set O^{com} of the combined train departure order
  Create set O^{uni} of the combined train arrival order
  For each combined train $p \in T^{com}$
    For each combined train $p' \in T^{com}$
      If $\mathbf{g1}(p) < \mathbf{g1}(p')$
        Let $o_{p,p'} = 1$ and add $o_{p,p'}$ to O^{com}
    Else
        Let $o_{p,p'} = 0$ and add $o_{p,p'}$ to O^{com}
  For each unit train $q \in T^{uni}$
    For each combined train $q' \in T^{uni}$
      If $a_q < a_{q'}$
        Let $o_{q,q'} = 1$ and add $o_{q,q'}$ to O^{uni}
    Else
        Let $o_{q,q'} = 0$ and add $o_{q,q'}$ to O^{uni}

After the train operation sequence is determined, the solution space of the original problem is greatly reduced. Using parameters $O^{com}(p, p')$, $O^{uni}(q, q')$, and a large positive number M, we can rewrite the absolute value inequality in constraints (12) and (13). The operation time of the combined trains and unit trains in the CBS can be easily solved. We rewrite this part as the combination planning problem (CPP). The objective function of the CPP is the same as TTP, that is, to minimize the total operation time of the CBS.

Additionally, the operating time of the trains in the CBS needs to meet the station operating time requirements.

To sum up, this program's objective function and constraints are shown in (16) and (17).

$$\min z_3 = \sum_q \left[\sum_p d_p(s_z, j)\theta_{p,q} - a_q(i, s_z) \right], \forall p \in T^{com}, \forall q \in T^{uni} \tag{16}$$

$$s.t. \begin{cases} d_p(s_z, j) - a_q(i, s_z) \geq \theta_{p,q} \sum_k p_k t_k^{tec} + (\theta_{p,q} - 1)M, \forall p \in T^{com}, \forall q \in T^{uni} \\ d_{p'}(s_z, j) - d_p(s_z, j) + (1 - o_{p,p'})M \geq \sum_k p_k I_k, \forall p, p' \in T^{com} \\ a_{q'}(i, s_z) - a_q(i, s_z) + (1 - o_{q,q'})M \geq \sum_u q_u I_u, \forall q, q' \in T^{uni} \\ d_p(s_z, j) \in \left[D^{start}, D^{end} \right], \forall p \in T^{com} \\ a_q(i, s_z) \in \left[A^{start}, A^{end} \right], \forall q \in T^{uni} \end{cases} \tag{17}$$

We will call the commercial solver Gurobi to solve the CPP.

Finally, calculate the train timetable based on the solution of CPP. In this paper, we assume that all trains run at the same speed class and that there is no train overtaking that occurs when running on a heavy-haul railway. Therefore, once the time of a combined train departing from the CBS is determined, its arrival time at the unloading station and stop time at the adjacent stations can also be determined. Same as for combined trains, the schedule of a unit train can also be calculated once its arrival time at the CBS is determined. Formulas for calculating the arrival and departure times of other stations are shown in (18).

$$\begin{cases} a_p(i, j) - d_p(i, j) = r(i, j)p(i, j), \forall (i, j) \in V, \forall p \in T^{com} \\ a_q(i, j) - d_q(i, j) = r(i, j)q(i, j), \forall (i, j) \in V, \forall q \in T^{uni} \end{cases} \tag{18}$$

4.4. Fitness Function

The fitness function is shown in (19), where the denominator represents the total operating time of the trains in the CBS. Its reciprocal is good for evaluating the quality of the solution: a larger fitness function value indicates a better solution.

$$f = \frac{1}{\sum_q \left[\sum_p d_p(i, j)\theta_{p,q} - a_q(i, j) \right]} \tag{19}$$

4.5. Crossover

Crossover is one of the genetic operations that combines two chromosomes to generate a new chromosome. First, select two chromosomes from the current population with probability P_s. P_s is a set of probability, indicating the possibility of each individual being selected. The selection probability of each individual is proportional to its fitness value. The chosen two chromosomes will crossover with the probability of P_c. Then, randomly select a crossover point for gene fragment I and gene fragment . After crossover, it should avoid generating infeasible solutions [38]. For the newly generated chromosomes that may not satisfy the constraint, we use the adjustment strategy proposed in Algorithm 1 to ensure the feasibility of the newly generated chromosomes.

4.6. Mutation

The mutation is another genetic operation to evolve the chromosomes. The randomness mutation of the population can allow the search process to jump out of the local optimal solution and search for the global optimal solution. However, this randomness does not necessarily mean that the population will evolve in a better direction. A high mutation probability may make the population mixed with poor individuals and the results experience difficulty in converging. However, when the mutation probability is too low, the population may fail to evolve for many generations and become trapped in the local optimal solution. To solve this problem, Glover [39] tried to use a heuristic search in the mutation process to improve the performance.

Therefore, we introduced a neighborhood search into the mutation process to ensure that the population evolves in a better direction. The select chromosomes will be mutated with probability P_m. We separately apply the following mutation procedure for each gene fragment.

For gene fragment I, randomly select two chromosomal sites and exchange the genes of the two sites. For gene fragment II, randomly select a chromosomal site and change the value of the gene. The gene of fragment II is a 0–1 binary variable, that is, to change 0 to 1 or 1 to 0.

During each mutation, we generate a neighborhood set for the chromosome. Then, we use the fitness function to evaluate the quality of chromosomes in the neighborhood set and choose the best mutant chromosome as the newly generated chromosome.

4.7. Termination Conditions

The algorithm ends and outputs the result when the optimal objective value does not change continuously or when the number of iterations reaches the predetermined value.

4.8. Algorithm

The framework of the algorithmic procedure is summarized below in Figure 7.

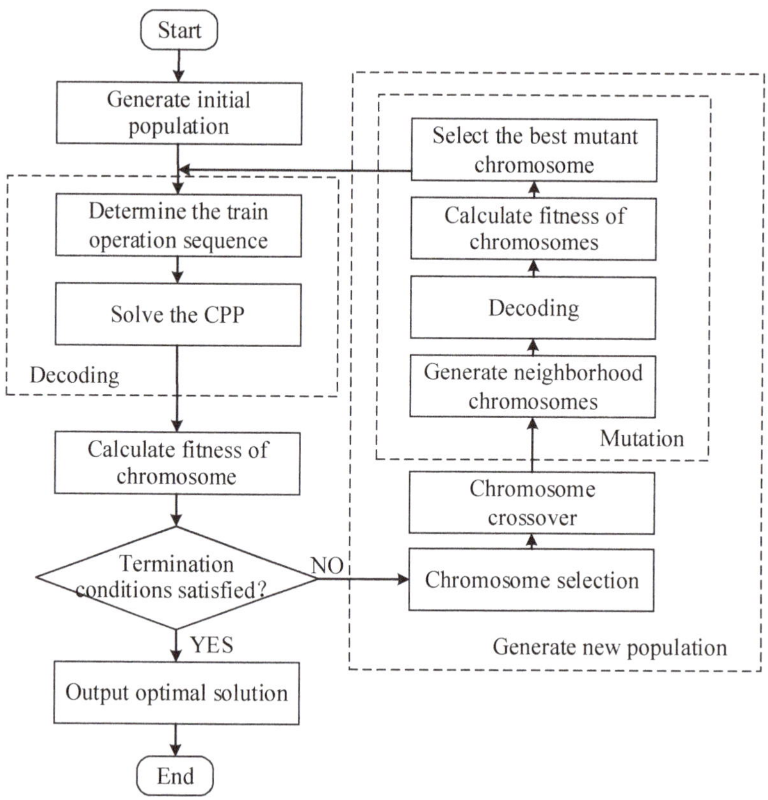

Figure 7. Flow chart of the solution approach.

5. Case Study

This section provides several numerical experiments on the proposed heavy-haul railway train timetabling problem in order to prove the effectiveness of our models and algorithms.

5.1. A Small Case

In this section, we use a small case to illustrate the effectiveness of the proposed model and hybrid algorithm.

5.1.1. Operation Data

The corresponding schematic diagram of a small heavy-haul railway network is presented in Figure 8. As shown in the figure, this small heavy-haul railway consists of three loading stations (station *a*, station *b*, and station *c*), three unloading stations (station *d*, station *e*, and station *f*), and a CBS. The loading capacity of station *a*, station *b*, and station *c* is each 550 cars per day. The unloading demand of station *d*, station *e*, and station *f* is 360 cars/day, 720 cars/day, and 540 cars/day, respectively. Lastly, the unloading capacity of station *d*, station *e*, and station *f* can meet the unloading requirements.

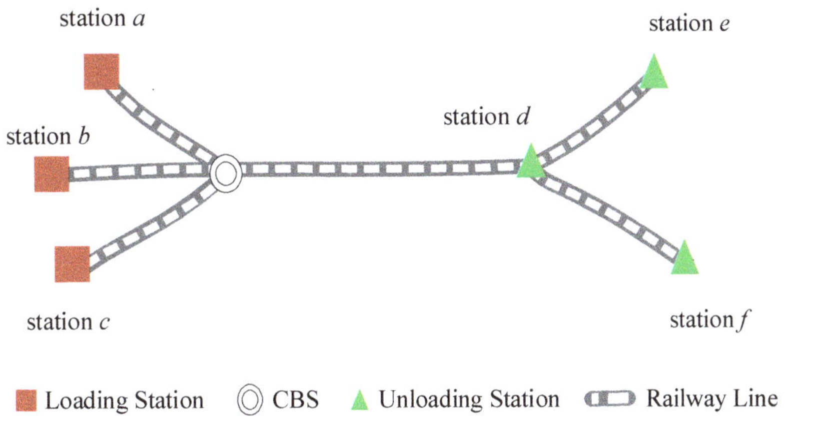

Figure 8. Simplified heavy-haul railway network of the small case.

In this small heavy-haul railway, the unit train types include 5 kt unit trains and 10 kt unit trains that contain 60 railcars and 120 railcars, respectively. The combined train types include 10 kt (2 × 5 kt) combined trains, 15 kt (3 × 5 kt) combined trans, and 20 kt (4 × 5 kt) combined trans. The corresponding combination rule and operation cost of these combined trans are 10 kt (2 × 5 kt) combined trans (two 5 kt unit trains, cost = 0.9), 15 kt (3 × 5 kt) combined trans (three 5 kt unit trains, cost = 1), and 20 kt (4 × 5 kt) combined trans (four 5 kt unit trains, cost = 1; or two 10 kt unit trains, cost = 1.8). The combined operation in the CBS should be during 0:00–8:00 and the combined train needs to depart from the CBS during 6:00–8:00.

5.1.2. Optimization Results

By solving the TSPP, we derive the type and number of unit trains that need to be loaded at each loading station, as well as the type and number of combined trains that need to be unloaded at each unloading station. The solution of the TSPP is shown in Table 4. The TTP will be further solved based on this solution.

Table 4. Solution of the TSPP in the small case.

Unloading Stations	Unit Train Type	Number of Trains	Loading Stations	Combined Train Type	Number of Trains
station a	5 kt	9	station d	10 kt (2 × 5 kt)	1
station b	5 kt	9	station d	20 kt (4 × 5 kt)	1
station c	5 kt	9	station e	20 kt (4 × 5 kt)	3
			station f	10 kt (2 × 5 kt)	1
			station f	15 kt (3 × 5 kt)	1
			station f	20 kt (4 × 5 kt)	1

Then, we apply the proposed HGA to obtain the optimal combination scheme and train timetable to minimize the total dwell time of all heavy-haul trains. Specifically, some parameters of the HGA are set as follows:

- The crossover rate is set to 0.85;
- The mutation rate is set to 0.15 and the number of elements in the neighborhood set of mutation chromosomes is set as 20;
- The larger positive number M is set to 10,000; and
- When the optimal value exceeds 50 iterations without optimization, the iteration reaches the termination condition.

The corresponding results and the solution searching process of the small case are plotted in Figure 9. By generating and optimizing the combined operation plan of heavy-haul trains at the CBS, the total combined operation time in the system is 4664 car–hours.

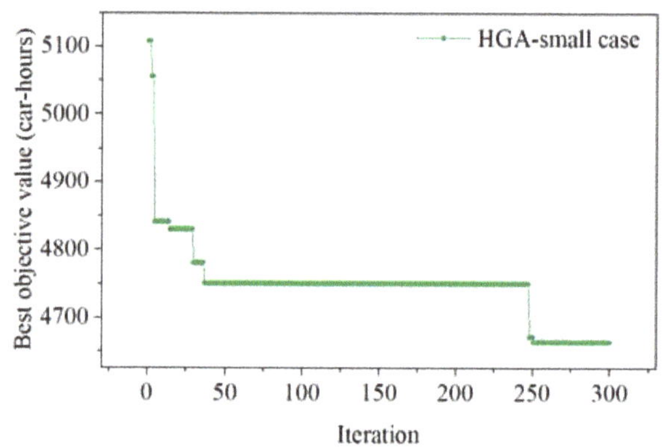

Figure 9. The solution searching process of the small case using the HGA.

In order to verify the effectiveness of the proposed model, we compare the results with the timetable obtained by the manual scheduling operation. To this end, we designed a simulation process to imitate the manual scheduling operation approaches of the heavy-haul railway scheduling system. The simulation steps are as follows:

Step 1: Assign a combination scheme of combined trains and unit trains based on experience.

Step 2: Randomly assign a series of the departure times of the combined trains in the CBS.

Step 3: Determine the arrival time of the unit trains at the CBS backwards according to the departure time of the combined trains and the required dwell time.

Step 4: Calculate trains' arrival and departure times in each section according to the section running time.

Step 5: Check for conflicts and adjust the train timetable.

Next, the simulation method is applied to the small case to obtain a solution. The total dwell time obtained by the simulation is 7876 car–hours. Compared to the proposed model's optimization results, the simulation method's operation time increases by 3212 car–hours, resulting in a time waste of about 40.8%.

Figure 10 shows the combination scheme's diagram using the HGA and the simulation. The horizontal axis shows the time and the vertical axis shows the position of the trains. These combination scheme diagrams can be divided into three parts. The upper oblique lines represent the unit trains running between the loading stations and the CBS. The lower oblique lines represent the combined trains running between the CBS and the unloading stations. The horizontal and vertical lines in the middle represent the combination relationship and the dwell time of each train. As shown in the figure, compared to the combination scheme obtained by the simulation, the solution obtained by the HGA is more compact and uses less time in the CBS.

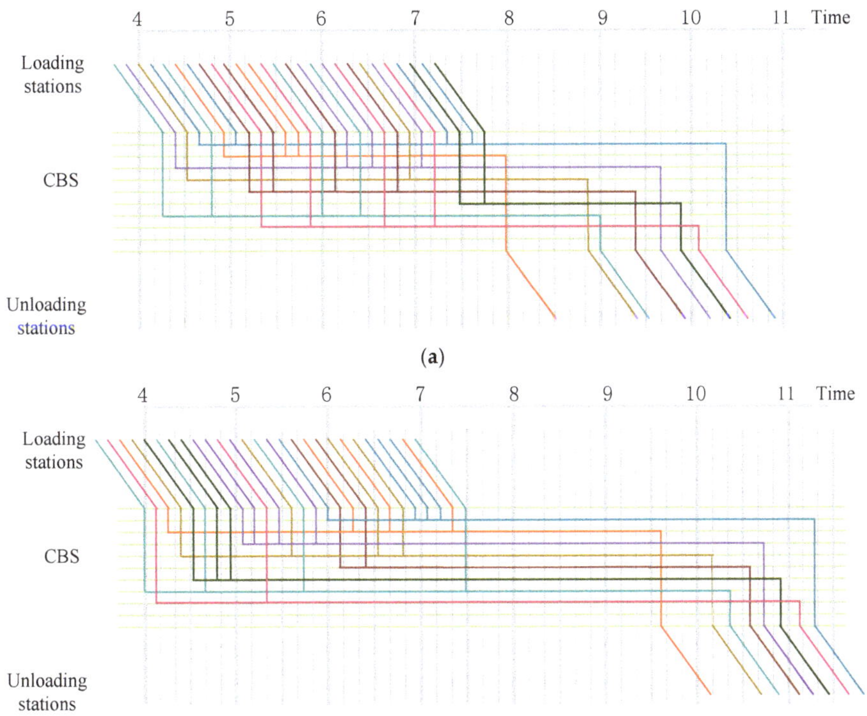

Figure 10. The comparison of the results of the two methods: (**a**) the diagram of the combination scheme using the HGA and (**b**) the diagram of the combined combination scheme using the simulation.

To further compare the two methods, we calculate the waiting time of each unit train that waits for the combination operation in the two combination schemes. The waiting time is the period that lasts from when the unit train enters the CBS to when it starts the combination operation. The waiting time of each unit train in the CBS is shown in (20). We define the unit train whose waiting time in the CBS is less than 90 min as the efficient turnover train (ETT).

$$w_p = \sum\nolimits_{p \in T^{com}} d_p(s_z, j)\theta_{p,q} - a_q(i, s_z), \forall q \in T^{uni} \tag{20}$$

Figure 11 plots the waiting time distribution of unit trains in the two solutions. The horizontal axis represents the index of the unit train and the vertical axis represents the waiting time of the corresponding unit train in the CBS. The green dotted line indicates that the waiting time is 90 min and the points below the green dotted line are the ETT. As shown in Figure 11, 18 unit trains in the combination scheme obtained by the TTP belong to the ETT, while only 7 ETT are in the combination scheme obtained by the simulation. Therefore, the combination scheme solved by the TTP can effectively reduce the waiting time of unit trains in the CBS, thus reducing the service load of stations and improving train turnover efficiency.

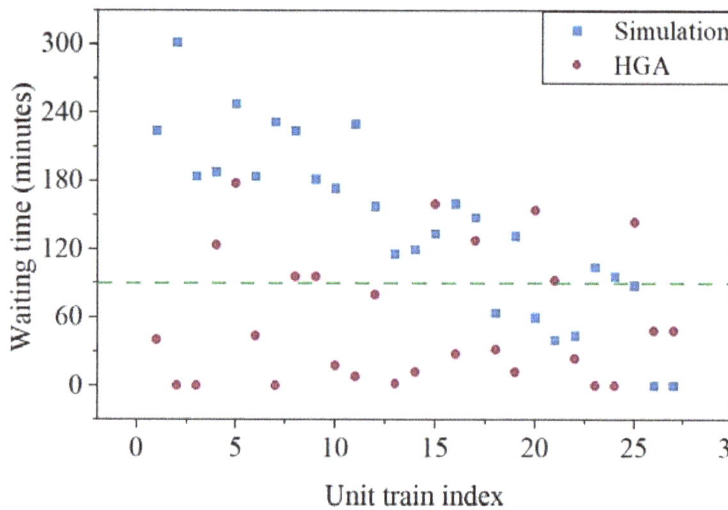

Figure 11. The distribution of the waiting time for unit trains in the two solutions.

By comparing these two methods of scheduling the combination schemes, the efficiency of the TTP is verified.

5.1.3. Effectiveness of the Algorithm

In order to verify the effectiveness of the algorithm, we introduce an unimproved genetic algorithm (GA) and a tabu search (TS) to solve the optimization problem.

In experiments of GA, we still adopt the same representation scheme, population initialization operation, and chromosome-crossing operation as the method in this paper. The fitness function still adopts the reciprocal of the objective function. The difference is that the CPP is not called during the decoding process to find the optimized combination scheme. Meanwhile, the mutation operation uses single random mutation instead of neighborhood search.

As for the TS, it is another widely used stochastic search method designed to find the optimal solution. The tabu search algorithm imitates human memory and uses a tabu list to forbid certain moves to avoid cycling search [40]. The tabu search adopts neighborhood optimization. In order to transcend the local optimal solution, the algorithm can accept inferior solutions. In this algorithm, we use the same representation scheme in the HGA and we employ the neighborhood generation method of the mutation operator in the HGA. The objective function of the TTP is used as the evaluation function to evaluate the quality of neighbor solutions.

Then, we conducted five experiments with different parameters. Due to the randomness of the search process, we ran each experiment 10 times and calculated the average value of the best five solutions as the final result. The experimental parameters and results are shown in Tables 5–7.

Table 5. The parameters and results of the GA.

Index	Population Size	Generations	Crossover Rate	Mutation Rate	Total Operation Time (car–hours)
1	30	150	0.90	0.05	5735
2	50	200	0.80	0.1	5531
3	50	150	0.80	0.15	5623
4	80	250	0.75	0.15	5703
5	100	250	0.85	0.15	5537

Table 6. The parameters and results of the HGA.

Index	Population Size	Generations	Crossover Rate	Mutation Rate	Total Operation Time (car–hours)
1	30	150	0.90	0.05	4867
2	50	200	0.80	0.1	4855
3	50	150	0.80	0.15	4818
4	80	250	0.75	0.15	4770
5	100	250	0.85	0.15	4727

Table 7. The parameters and results of the TS.

Index	Number of Neighbors	Tabu Length	Iterations	Total Operation Time (car–hours)
1	20	15	200	5032
2	40	15	200	4993
3	40	10	200	5057

The search process of the two algorithms is shown in Figures 12–14. All algorithms have made some progress in reducing the objective value. In the experimental group of the GA, experiment GA-3 obtained the best objective value, which was 5531 car–hours. In the experimental group of the HGA, the best objective function value was obtained in experiment HGA-4, which was 4727 car–hours. In the experimental group of the TS, the best objective function value was obtained in experiment TS-2, which was 4993 car–hours. In comparing Figure 12 with Figure 13, it is obvious that the HGA achieves better optimization values than the GA on the whole. As shown in Figure 14, compared with the TS, the HGA obtains a better initial solution and has a high convergence speed. The above experiments showed that, compared to the GA and TS, the HGA obtains a better initial solution because we solve the CPP model to improve the initial solution in the HGA. Additionally, the HGA achieves a high convergence speed. These experiments demonstrate the effectiveness of the designed HGA.

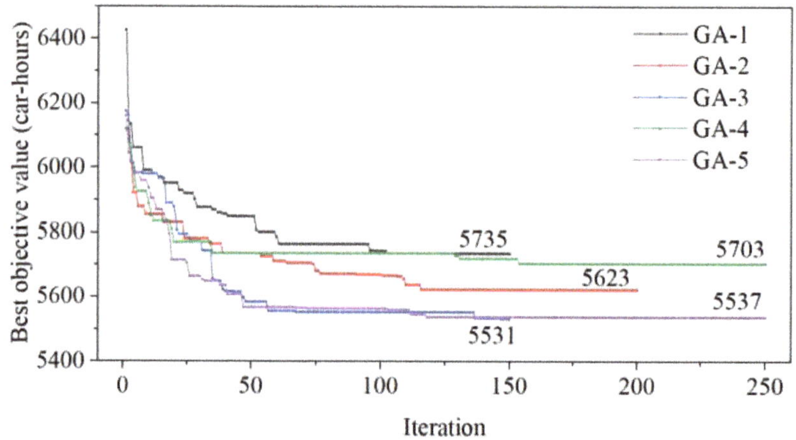

Figure 12. The solution searching process of the GA.

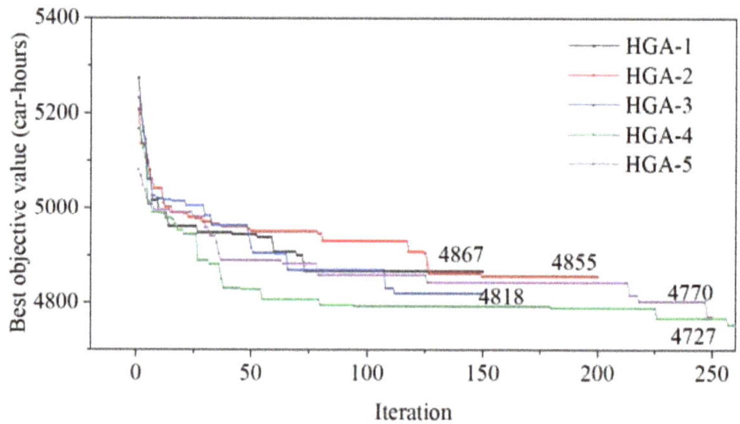

Figure 13. The solution searching process of the HGA.

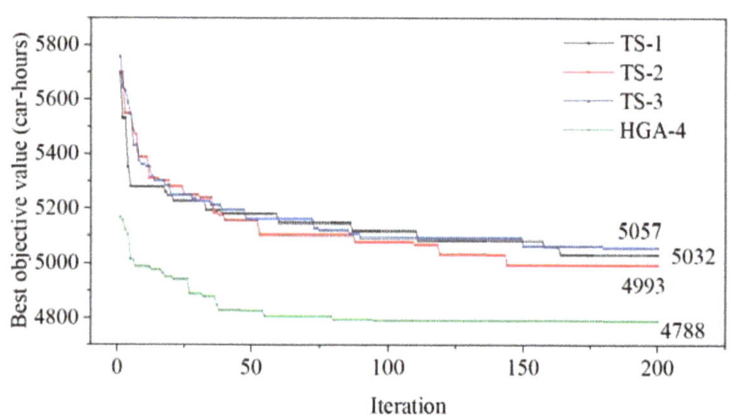

Figure 14. The solution searching process of the TS.

5.2. Large-Scale Experiments

5.2.1. Description of the Experimental Setting

To further show the effectiveness of the proposed approach, we select the Datong-Qinhuangdao Heavy-haul Railway (DQHR) as a large-scale study case. The DQHR, with a total length of 653 km, is China's first double-line electrified heavy-haul railway and it is an important coal transportation corridor from Shanxi, Shaanxi, and western Inner Mongolia [41].

In these experiments, the DQHR contains 6 loading stations and 11 unloading stations. A simplified schematic is shown in Figure 15. Some important train operation parameters are listed in Tables 8 and 9. Table 8 shows the capacity and demand of the loading and unloading stations in the DQHR. Table 8 shows the running cost and operating time of each type of combined train.

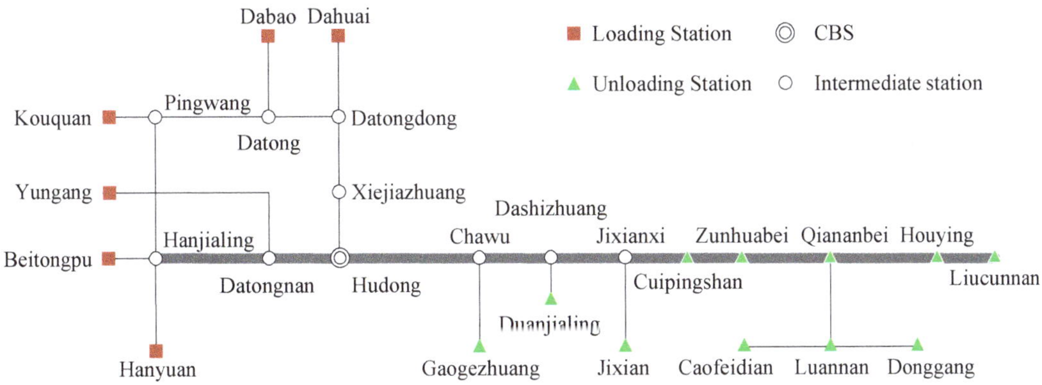

Figure 15. Simplified Datong-Qinhuangdao Heavy-haul Railway.

Table 8. Parameters of the loading stations and unloading stations.

Parameters of Loading Stations		**Parameters of Unloading Stations**		
Loading Station	**Daily Loading Capacity (cars/day)**	**Unloading Station**	**Daily Unloading Capacity (cars/day)**	**Cargo Demand (cars/day)**
Hanyuan	15,058	Gaogezhuang	382	300
Beitongpu	18,888	Duanjialing	870	500
Yungang	2365	Jixian	403	300
Kouquan	1134	Cuipingshan	960	600
Dabao	6845	Zunhuabei	1296	1000
Dahuai	7289	Qiananbei	480	300
		Caofeidian	4992	3000
		Donggang	1944	1500
		Luannan	4584	3500
		Houying	2544	1500
		Liucunnan	6720	4000

Table 9. The running cost and operating time of each type of combined train.

Index	Type of Combined Trains	Type and Number of Unit Trains	Cost of Train	Operating Time of Combination (min)
1	10 kt combined train	2 × 5 kt unit train	1.3	126
2	10 kt combined train	1 × 10 kt unit train	1	93
3	15 kt combined train	3 × 5 kt unit train	1.4	152
4	20 kt combined train	2 × 10 kt unit train	1.5	135
5	20 kt combined train	4 × 5 kt unit train	1.8	182
6	30 kt combined train	3 × 10 kt unit train	2	168

5.2.2. Computational Results of the Large-Scale Experiment

The solution of the TSPP in the large-scale experiment is shown in Table 10. Subsequent TTP will be further solved based on this solution.

Table 10. Solution of the TSPP in the large-scale experiment.

Unloading Stations	Unit Train Type	Number of Trains	Loading Stations	Combined Train Type	Number of Trains
Hanyuan	10 kt	24	Chawu	30 kt (3 × 10 kt)	1
Beitongpu	5 kt	114	Duanjialing	15 kt (3 × 5 kt)	3
Dahuai	5 kt	121	Jixian	30 kt (3 × 10 kt)	1
			Zunhua	15 kt (3 × 5 kt)	3
			Zunhua	20 kt (4 × 5 kt)	2
			Cuipingshan	15 kt (3 × 5 kt)	2
			Cuipingshan	20 kt (4 × 5 kt)	1
			Qianan	30 kt (3 × 10 kt)	1
			Caofeidian	15 kt (3 × 5 kt)	14
			Caofeidian	20 kt (4 × 5 kt)	2
			Donggang	15 kt (3 × 5 kt)	3
			Donggang	20 kt (4 × 5 kt)	4
			Luannan	15 kt (3 × 5 kt)	16
			Luannan	30 kt (3 × 10 kt)	2
			Houying	15 kt (3 × 5 kt)	7
			Houying	20 kt (4 × 5 kt)	1
			Liucunnan	15 kt (3 × 5 kt)	5
			Liucunnan	20 kt (4 × 5 kt)	9
			Liucunnan	30 kt (3 × 10 kt)	3

In this large-scale experiment, the crossover rate was set to 0.85 and the mutation rate was set to 0.15. The number of elements in the neighborhood set of mutation chromosomes was set as 20. The maximum number of iterations was set to 200.

Figure 16 shows the iterative process of the HGA when used to solve the TTP in the large-scale experiment. As shown in Figure 16, the total combined operation time in the system was 98,848 car–hours. This example provides an illustration of the usefulness and application of our proposed approach.

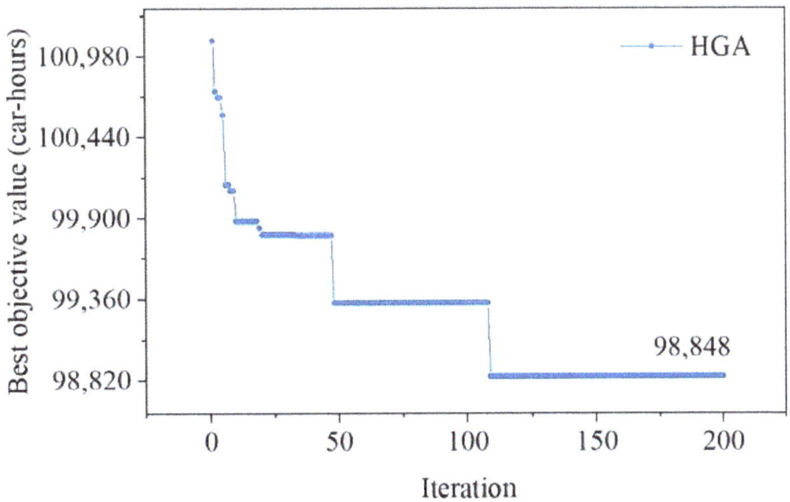

Figure 16. The solution searching process of the large-scale experiment using the HGA.

6. Conclusions

Based on the actual demand of the heavy-haul railway, this paper proposes a scheduling approach to optimize the heavy-haul railway transportation plan, including the combination scheme and train timetable. This approach can satisfy the freight transportation demands, improve the utilization of the CBS operation, and reduce the cost of the whole transportation process, thus improving rail operators' profits.

First, the train service plan problem (TSPP) model was proposed based on the loading stations' capacity and the unloading stations' demands. We solved the TSPP to determine the type and number of unit trains that need to be loaded at each loading station, as well as the type and number of combined trains that need to arrive and unload at unloading stations. On this basis, we put forward the train timetabling problem (TTP) model. The TTP model can solve the combination scheme in the CBS and the train timetable of the heavy-haul railway. Then, we applied a hybrid genetic algorithm (HGA) to solve the TTP model. In a small case study using the TTP model, we obtained a solution that reduces the total dwell time of unit trains by 40.8% compared to a manual scheduling simulation method. The comparison experiments with the unimproved genetic algorithm (GA) and tabu search (TS) demonstrate that the HGA can obtain better solutions and achieve a high convergence speed. By applying the approach into a large-scale case of DQHR, we demonstrated the practicability of the method in the real world.

In this paper, we assumed that the combined capacity of the CBS is large enough to deal with the unit trains. However, in practice, when the freight volume of the heavy-haul railway is too big, the CBS may not be able to combine many unit trains at once. In future studies, we will consider the combined capacity of the CBS as one of the constraints to determine a more feasible combination scheme. At the same time, the return and decomposed operation of the empty trains after the unloading operation is also a critical part of trains circulating in the heavy-haul railway system. Thus, the transportation organization of the empty trains will also be considered in future research.

Author Contributions: Conceptualization, H.Z. and B.G.; methodology, H.Z. and L.Z.; software, H.Z.; investigation, Z.B.; data curation, L.Y. and Z.W.; writing—original draft preparation, H.Z.; writing—review and editing, H.Z. and B.G. All authors have read and agreed to the published version of the manuscript.

Funding: This research study was funded by the High-speed Rail Joint Fund of the National Natural Science Foundation of China, grant number U1934216; Fundamental Research Funds for the Central Universities, grant number 2018RC012; and High-speed Rail Joint Fund of the National Natural Science Foundation of China, grant number U1834211.

Institutional Review Board Statement: Not applicable.

Informed Consent Statement: Not applicable.

Data Availability Statement: Not applicable.

Conflicts of Interest: The authors declare no conflict of interest.

References

1. Fu, J.; Chen, J. A Green Transportation Planning Approach for Coal Heavy-Haul Railway System by Simultaneously Optimizing Energy Consumption and Capacity Utilization. *Sustainability* **2021**, *13*, 4173. [CrossRef]
2. Zong, Y. Research on development countermeasures of railway coal transportation in China. *Railw. Freight* **2020**, *38*, 11–15.
3. Fan, Z. *Research on Car Flow Attractive Region of Heavy-Haul Channels and Direct Heavy-Haul Transportation*; Beijing Jiaotong University: Beijing, China, 2008.
4. Yang, Y.; Jia, C.; Hu, S. Relationship Determination with Train Weight, Speed, and Density of Heavy Haul Railway Line in China. *J. Beijing Jiaotong Univ.* **2006**, *30*, 1–4+9.
5. Ma, H. Discussion on influencing factors of railway transportation capacity of Da-qin railway. *J. Railw. Freight* **2014**, *8*, 40–43.
6. Sun, J.; Kong, L.; Pan, H. Develop Heavy-Loading Transportation and Increase Traffic Capacity of Baotou-Shenmu Railway. *Transp. Logist.* **2011**, *20*, 172–176.
7. Zhou, F. *Research on Key Technologies of Baoshen Railway Heavy Haul Transportation Organization*; Lanzhou Jiaotong University: Lanzhou, China, 2019.
8. Xue, F.; Zhao, L.; Fan, Q. Study on coupling and adjustment of marshalling yard stage plan and dynamic traffic flow. *J. China Railw. Soc.* **2020**, *42*, 18–26.
9. Wang, C. Application of goal programming in railway heavy haul transportation. *J. Southwest Jiaotong Univ.* **2009**, *44*, 392–395.
10. Zhao, P.; Zhang, J.; Tang, B. Study on the Optimization Model of Car Flow Organization in the Loading Area of Heavy Haul Railway Based on the Combined Trains. *China Railw. Sci.* **2010**, *31*, 116–121.
11. Tang, B. *Study on the Organization of Wagon Flow in Heavy Haul Loading Area*; Beijing Jiaotong University: Beijing, China, 2009.
12. Wang, W. *Research on the Organization of Direct Wagon Flow in Heavy-Haul Loading Area*; Southwest Jiaotong University: Chengdu, China, 2012.
13. Jing, L.; LI, G. Research on traffic organization optimization of heavy haul transportation based on improved fruit fly algorithm. *Railw. Transp. Econ.* **2018**, *40*, 30–35.
14. Xiao, J.; Lin, B.; Wang, J. Solving the train formation plan network problem of the single-block train and two-block train using a hybrid algorithm of genetic algorithm and tabu search. *Transp. Res. Part C Emerg. Technol.* **2018**, *86*, 124–146. [CrossRef]
15. China Railway Publishing House. Specification for the design of heavy-haul railways: TB 10625—2017[S]. In *China, National Railway Administration of the People's Republic of China*; China Railway Publishing House: Beijing, China, 2017.
16. Zhiye, W.; Bing, Z. Study of Schematic Design of Technical Operation Stations for Heavy Haul Railway. *China Railw.* **2019**, *06*, 71–76.
17. Wei, Y.; Zhang, H.; Yang, H. Impact of Turnout Type Selection on Station Carrying Capacity of Heavy Haul Railway. *China Railw. Sci.* **2013**, *34*, 95–98.
18. Ye, J.; Wang, H. Practice and development of 20,000 T combined train in Hudong Station. *China Railw.* **2009**, *6*, 22–25.
19. Liang, Q.; Han, D.; Zhang, J. The influence of 20,000 T heavy haul train on the arrival and departure of Hudong Marshalling Station of Daqin Line. *China Railw. Sci.* **2005**, *26*, 1–6.
20. Han, X.; Guo, H.; Peng, Q. Methods for optimization of combination schemes of combination stations in heavy-haul transportation. *J. China Railw. Soc.* **2012**, *34*, 1–7.
21. Ma, M.; Ding, D.; Ren, Z. Optimal Utilization of Arrival and Departure Track in Heavy Haul Railways. *Railw. Investig.* **2019**, *45*, 110–115.
22. Yaghini, M.; Momeni, M.; Sarmadi, M. A hybrid solution method for fuzzy train formation planning. *Appl. Soft Comput.* **2015**, *31*, 257–265. [CrossRef]
23. Martinelli, D.R.; Teng, H. Optimization of railway operations using neural networks. *Transp. Res. Part C Emerg. Technol.* **1996**, *4*, 33–49. [CrossRef]
24. Kozachenko, D.; Bobrovskiy, V.; Gera, B.; Skovron, I.; Gorbova, A. An optimization method of the multi-group train formation at flat yards. *Int. J. Rail Transp.* **2020**, *9*, 61–78. [CrossRef]
25. Lin, B.-L.; Wang, Z.-M.; Ji, L.-J.; Tian, Y.-M.; Zhou, G.-Q. Optimizing the freight train connection service network of a large-scale rail system. *Transp. Res. Part B Methodol.* **2012**, *46*, 649–667. [CrossRef]
26. Lin, B.; Zhao, Y. The Systematic Optimization of Train Formation in Loading Stations. *Symmetry* **2019**, *11*, 1238. [CrossRef]

27. Yaghini, M.; Momeni, M.; Sarmadi, M. Solving train formation problem using simulated annealing algorithm in a simplex framework. *J. Adv. Transp.* **2014**, *48*, 402–416. [CrossRef]
28. Murali, P.; Ordóñez, F.; Dessouky, M.M. Modeling strategies for effectively routing freight trains through complex networks. *Transp. Res. Part C Emerg. Technol.* **2016**, *70*, 197–213. [CrossRef]
29. Lazarev, A.A.; Musatova, E.G. The problem of trains formation and scheduling: Integer statements. *Autom. Remote Control* **2013**, *74*, 2064–2068. [CrossRef]
30. Ma, X. *Research on the Organization Mode of Railway Heavy Load Transportation and Related Problems*; Southwest Jiaotong University: Chengdu, China, 2006.
31. Tong, L.; Zhou, L.; Liu, J.; Zhou, X. Customized bus service design for jointly optimizing passenger-to-vehicle assignment and vehicle routing. *Transport. Res. C-Emerg. Technol.* **2017**, *85*, 451–475. [CrossRef]
32. Yue, Y.; Wang, S.; Zhou, L.; Tong, L.; Saat, M.R. Optimizing train stopping patterns and schedules for high-speed passenger rail corridors. *Transport. Res. C-Emerg. Technol.* **2016**, *63*, 126–146. [CrossRef]
33. Zhang, Y.; D'Ariano, A.; He, B.; Peng, Q. Microscopic optimization model and algorithm for integrating train timetabling and track maintenance task scheduling. *Transp. Res. Part B Methodol.* **2019**, *127*, 237–278. [CrossRef]
34. Oroojlooy Jadid, A.; Eshghi, K. Train Timetabling on double track and multiple station capacity railway with useful upper and lower bounds. *Sci. Iran* **2017**, *24*, 3324–3344. [CrossRef]
35. Holland, J.H. *Adaption in Natural and Artificial Systems*; The MIT Press: Cambridge, CA, USA, 1975; Volume 6, pp. 126–137.
36. Man, K.F.; Tang, K.S.; Kwong, S. Genetic algorithms: Concepts and applications. *IEEE Trans. Ind. Electron.* **1996**, *43*, 519–534. [CrossRef]
37. Coley, D.A. Genetic algorithms. *Contemp. Phys.* **1996**, *37*, 145–154. [CrossRef]
38. Zhou, Y.; Zhou, L.; Wang, Y.; Yang, Z.; Wu, J. Application of Multiple-Population Genetic Algorithm in Optimizing the Train-Set Circulation Plan Problem. *Complexity* **2017**, *2017*, 3717654. [CrossRef]
39. Glover, F.; Kelly, J.P.; Laguna, M. Genetic algorithms and tabu search: Hybrids for optimization. *Comput. Oper. Res.* **1995**, *22*, 111–134. [CrossRef]
40. Glover, F. Future paths for integer programming and links to artificial intelligence. *Comput. Oper. Res.* **1986**, *13*, 533–549. [CrossRef]
41. Ma, J. Research on Countermeasures for Improving Freight Capacity of Daqin Railway line. *Railw. Freight* **2021**, *39*, 1–6.

mathematics

MDPI

Article

Impact of Trapezoidal Demand and Deteriorating Preventing Technology in an Inventory Model in Interval Uncertainty under Backlogging Situation

Rajan Mondal [1], Ali Akbar Shaikh [1], Asoke Kumar Bhunia [1], Ibrahim M. Hezam [2,*] and Ripon K. Chakrabortty [3]

1 Department of Mathematics, The University of Burdwan, Burdwan 713104, India; rmondal@scholar.buruniv.ac.in (R.M.); aliashaikh@math.buruniv.ac.in (A.A.S.); akbhunia@math.buruniv.ac.in (A.K.B.)

2 Department of Statistics and Operations Research, College of Science, King Saud University, Riyadh 11451, Saudi Arabia

3 Capability Systems Centre, School of Engineering and IT, UNSW Canberra, Campbell, ACT 2612, Australia; r.chakrabortty@adfa.edu.au

* Correspondence: ialmishnanah@ksu.edu.sa

Abstract: The demand for a product is one of the important components of inventory management. In most cases, it is not constant; it may vary from time to time depending upon several factors which cannot be ignored. For any seasonal product, it is observed that at the beginning of the season, demand escalates over time, then it is stable and after that, it decreases. This type of demand is known as the trapezoidal type. Also, due to the uncertainty of customers' behavior, inventory parameters are not always fixed. Combining these two concepts together, an inventory model is formulated for decaying items in an interval environment. Preservative technology is incorporated to preserve the product from deterioration. The corresponding mathematical formulation is derived in such a way that the profit of the inventory system is maximized. Consequently, the corresponding optimization problem is converted into an interval optimization problem. To solve the same, different variants of quantum-behaved particle swarm optimization (QPSO) techniques are employed to determine the duration of stock-in time and preservation technology cost. To illustrate and also to validate the model, three numerical examples are considered and solved. Then the computational results are compared. Thereafter, to study the impact of different parameters of the proposed model on the best found (optimal or very close to optimal) solution, sensitivity analysis are performed graphically.

Keywords: trapezoidal type demand; interval-valued inventory costs; deterioration; preservation technology; QPSO algorithms

Citation: Mondal, R.; Shaikh, A.A.; Bhunia, A.K.; Hezam, I.M.; Chakrabortty, R.K. Impact of Trapezoidal Demand and Deteriorating Preventing Technology in an Inventory Model in Interval Uncertainty under Backlogging Situation. *Mathematics* **2022**, *10*, 78. https://doi.org/10.3390/math10010078

Academic Editors: Humberto Rocha and Ana Maria Rocha

Received: 1 November 2021
Accepted: 22 December 2021
Published: 27 December 2021

1. Introduction

In the literature of inventory, it is observed that several investigators drew their attention to investigate the impact of trapezoidal type demand rate on the different inventory systems. To the best of our knowledge, Cheng and Wang [1] first proposed the idea of trapezoidal type demand in the modeling of an inventory control problem. Cheng et al. [2] expanded the model of Cheng and Wang [1] with the help of partially backlogged shortages and also the effect of deterioration. Then, Lin [3] developed an inventory model considering the demand which follows the trapezoidal pattern. After that, Chuang et al. [4] and Singh & Pattanayak [5] investigated inventory models considering trapezoidal type demand for deteriorating items. Lin et al. [6] wrote a note on Cheng et al. [2] based on modeling and solutions. Mishra [7] introduced a deteriorating inventory model considering deterioration prevention technology and trapezoidal demand. Wu et al. [8] developed two inventory models with trapezoidal demand, time-dependent deterioration, and completely backlogged shortages. Recently, Vandana and Srivastava [9] developed an inventory model for ameliorating items with trapezoidal demand and complete shortages under inflation conditions. Wu et al. [10] formulated an inventory model with trapezoidal demand and the rate of decaying is dependent on the maximum lifetime of an item along with trade credit facilities. Garai et al. [11] proposed a fuzzy inventory model with time-varying holding cost under price-dependent demand. Xu et al. [12] studied an inventory model for nonperishable items with trapezoidal type demand and partial backlogging shortages. Kumar [13] investigated a fuzzy inventory model with trapezoidal demand and time-varying holding costs under shortages.

Usually, the selling price of an item is not always fixed. It may vary from time to time within a certain range. In this connection, different types of costs, like ordering cost, carrying cost, shortage cost, etc. may also vary. So, the authors should give attention to the flexible nature of the system parameters in the formulation of the inventory model. The impreciseness of inventory parameters can be represented with the help of fuzzy, probabilistic, and interval approaches. In this connection, one may refer to the works of Kazemi et al. [14], De and Sana [15], Mondal et al. [16], Mondal et al. [17], and De et al. [18] in which the imprecise parameters are represented by either fuzzy sets or fuzzy numbers. Representing the impreciseness by random variables, the works of Pulido-Rojano [19] and Adak and Mahapatra [20] are worth mentioning. Using the interval approach, Dutta and Kumar [21], Bhunia and Shaikh [22], and Bhunia et al. [23] proposed several inventory models. Over the last two decades, several researchers applied the concept of interval uncertainty in inventory control theory and formulated several inventory models. To the best of our knowledge, Gupta et al. [24] first applied this concept in the formulation of their inventory model and solved the corresponding interval optimization problem by using a modified genetic algorithm. Then, Gupta et al. [25] proposed another inventory model and solved it with the help of a genetic algorithm. They considered the concept of the advance payment and assumed the inventory costs as interval-valued. Chakrabortty et al. [26] developed an algorithm for solving an inventory problem under an interval environment. Dutta and Kumar [21] developed a deteriorating inventory model along with time-varying holding cost and demand. Bhunia and Shaikh [22] proposed two warehouse inventory models under inflation in an interval environment. Bhunia et al. [23] formulated a partially integrated production model with variable demand and reliability of the product in an interval environment. Mondal et al. [27] introduced an ameliorating inventory model for deteriorating items in crisp and interval environments. They have solved the corresponding optimization problem with the help of different variants of quantum-behaved particle swarm optimization techniques. Shaikh et al. [28] studied an inventory problem of a two-warehouse system for non-instantaneous deteriorating items in an interval environment. Rahman et al. [29] proposed a parametric approach of interval in formulating an inventory model with price-dependent demand. Ruidas, et al. [30] developed an interval-valued production inventory model with price-sensitive demand under interval-valued carbon emission.

The products like pharmaceuticals, blood, food items, chemicals, and radioactive chemicals deteriorate very fast with time. Various factors like heat, worm effect, vaporization, dryness, perishability, spoilage, lack of preservation facility, etc. are responsible for this deterioration. The loss that occurs due to the effect of deterioration cannot be neglected in the inventory analysis. In 1963, Ghare and Schrader [31] first proposed the concept of decaying of the product in the modeling of an inventory control problem and they formulated an inventory model with an exponentially decaying rate. Covert and Phillip [32] extended the work of Ghare and Schrader [31] by considering Weibull distributed deterioration rate. After that, several works were developed assuming fixed or variable deterioration rates. Mahapatra et al. [33] proposed an inventory model with reliability-dependent demand for deteriorating items. Shaikh et al. [34] developed a stock and price-related inventory model for non-instantaneous decaying items. Shah and Naik [35] proposed a non-instantaneous decaying model assuming price-sensitive demand and considering learning effects. Chen et al. [36] developed an optimal pricing inventory model by taking stock-level, price, and time-dependent demand for decaying items. Mahmoodi [37] introduced the concept of duopoly retailers and formulated a deteriorating inventory model with a linear trend in demand. Saha and Sen [38] proposed a price-dependent inventory model for deteriorating items considering shortages. Khakzad and Gholamian [39] introduced an advance payment-related inventory model with the effect of the inspection rate of deterioration. Khan et al. [40] discussed the effect of non-instantaneous deterioration in a two-warehouse system under advance payment and shortages. Xu et al. [41] studied the strategy of inventory control for deteriorating items with time-varying demand and carbon emission regulations. A comparative study between the proposed work and the related works reported in the existing literature is shown in Table 1. To preserve the product in store room, a preservation technology cost is required. This cost undoubtedly affects inventory control optimization. Dye [42] proposed a non-instantaneous decaying inventory model considering preservation facility.

Table 1. Literature review related to the proposed model.

Reported Articles	Type of Model	Deterioration	Backlog-ging Situation	Demand Type	Preventing Technology	Uncertainty	Solution Procedure
Wahab et al. [43]	Purchase model	×	×	-	×	Not considered	Gradient best numerical method
Cheng et al. [2]	Purchase model	√	√	Trapezoidal demand	×	Not considered	Gradient best numerical method
Zhao, L. [44]	Purchase model	√	√	Trapezoidal demand	×	Not considered	Gradient best numerical method
Wu et al. [8]	Purchase model	√	√	Trapezoidal demand	×	Not considered	Gradient best numerical method
Bhunia and Shaikh [22]	Purchase model	√	√	-	×	Interval	Different variants of PSO
Taleizadeh et al. [45]	Purchase model	×	×	-	×	Not considered	Gradient best numerical method
Wu et al. [10]	Purchase model	√	√	Trapezoidal demand	×	Not considered	Gradient best numerical method
Mondal et al. [27]	Purchase model	√	×	Price dependent	×	Crisp and Interval	Different variants of QPSO techniques
Rahman et al. [46]	Purchase model	√	×	Known and constant	√	Interval-valued	Different variants of QPSO techniques
Dey et al. [47]	Supply chain	×	×	Advertisement dependent demand	×	Not considered	Gradient best numerical method
Shaikh et al. [48]	Purchase model	√	√	Price dependent	×	Crisp	Multi-section Method
Jabbarzadeh et al. [49]	Purchase model	Shaikh et al. [48] ×	√	-	×	Crisp	Signomial Geometric Programming

Table 1. *Cont.*

Reported Articles	Type of Model	Deterioration	Backlog-ging Situation	Demand Type	Preventing Technology	Uncertainty	Solution Procedure
This work	inventory model	√	√	trapezoidal demand	√	Interval	Different variants of QPSO techniques

Singh and Rathore [50] proposed a trade credit policy-oriented inventory model for deteriorating items under preservation facility. Tayal et al. [51] investigated a production inventory model with a preservation facility for a deteriorating item. Mishra et al. [52] formulated a deteriorated inventory model taking the impact of decaying reduction technology investment. They considered stock and price-dependent demand with shortages in their model. Mishra et al. [53] proposed an inventory model with a preservation facility for the deteriorating item under a trade credit facility. Bardhan et al. [54] applied the concept of reduction technology in the modeling of inventory control Shah et al. [55] proposed an inventory model with preservation investment Das et al. [56] introduced an inventory model with price dependent demand under preservation investment and backlogging. Khanna and Jaggi [57] formulated an inventory model with a preservation facility considering the price and stock-dependent demand.

In the existing literature, several research works are available for solving the interval-valued optimization problem. Bhunia and Shaikh [22] developed a two-warehouse inventory model for the deteriorating item under inflation with interval-valued inventory cost. Bhunia et al. [23] introduced a production inventory model with a reliability factor of the product in an interval environment. Shaikh et al. [28] proposed an inventory model for stock-dependent demand with inventory costs as interval-valued. Rahman et al. [58] studied an inventory model in interval environment with parametric approach of interval. To the best of our knowledge, no one solved the inventory model with trapezoidal demand for deteriorating items considering preservation facility, partially backlogged shortages along interval-valued inventory costs. The proposed work is developed for decaying items considering trapezoidal type demand, preservation technology, and completely backlogged shortages. Also, the cost of inventory parameters is considered interval-valued. Due to the consideration of interval-valued inventory cost parameters, the corresponding optimization problem is converted into an interval-valued optimization problem. Also, this optimization problem is highly nonlinear in nature. So, it cannot be solved with the help of classical and numerical gradient-based optimization techniques. Due to this limitation, interval order relation and different variants of the quantum-behaved particle swarm optimization technique (QPSO) are used. These techniques are modified with the interval fitness to solve the interval-valued optimization problem. Finally, sensitivity analyses are presented graphically for Example 3 to show the impact on the best found (optimal) policies.

The remaining paper is organized in the following ways: Section 2 represents notations. In Section 3, assumptions of the proposed model are mentioned. Mathematical formulations are derived in Section 4. Section 5 represents the numerical solution of the proposed model. A sensitivity analysis is performed in Section 6. Section 7 represents some managerial insight into the proposed model. Finally, a conclusion is made in Section 8.

2. Notation

S	Initial inventory level
$D(t)$	Time-dependent trapezoidal demand rate
a, c, b, e_1	Constant demand parameters
θ	Constant deterioration rate
D'	Total deteriorated units throughout the business period
$[C_{pL}, C_{pU}]$	Purchasing cost (\$)/unit.
$[p'_L, p'_U]$	Interval valued salvage value (\$)/unit $(p'_U < C_{pL})$
ξ	Preservation cost (\$)/unit/unit time
$m(\xi)$	Preservation technology function
$[C_{OL}, C_{OU}]$	Replenishment cost (\$)
$[C_{hL}, C_{hU}]$	Interval valued inventory holding cost (\$)/unit/unit time
p	Selling price (\$)/unit
t_1	Stock-in period
T	Cycle length
R	Maximum shortage level
γ_1, γ_2	Time points of trapezoidal demand in which the demand becomes constant during the time period $[\gamma_1, \gamma_2]$
SR	Sales revenue
$[c_{bL}, c_{bU}]$	Interval valued shortage cost/unit/unit time
$[c_{lL}, c_{lU}]$	Interval-valued lost-sale cost
δ	Backlogging rate
TC	Crisp valued total system cost (\$)
$[TC_L, TC_U]$	Interval-valued total cost of the system (\$)
$Z/[Z_L, Z_U]$	Crisp/ Interval valued average profit (\$)

3. Assumptions

Basically, the proposed model is developed based on trapezoidal type demand, deterioration, preservation facility, backlogged shortage, and interval-valued inventory costs. The following assumptions are considered before developing this type of particular inventory model.

(i) The replenishment rate is infinite.

(ii) The demand pattern is following a trapezoidal function of time whose mathematical form is as follows (the pictorial view is shown in Figure 1):

$$D(t) = \begin{cases} a + bt, & t \leq \gamma_1 \\ c, & \gamma_1 < t \leq \gamma_2 \\ c - e_1(t - \gamma_2), & \gamma_2 < t \leq T \end{cases}$$

(iii) The deterioration rate $\theta(0 < \theta << 1)$ is constant.

(iv) To prevent the decaying rate, deterioration reduction technology is incorporated on the item, and a preservation technology functions $m(\xi) = \frac{a_1\xi}{1+a_1\xi}$, $a_1 > 0$ or $m(\xi) = 1 - \exp(-a_2\xi)$, $a_2 > 0$ is considered Hsu et al. [59], Hasan et al. [60], Masud et al. [61], Dye [42], Yang et al. [62], and Das et al. [56,63]. It should be noted that $m(\xi)$ is an increasing function with $m''(\xi) < 0$.

(v) Various costs related to inventory, like purchasing cost, holding cost, ordering cost are known and interval types due to the uncertainty of marketing price.

(vi) Shortages are allowed and it is completely backlogged.

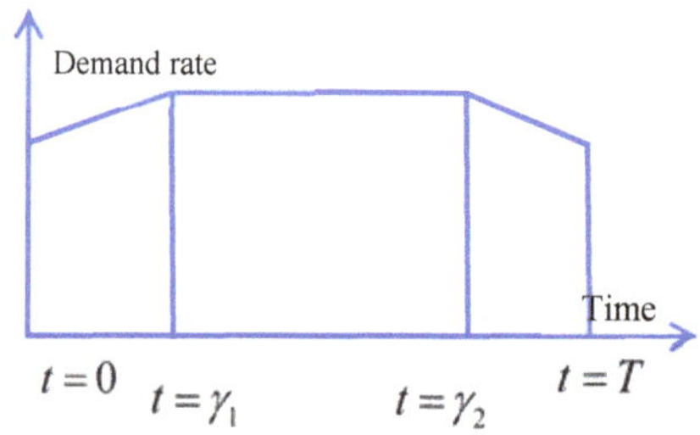

Figure 1. Pictorial representation of trapezoidal demand pattern function of the demand rate.

4. Mathematical Formulation

It is assumed that before the beginning of an inventory cycle, an enterprise makes an order of $(S + R)$ units of a perishable item. After receiving the lot at the beginning ($t = 0$), R units are utilized to satisfy the backlogged quantities of an earlier cycle and the remaining stock becomes S units. After that, the level of inventory gradually decreases due to the combined effects of deterioration and customers' requirements. Finally, at the time point, $t = t_1$, the level of inventory reaches zero. Thereafter, the stock-out situation occurs and at the end of the cycle i.e., at time point $t = T$ along with the maximum shortage R units. Then the entire cycle is repeating itself.

According to the assumptions, the behavior of inventory level at any time t can be presented with the help of the following differential Equations:

$$q'(t) = -D(t) - \theta\{1 - m(\xi)\}q(t), \; 0 \leq t \leq t_1 \quad (1)$$

$$q'(t) = -D(t), \;\; t_1 < t \leq T \quad (2)$$

with $q(t_1) = 0$.

From the demand function, one can easily obtain the relations $\gamma_1 = (c - a)/b$, and $\gamma_2 = T - (c - a)/e_1$. Depending upon the time t_1, γ_1 and γ_2, three cases may arise:

Case-I: $0 \leq t_1 \leq \gamma_1$

Case-II: $\gamma_1 < t_1 \leq \gamma_2$

Case-III:$\gamma_2 < t_1 \leq T$

Now, all the cases are discussed in detail.

Case-I: $0 \leq t_1 \leq \gamma_1$

The level of inventory depletes due to the trapezoidal type of demand and constant decaying rate with preservation technology during the time period $[0, t_1]$ and it becomes empty at time $t = t_1$ (see Figure 2). Then from Equations (1) and (2), one can write

$$q'(t) = -(a + bt) - kq(t), \; 0 < t \leq t_1 \quad (3)$$

$$q'(t) = -\delta(a + bt), \;\; t_1 < t \leq \gamma_1 \quad (4)$$

$$q'(t) = -\delta c, \;\; \gamma_1 < t \leq \gamma_2 \quad (5)$$

and

$$q'(t) = -\delta\{c - e_1(t - \gamma_2)\}, \;\; \gamma_2 < t \leq T \quad (6)$$

with the conditions

$$q(0) = S, q(t_1) = 0 \text{ and } q(T) = -R \quad (7)$$

where $k = \theta\{1 - m(\xi)\}$.

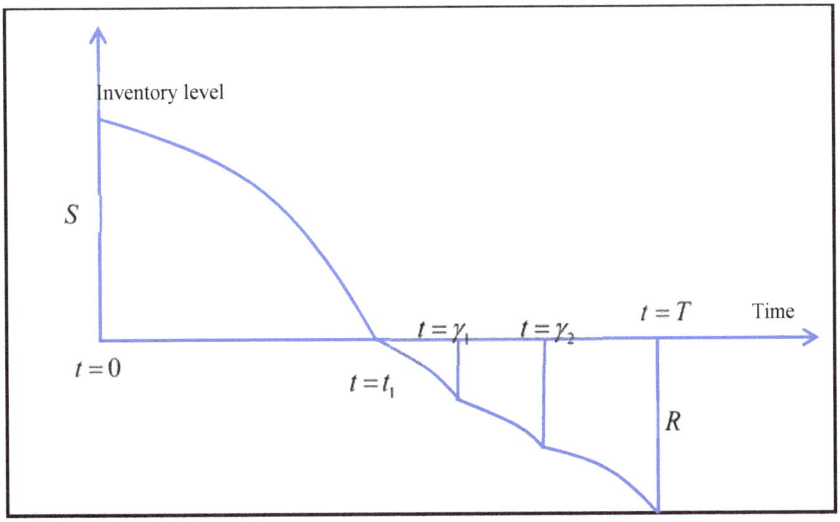

Figure 2. Pictorial representation of inventory situation under Case-I.

The solutions of the differential Equations (3)–(6) with the condition (7) are given by

$$q(t) = -\left(\frac{a+bt}{k} - \frac{b}{k^2}\right) + \left\{\frac{a+bt_1}{k} - \frac{b}{k^2}\right\}\exp\{k(t_1 - t)\},\ 0 \le t \le t_1 \tag{8}$$

$$q(t) = \delta\left\{a(t_1 - t) + \frac{b}{2}(t_1^2 - t^2)\right\},\ t_1 < t \le \gamma_1 \tag{9}$$

$$q(t) = \delta\{c(\gamma_1 - t) + k_1\},\ \gamma_1 < t \le \gamma_2 \tag{10}$$

$$q(t) = \delta\left\{k_2 + c\gamma_2 + \frac{e_1}{2}\gamma_2^2 - ct + e_1\left(\frac{t^2}{2} - \gamma_2 t\right)\right\},\ \gamma_2 < t \le T \tag{11}$$

where $k = \theta\{1 - m(\xi)\}$

$$k_1 = q(\gamma_1) = a(t_1 - \gamma_1) + \frac{b}{2}\left(t_1^2 - \gamma_1^2\right)$$

$$k_2 = q(\gamma_2) = c(\gamma_1 - \gamma_2) + k_1$$

$$k_3 = c\gamma_2 + \frac{e_1}{2}\gamma_2^2 + k_2$$

Again, condition (7) implies

$$S = q(0) = \left(\frac{a+bt_1}{k} - \frac{b}{k^2}\right)\exp(kt_1) - \left(\frac{a}{k} - \frac{b}{k^2}\right) \tag{12}$$

As at the time $t = T$, $q(t) = -R$, so the maximum shortage level R is given by

$$R = \delta\left\{cT - e_1\left(\frac{T^2}{2} - \gamma_2 T\right) - k_2 - c\gamma_2 - \frac{e_1}{2}\gamma_2^2\right\}$$

The number of units that deteriorated during the time $[0, t_1]$ is given by

$$\begin{aligned} D' &= S - \int_0^{t_1} D(t)dt = S - \int_0^{t_1} (a+bt)dt \\ &= S - at_1 - \tfrac{b}{2}t_1^2 \end{aligned} \tag{13}$$

The total salvage value throughout the cycle T is given by $[p'_L, p'_U]D'$.
The bounds of the carrying cost are $C_{hL}H_1$ and $C_{hU}H_1$, where

$$H_1 = \int_0^{t_1} q(t)dt = \left(\frac{a+bt_1}{k^2} - \frac{b}{k^3}\right)\{\exp(kt_1) - 1\} - \frac{at_1}{k} + \frac{bt_1}{k^2} + \frac{bt_1^2}{2k}$$

Again, the total shortage of units throughout the entire cycle $[t_1, T]$ are given by

$$SC = -\int_{t_1}^{T} q(t)dt = -\int_{t_1}^{\gamma_1} q(t)dt - \int_{\gamma_1}^{\gamma_2} q(t)dt - \int_{\gamma_2}^{T} q(t)dt$$
$$= \delta\left\{a\gamma_1\left(\tfrac{\gamma_1}{2} - t_1\right) + \tfrac{b\gamma_1}{2}\left(\tfrac{\gamma_1^2}{3} - t_1^2\right) + a\tfrac{t_1^2}{2} + \tfrac{b}{3}t_1^3\right\} + \delta\left\{c\gamma_2\left(\tfrac{\gamma_2}{2} - \gamma_1\right) - k_1\gamma_2 + \tfrac{c}{2}\gamma_1^2 + k_1\gamma_1\right\}$$
$$+\delta\left\{\tfrac{cT^2}{2} - e_1T^2\left(\tfrac{T}{6} - \tfrac{\gamma_2}{2}\right) - k_3T - \tfrac{c}{2}\gamma_2^2 - e_1\tfrac{\gamma_2^3}{3} + k_3\gamma_2\right\}$$

The total shortage cost for the entire cycle is $[c_{bL}, c_{bU}]SC$.
Lost sale cost,

$$[\mathrm{LSC}_L,\ LSC_U] = [c_{lL}, c_{lU}](1-\delta)\left[\int_{t_1}^{\gamma_1}(a+bt)\,dt + \int_{\gamma_1}^{\gamma_2} c\,dt + \int_{\gamma_2}^{T}\{c - e_1(t-\gamma_2)\}\,dt\right]$$
$$= [c_{lL}, c_{lU}](1-\delta)\left[a(\gamma_1 - t_1) + \tfrac{b}{2}\left(\gamma_1^2 - t_1^2\right) + c(T-\gamma_1) - \tfrac{e_1}{2}(T-\gamma_2)^2\right]$$

Thus the bounds of lost sale cost are

$$\mathrm{LSC}_L = c_{lL}(1-\delta)\left[a(\gamma_1 - t_1) + \frac{b}{2}\left(\gamma_1^2 - t_1^2\right) + c(T-\gamma_1) - \frac{e_1}{2}(T-\gamma_2)^2\right]$$

and

$$\mathrm{LSC}_U = c_{lU}(1-\delta)\left[a(\gamma_1 - t_1) + \frac{b}{2}\left(\gamma_1^2 - t_1^2\right) + c(T-\gamma_1) - \frac{e_1}{2}(T-\gamma_2)^2\right]$$

Preservation cost $= \zeta T$ per cycle
Ordering cost $= [C_{OL}, C_{OU}]$ per cycle

$$\text{Sales revenue } (SR) = p\int_0^{t_1} D(t)dt + pR = p\int_0^{t_1}(a+bt)dt + pR = p\left(at_1 + \frac{bt_1^2}{2}\right) + pR$$

System Cost:
The total system cost is given by

$$[TC_L, TC_U]$$

where $TC_L = C_{OL} + C_{pL}(S+R) + C_{hL}H_1 + c_{bL}SC + \mathrm{LSC}_L + \zeta T$, and $TC_U = C_{OU} + C_{pU}(S+R) + C_{hU}H_1 + c_{bU}SC + \mathrm{LSC}_U + \zeta T$.

Profit Function:

So, the profit function per unit time Z with respect to two variables t_1 and ξ.

Hence, the profit per unit time is given by $[Z_L(t_1,\xi),\ Z_U(t_1,\xi)]$, where $Z_L(t_1,\xi) = \frac{1}{T}[SR + p'_L D' - TC_U]$ and $Z_U(t_1,\xi) = \frac{1}{T}[SR + p'_U D' - TC_L]$.

Therefore, the related optimization problem can be written as:

Maximize $[Z_L(t_1,\xi),\ Z_U(t_1,\xi)]$,

With the conditions $t_1,\ \xi > 0$.

Case-II: $\gamma_1 < t_1 \le \gamma_2$

In this case, from the starting point of the cycle, the level of inventory depletes due to the trapezoidal type demand and constant deterioration rate with preservation technology throughout the time period $[0, t_1]$ and it reaches zero level at the time $t = t_1$ (see Figure 3). Then, from Equations (1) and (2), we have

$$q'(t) = -(a+bt) - kq(t),\ 0 \le t \le \gamma_1 \tag{14}$$

$$q'(t) = -c - kq(t),\ \gamma_1 < t \le t_1 \tag{15}$$

$$q'(t) = -\delta c,\ t_1 < t \le \gamma_2 \tag{16}$$

and

$$q'(t) = -\delta\{c - e_1(t-\gamma_2)\},\ \gamma_2 < t \le T \tag{17}$$

with

$$q(t) = S \text{ at } t = 0 \text{ and } q(t) = 0 \text{ at} \tag{18}$$

where $k = \theta\{1 - m(\xi)\}$.

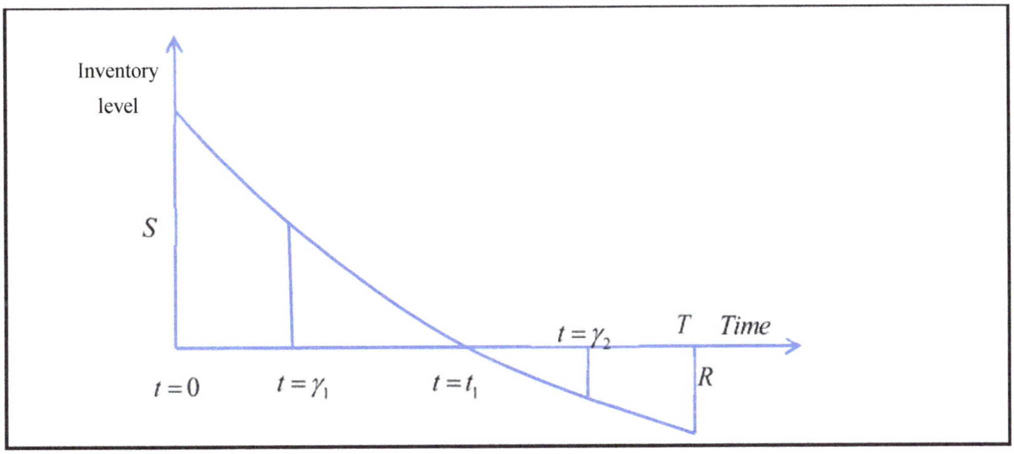

Figure 3. Pictorial representation of inventory situation under Case-II.

The solutions of the differential Equations (14)–(17) with the condition (18) are given by

$$q(t) = -\frac{a}{k} - \frac{bt}{k} + \frac{b}{k^2} + \left\{S + \frac{a}{k} - \frac{b}{k^2}\right\}\exp(-kt),\ 0 < t \le \gamma_1 \tag{19}$$

$$q(t) = \frac{c}{k}[\exp\{k(t_1 - t)\} - 1],\ \gamma_1 < t \le t_1 \tag{20}$$

$$q(t) = c\delta(t_1 - t),\ t_1 < t \le \gamma_2 \tag{21}$$

$$q(t) = \delta\left\{-ct + e_1\left(\frac{t^2}{2} - \gamma_2 t\right) + k_3\right\},\ \gamma_2 < t \le T \tag{22}$$

where $k = \theta\{1 - m(\xi)\}$

$$k_1 = \left[\tfrac{c}{k}[\exp\{k(t_1-\gamma_1)\} - 1] + \tfrac{a}{k} + \tfrac{b\gamma_1}{k} - \tfrac{b}{k^2}\right]\exp(k\gamma_1)$$
$$k_2 = q(\gamma_2) = c(t_1 - \gamma_2)$$
$$\text{and } k_3 = k_2 + \left\{c\gamma_2 + e_1\tfrac{\gamma_2^2}{2})\right\}$$

Now, $q(t) = S$ at $t = 0$ implies

$$S = q(0) = \frac{b}{k^2} - \frac{a}{k} + k_1 \tag{23}$$

At the time $t = T$, $q(t) = -R$, so the highest shortage level R is given by

$$R = \delta\left\{cT - e_1\left(\frac{T^2}{2} - \gamma_2 T\right) - k_3\right\}$$

The total number of units that deteriorate throughout the period $[0, t_1]$ is given by

$$\begin{aligned} D' &= S - \int_0^{t_1} D(t)dt = S - \int_0^{\gamma_1} (a+bt)dt - \int_{\gamma_1}^{t_1} c\,dt \\ &= S - a\gamma_1 - \tfrac{b}{2}\gamma_1^2 - c(t_1 - \gamma_1) \end{aligned} \tag{24}$$

The total salvage value throughout the cycle time is $[p'_L, p'_U]D'$.
The bounds of the carrying cost of the system are $C_{hL}H_2$ and $C_{hU}H_2$ where

$$\begin{aligned} H_2 &= \int_0^{t_1} q(t)dt = \int_0^{\gamma_1} q(t)dt + \int_{\gamma_1}^{t_1} q(t)dt \\ &= \left[-\tfrac{a\gamma_1}{k} - \tfrac{b}{2k}\gamma_1^2 + \tfrac{b\gamma_1}{k^2} + \tfrac{1}{k}\left(S + \tfrac{a}{k} - \tfrac{b}{k^2}\right)\{1 - \exp(-k\gamma_1)\}\right] + \tfrac{c}{k^2}[\exp\{k(t_1 - \gamma_1)\} - k(t_1 - \gamma_1) - 1] \end{aligned} \tag{25}$$

The total shortage unit throughout the period $[t_1, T]$ is given by

$$\begin{aligned} SC &= -\int_{t_1}^{T} q(t)dt = -\int_{t_1}^{\gamma_2} q(t)dt - \int_{\gamma_2}^{T} q(t)dt \\ &= \tfrac{c\delta}{2}(\gamma_2 - t_1)^2 + \delta\left\{\tfrac{cT^2}{2} - e_1\left(\tfrac{T^3}{6} - \tfrac{\gamma_2 T^2}{2}\right) - k_3 T\right\} - \delta\left\{\tfrac{c\gamma_2^2}{2} - e_1\tfrac{\gamma_2^3}{6} - k_3\gamma_2\right\} \end{aligned} \tag{26}$$

Preservation cost $= \zeta T$ per cycle.
Ordering cost $= [C_{OL}, C_{OU}]$ per cycle.
The total shortage cost for the entire cycle T is $[c_{bL}, c_{bU}]SC$.
Lost sale cost, $[\mathrm{LSC}_L, \mathrm{LSC}_U] = [c_{lL}, c_{lU}](1-\delta)\left[\int_{t_1}^{\gamma_2} c\,dt + \int_{\gamma_2}^{T}\{c - e_1(t - \gamma_2)\}\,dt\right]$
$= [c_{lL}, c_{lU}](1-\delta)\left[c(T - t_1) - \frac{e_1}{2}(T - \gamma_2)^2\right]$.
Thus, the bounds of lost sale cost are

$$[\mathrm{LSC}_L, \mathrm{LSC}_U] = \left[c_{lL}(1-\delta)\left\{c(T-t_1) - \frac{e_1}{2}(T-\gamma_2)^2\right\}, c_{lU}(1-\delta)\left\{c(T-t_1) - \frac{e_1}{2}(T-\gamma_2)^2\right\}\right]$$

$$\text{Sales revenue } (SR) = p\int_0^{t_1} D(t)dt + pR = p\left\{\int_0^{\gamma_1}(a+bt)dt + \int_{\gamma_1}^{t_1} c\,dt\right\} + pR = p\left\{a\gamma_1 + \frac{b\gamma_1^2}{2} + c(t_1 - \gamma_1)\right\} + pR$$

System Cost:
The total system cost is given by $[TC_L, TC_U]$, where $TC_L = C_{OL} + C_{pL}(S+R) + C_{hL}H_2 + c_{bL}SC + \mathrm{LSC}_L + \zeta T$ and $TC_U = C_{OU} + C_{pU}(S+R) + C_{hU}H_2 + c_{bU}SC + \mathrm{LSC}_U + \zeta T$.
Profit Function:
So, the profit function Z is a function of two variables t_1 and ξ.
Hence, the profit function can be written as $[Z_L(t_1,\xi), Z_U(t_1,\xi)]$, where $Z_L(t_1,\xi) = \frac{1}{T}[SR + p'_L D' - TC_U]$ and $Z_U(t_1,\xi) = \frac{1}{T}[SR + p'_U D' - TC_L]$.
Again, the corresponding optimization problem is given by
Maximize $[Z_L(t_1,\xi), Z_U(t_1,\xi)]$,
Subject to t_1, $\xi > 0$.
Case-III: $\gamma_2 < t_1 \le T$
In this case, from the starting of the entire cycle, the inventory level depletes due to the combined effect of constant decaying rate with preservation technology and demand

of an item throughout the time period $[0, t_1]$. Finally, it reaches to empty level at the time $t = t_1$ (see Figure 4). Then from Equations (1) and (2), we have

$$q'(t) = -(a+bt) - kq(t),\ 0 < t \le \gamma_1 \tag{27}$$

$$q'(t) = -c - kq(t),\ \gamma_1 < t \le \gamma_2 \tag{28}$$

$$q'(t) = -\{c - e_1(t-\gamma_2)\} - kq(t),\ \gamma_2 < t \le t_1 \tag{29}$$

$$\text{and } q'(t) = -\delta\{c - e_1(t-\gamma_2)\},\ t_1 < t \le T \tag{30}$$

with

$$q(0) = S,\ q(t_1) = 0 \text{ and } q(T) = -R. \tag{31}$$

$q(t)$ is also continuous at $t = \gamma_1$ and γ_2.

Solving the differential Equations (27)–(30) with the conditions (31) are given by

$$q(t) = -\frac{a}{k} - \frac{bt}{k} + \frac{b}{k^2} + (S + \frac{a}{k} - \frac{b}{k^2})\exp(-kt),\ 0 < t \le \gamma_1 \tag{32}$$

$$q(t) = -\frac{c}{k} + k_4\exp\{k(\gamma_2 - t)\},\ \gamma_1 < t \le \gamma_2 \tag{33}$$

$$q(t) = -\left[\frac{c}{k} - \frac{e_1(t-\gamma_2)}{k} + \frac{e_1}{k^2}\right] + k_1\exp\{k(t_1 - t)\},\ \gamma_2 < t \le t_1 \tag{34}$$

$$q(t) = \delta\left\{-ct + e_1\left(\frac{t^2}{2} - \gamma_2 t\right) + k_7 - R\right\},\ t_1 < t \le T \tag{35}$$

where $k = \theta\{1 - m(\xi)\}$,

$$\begin{aligned}
k_1 &= \tfrac{c}{k} - \tfrac{e_1(t_1-\gamma_2)}{k} + \tfrac{e_1}{k^2},\\
k_2 &= \tfrac{c}{k} + \tfrac{e_1}{k^2},\\
k_3 &= -k_2 + k_1\exp\{k(t_1-\gamma_2)\},\\
k_4 &= k_3 + \tfrac{c}{k},\\
k_5 &= -\tfrac{c}{k} + k_4\exp\{k(\gamma_2-\gamma_1)\},\\
k_6 &= \tfrac{a}{k} + \tfrac{b\gamma_1}{k} - \tfrac{b}{k^2},\\
k_7 &= cT - e_1\left(\tfrac{T^2}{2} - \gamma_2 T\right).
\end{aligned}$$

Now, $q(0) = S$ implies

$$S = (k_5 + k_6)\exp(k\gamma_1) - \frac{a}{k} + \frac{b}{k^2} \tag{36}$$

At the time $t = T$, $q(t) = -R$, so the highest shortage level R is given by

$$R = \delta\left\{k_7 - ct_1 + e_1\left(\frac{t_1^2}{2} - \gamma_2 t_1\right)\right\}$$

The total number of units deteriorated throughout the time $t = 0$ to $t = t_1$ is

$$D' = S - \int_0^{t_1} D(t)dt = S - \left[a\gamma_1 + b\frac{\gamma_1^2}{2} + c(t_1-\gamma_1) - \frac{e_1}{2}(t_1-\gamma_2)^2\right] \tag{37}$$

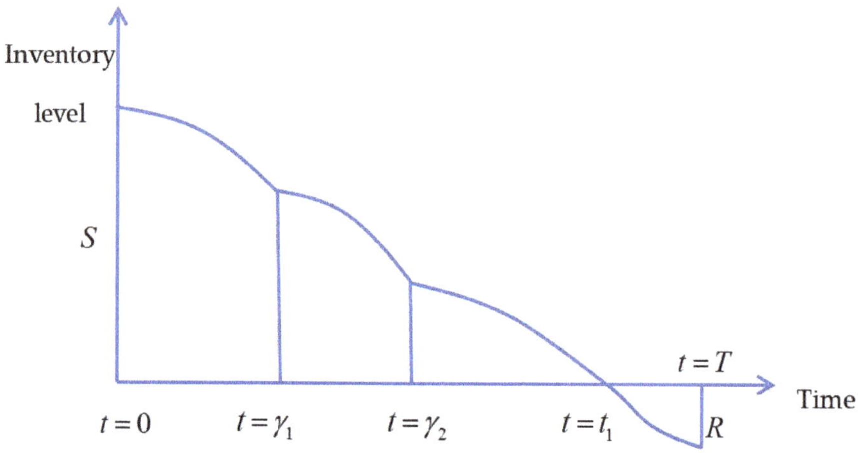

Figure 4. Pictorial representation of inventory situation under Case-III $\gamma_2 < t_1 \leq T$.

The total salvage value throughout the period is $[p'_L, p'_U]D'$.
The bounds of the carrying cost of the system are $C_{hL}H_3$ and $C_{hU}H_3$ where

$$H_3 = \int_0^{t_1} q(t)dt = \int_0^{\gamma_1} q(t)dt + \int_{\gamma_1}^{\gamma_2} q(t)dt + \int_{\gamma_2}^{t_1} q(t)dt$$
$$= \tfrac{1}{k}\left(S + \tfrac{a}{k} - \tfrac{b}{k^2}\right)\{1 - \exp(-k\gamma_1)\} - \left(\tfrac{a\gamma_1}{k} + \tfrac{b\gamma_1^2}{2k} - \tfrac{b\gamma_1}{k^2}\right) + \tfrac{k_4}{k}[\exp\{k(\gamma_2 - \gamma_1)\} - 1] - \tfrac{c}{k}((\gamma_2 - \gamma_1))$$
$$+ \tfrac{k_1}{k}[\exp\{k(t_1 - \gamma_2)\} - 1] - \left[\tfrac{c}{k}(t_1 - \gamma_2) - \tfrac{e_1}{2k}(t_1 - \gamma_2)^2 + \tfrac{e_1}{k^2}(t_1 - \gamma_2)\right]$$

The total shortage of units throughout the period $[t_1, T]$ is given by

$$SC = -\int_{t_1}^{T} q(t)dt = \delta\left\{(R - k_7)(T - t_1) + \frac{1}{2}(c - \gamma_2)\left(T^2 - t_1^2\right) - \frac{e_1}{6}\left(T^3 - t_1^3\right)\right\}$$

Preservation cost $= \zeta T$ per cycle,
Ordering cost $= [C_{0L}, C_{0U}]$ per cycle,
The total shortage cost for the entire cycle T is $[c_{bL}, c_{bU}]SC$.
The lost sale cost is given by

$$[\mathrm{LSC}_L, \mathrm{LSC}_U] = [c_{lL}, c_{lU}](1 - \delta)\left[\int_{t_1}^{T} \{c - e_1(t - \gamma_2)\}\,dt\right] = [c_{lL}, c_{lU}](1 - \delta)\left[c(T - t_1) - \frac{e_1}{2}\left\{(T - \gamma_2)^2 - (t_1 - \gamma_2)^2\right\}\right]$$

Thus, the bounds of lost sale cost are

$$\mathrm{LSC}_L = c_{lL}(1 - \delta)\left[c(T - t_1) - \frac{e_1}{2}\left\{(T - \gamma_2)^2 - (t_1 - \gamma_2)^2\right\}\right]$$

and $\mathrm{LSC}_U = c_{lU}(1 - \delta)\left[c(T - t_1) - \frac{e_1}{2}\left\{(T - \gamma_2)^2 - (t_1 - \gamma_2)^2\right\}\right]$.

Sales revenue (SR) $= p\int_0^{t_1} D(t)dt + pR = p\left[a\gamma_1 + \frac{b\gamma_1^2}{2} + c(t_1 - \gamma_1) - \frac{e_1}{2}(t_1 - \gamma_2)^2\right] + pR$

System Cost:

The total system cost is given by $[TC_L, TC_U]$, where $TC_L = C_{0L} + C_{pL}(S + R) + C_{hL}H_3 + c_{bL}SC + \mathrm{LSC}_L + \zeta T$ and $TC_U = C_{0U} + C_{pU}(S + R) + C_{hU}H_3 + c_{bU}SC + \mathrm{LSC}_U + \zeta T$.

Profit Function:

So, the profit function Z is a function concerning two variables t_1 and ξ.

Hence, the profit per unit time can be written as $[Z_L(t_1,\xi),\, Z_U(t_1,\xi)]$, where $Z_L(t_1,\xi) = \frac{1}{T}[SR + p'_L D' - TC_U]$ and $Z_U(t_1,\xi) = \frac{1}{T}[SR + p'_U D' - TC_L]$.

Again, the related optimization problem can be written as

$$\text{Maximize } [Z_L(t_1,\xi),\, Z_U(t_1,\xi)] \tag{38}$$

subject to $t_1,\ \gamma_2,\ \xi > 0$ and $t_1 > \gamma_2$.

5. Numerical Illustration

To validate and also to illustrate the proposed models, three numerical examples are considered and solved. The best-found solutions for the feasible cases of each example are shown in Tables 2–10. To solve each optimization problem of the hypothetical inventory model, different variants of QPSO, viz. GQPSO, AQPSO, and WQPSO techniques are used and these algorithms are coded in C language. The corresponding computational works are performed on a laptop with the configuration Intel core i-3 with 2.40 GHz 7th generation processor in the Linux operating system. Every algorithm is run 50 times independently to solve each example. It is also to be mentioned that the obtained results are called best-found solutions which are either optimal or nearer to the optimal solution. The corresponding results of these computations are shown in Tables 2–4 for Example 1, Tables 5–7 for Example 2, and Tables 8–10 for Example 3.

Example 1. *The values of different parameters of the proposed models are as follows:*

$C_{pL} = \$12/unit$, $C_{pU} = \$14/unit$, $p = \$32/unit$, $a = 15$, $b = 2.8$, $c = 70$, $C_{0L} = \$95/order$, $C_{0U} = \$105$ /order, $e_1 = 15$, $C_{hL} = \$0.9/unit$, $C_{hU} = \$1.1/unit$, $\theta = 0.1$, $a_2 = 0.1$, $c_{bL} = \$1.5/unit/unit\ time$, $c_{bU} = \$2.5/unit/unit\ time$, $p'_L = \$7.5/unit$, $p'_U = \$9.5$, $c_{lL} = \$0.7$, $c_{lU} = \$0.9$, $\delta = 0.2$ *and* $T = 24$ *weeks.*

For the above hypothetical data, Case-I is feasible whereas the rest two cases are infeasible. It indicates that the other constraints are not satisfied with this particular example. The best-found, worst found solutions and statistical results are shown in Tables 2–4.

Table 2. Best found results for Case-I of Example 1.

Types of QPSO	Z_L (in \$)	Z_U (in \$)	Z_C (in \$)	Z_r	t_1 (Weeks)	γ_1 (Weeks)	γ_2 (Weeks)	ξ	S	R	Computational Time (s)
GQPSO	145.4559	289.9762	217.7161	72.26013	17.6825	19.6429	20.3333	35.5954	724.7227	67.2026	0.0308
AQPSO	145.456	289.9762	217.7161	72.26011	17.6825	19.6429	20.3333	35.5954	724.7224	67.2027	0.0221
WQPSO	145.456	289.9762	217.7161	72.2601	17.6825	19.6429	20.3333	35.5954	724.7223	67.2027	0.0542

Table 3. Worst found results for Case-I of Example 1.

Types of QPSO	Z_L (in \$)	Z_U (in \$)	Z_C (in \$)	Z_r	t_1 (Weeks)	γ_1 (Weeks)	γ_2 (Weeks)	ξ	S	R	Computational Time (s)
GQPSO	145.456	289.9762	217.7161	72.26011	17.6825	19.6429	20.3333	35.5954	724.7227	67.2027	0.037
AQPSO	145.456	289.9762	217.7161	72.26012	17.6825	19.6429	20.3333	35.5954	724.7225	67.2027	0.0287
WQPSO	145.456	289.9762	217.7161	72.2601	17.6825	19.6429	20.3333	35.5954	724.7224	67.2027	0.0656

Example 2. *The values of different parameters are given as follows:*

$C_{pL} = \$12/unit$, $C_{pU} = \$14/unit$, $p = \$32/unit$, $a = 15$, $b = 10$, $c = 65$, $C_{0L} = \$95/order$, $C_{0U} = \$105$ /order, $e_1 = 15$, $C_{hL} = \$0.9/unit$, $C_{hU} = \$1.1/unit$, $\theta = 0.1$, $a_2 = 0.1$,

$c_{bL} = \$1.5/unit/unit\ time$, $c_{bU} = \$2.5/unit/unit\ time$, $p'_L = \$7.5/unit$, $p'_U = \$9.5$, $c_{lL} = \$0.7$, $c_{lU} = \$0.9$, $\delta = 0.2$ *and* $T = 24$ *weeks.*

Table 4. Statistical Analysis for various types of QPSO for Case-I of Example 1.

Types of QPSO	Best Found Z_C (in \$)	Worst Found Z_C (in \$)	Mean of Z_C (in \$)	Standard Deviation
GQPSO	217.7161	217.7161	217.7161	0
AQPSO	217.7161	217.7161	217.7161	0
WQPSO	217.7161	217.7161	217.7161	0

For the above hypothetical data, Case-II is feasible whereas the rest two cases are infeasible. It indicates that the other constraints are not satisfied with this particular example. The best-found, worst found solutions and statistical results are shown in Tables 5–7.

Table 5. Best found solutions for Case-II of Example 2.

Types of QPSO	Z_L (in \$)	Z_U (in \$)	Z_C (in \$)	Z_r	γ_1 (Weeks)	t_1 (Weeks)	γ_2 (Weeks)	ξ	S	R	Computational Time (s)
GQPSO	271.0675	463.4415	367.2545	96.18704	5.0000	17.7892	20.6667	38.1136	1053.8824	64.0743	0.0087
AQPSO	271.0675	463.4415	367.2545	96.18699	5.0000	17.7891	20.6667	38.1136	1053.8817	64.0744	0.0054
WQPSO	271.0675	463.4415	367.2545	96.18699	5.0000	17.7891	20.6667	38.1136	1053.8818	64.0744	0.0123

Table 6. Worst found results for Case-II of Example 2.

Types of QPSO	Z_L (in \$)	Z_U (in \$)	Z_C (in \$)	Z_r	γ_1 (Weeks)	t_1 (Weeks)	γ_2 (Weeks)	ξ	S	R	Computational Time (s)
GQPSO	271.0675	463.4415	367.2545	96.18699	5.0000	17.7891	20.6667	38.1136	1053.8817	64.0744	0.0098
AQPSO	271.0675	463.4415	367.2545	96.187	5.0000	17.7892	20.6667	38.1136	1053.8818	64.0744	0.0062
WQPSO	271.0675	463.4415	367.2545	96.18699	5.0000	17.7891	20.6667	38.1136	1053.8818	64.0744	0.0163

Table 7. Statistical analysis for different types of QPSO for Case-II of Example 2.

Types of QPSO	Best Found Z_C (in \$)	Worst Found Z_C (in \$)	Mean of Z_C (in \$)	Standard Deviation
GQPSO	367.2545	367.2545	367.2545	0
AQPSO	367.2545	367.2545	367.2545	0
WQPSO	367.2545	367.2545	367.2545	0

Example 3. *The input values of different system parameters are given as follows:*

$C_{pL} = \$12/unit$, $C_{pU} = \$14/unit$, $p = \$32/unit$, $a = 15$, $b = 40$, $c = 85$, $C_{OL} = \$95/order$, $C_{OU} = \$105$ /order, $e_1 = 3.5$, $C_{hL} = \$0.9/unit$, $C_{hU} = \$1.1/unit$, $\theta = 0.1$, $a_2 = 0.1$, $c_{bL} = \$1.5/unit/unit\ time$, $c_{bU} = \$2.5/unit/unit\ time$, $p'_L = \$7.5/unit$, $p'_U = \$9.5$, $c_{lL} = \$0.7$, $c_{lU} = \$0.9$, $\delta = 0.2$ *and* $T = 24$ *weeks.*

For the above hypothetical data, Case-III is feasible whereas the rest two cases are infeasible. It indicates that the other constraints are not satisfied with this particular example. The best-found, worst found and statistical results are shown in Tables 8–10.

Table 8. Best found results for Case-III of Example 3.

Types of QPSO	Z_L (in $)	Z_U (in $)	Z_C (in $)	Z_r	γ_1 (Weeks)	γ_2 (Weeks)	t_1 (Weeks)	ξ	S	R	Computational Time (s)
GQPSO	345.2759	438.5276	391.9018	46.62583	1.7500	4.0000	6.5651	10.3321	703.4356	158.6968	0.0103
AQPSO	345.2928	438.5107	391.9017	46.60898	1.7500	4.0000	6.5637	9.8094	703.2428	158.7168	0.0054
WQPSO	345.2759	438.5276	391.9018	46.62583	1.7500	4.0000	6.5651	24.0148	703.4356	158.6968	0.0139

Table 9. Worst found results for Case- III of Example 3.

Types of QPSO	Z_L (in $)	Z_U (in $)	Z_C (in $)	Z_r	γ_1 (Weeks)	γ_2 (Weeks)	t_1 (Weeks)	ξ	S	R	Computational Time (s)
GQPSO	345.2759	438.5276	391.9018	46.62583	1.7500	4.0000	6.5651	14.5762	703.4356	158.6968	0.0124
AQPSO	338.219	443.7263	390.9726	52.75365	1.7500	4.0000	7.0334	15.4843	772.9438	151.6534	0.0054
WQPSO	345.2759	438.5276	391.9018	46.62583	1.7500	4.0000	6.5651	13.0511	703.4356	158.6968	0.019

Table 10. Statistical Analysis for different variants of QPSO for Case-III of Example 3.

Types of QPSO	Best Found Z_C (in $)	Worst Found Z_C (in $)	Mean of Z_C (in $)	Standard Deviation
GQPSO	391.9018	391.9018	391.9018	0
AQPSO	391.9017	390.9726	391.714238	0.270107162
WQPSO	391.9018	391.9018	391.9018	0

6. Discussions

- Tables 2 and 3, represent that the average profit/mid-value of the profit (Z_C) obtained by using GQPSO, AQPSO and WQPSO techniques be the same up to certain decimal places. Also, it should be noted that the AQPSO technique takes less computational time to find the best-found solution.
- It is clear from Tables 5 and 6 that the average profit (Z_C) obtained by using GQPSO, AQPSO and WQPSO techniques be the same up to certain decimal places. It is observed that the AQPSO technique takes less time to find the best-found solution. To solve this particular problem AQPSO is taking the least time. It does not give any guarantee that AQPSO always takes less time, it may vary from problem to problem.
- From Tables 9 and 10, it is also remarked that the average profit obtained by using GQPSO and WQPSO techniques be the same up to certain decimal places although it is different when applying the AQPSO technique. In this case, the AQPSO technique takes less time to find the best-found solution. From Tables 4 and 7, it is remarked that the statistical results assured that the GQPSO, AQPSO, and WQPSO algorithms equally perform and they are equally efficient to find the best-found solutions for Examples 1 and 2. From Table 10, it is also remarked that the statistical results assured that the GQPSO and WQPSO algorithms equally perform and they are equally efficient to find the best-found solutions for Examples 3.
- From Table 2, Table 5and Table 8, it is remarked that the best-found value of average profit (Z_C) of Examples 1, 2, and 3 lie in between the bounds of the best found (optimal) value of interval-valued average profit of Examples 1, 2, and 3. So, the study of the best found (optimal) policy in an interval environment is well validated.

7. Sensitivity Analysis

For Example 2, sensitivity analyses are performed to observe the effect of different parameters on the center of the average profit (Z_C), initial inventory level (S), maximum shortage level (R), stock-in period (t_1), and preservation technology cost (ξ). This experiment is performed by the GQPSO technique and it is obtained by changing each bound of a parameter by −20% to +20% keeping the values of the rest parameters as their original input values. For each problem, the best-found results are taken from 50 independent runs. The detailed analyses are depicted in Figures 5–10.

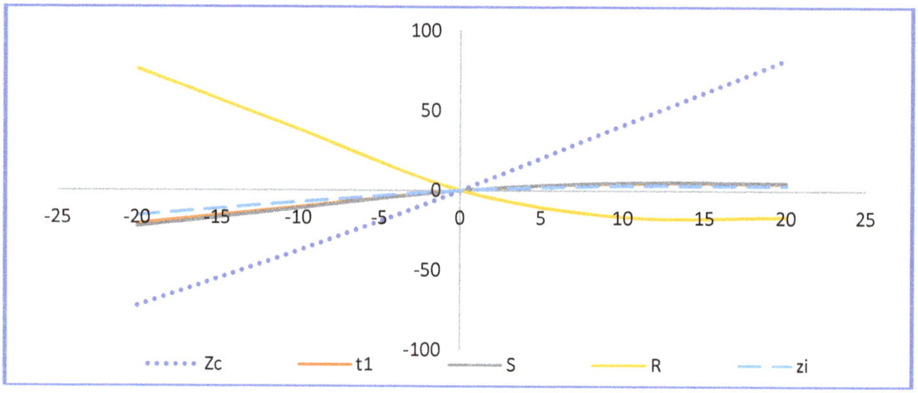

Figure 5. Impact of post optimality analysis of 'p' on Z_C, t_1, S, R, and ξ.

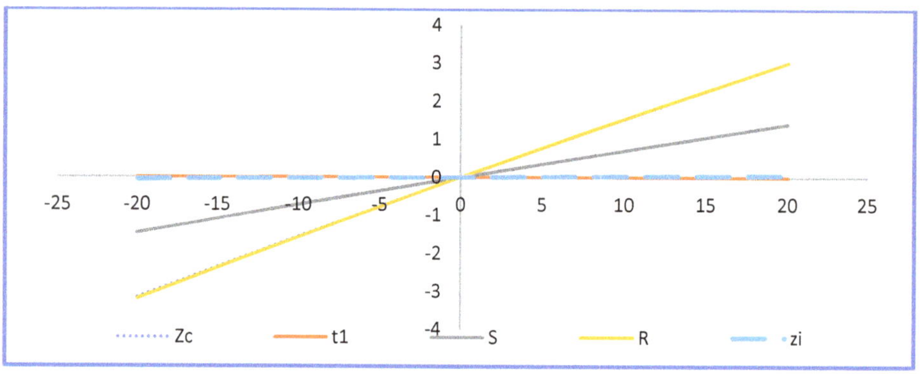

Figure 6. Impact of post optimality analysis of 'a' on Z_C, t_1, S, R, and ξ.

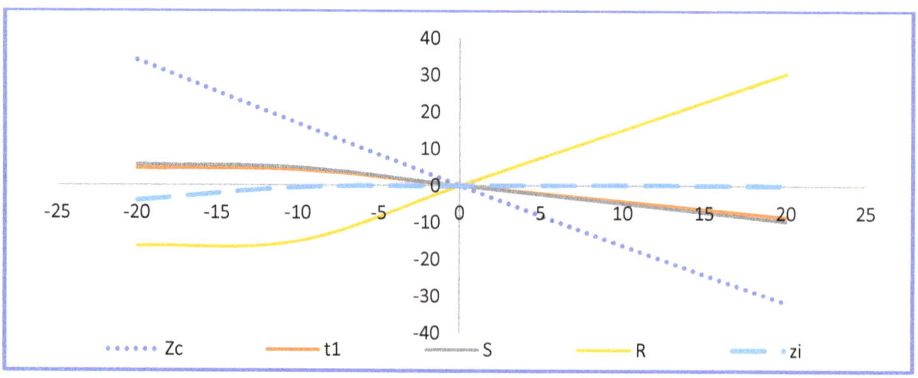

Figure 7. Impact of post optimality analysis of '$[C_{pL}, C_{pU}]$' on Z_C, t_1, S, R, and ξ.

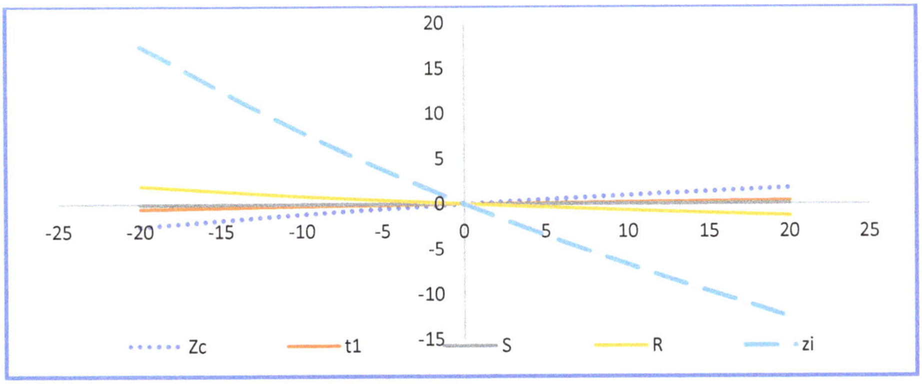

Figure 8. Impact of post optimality analysis of 'a_2' on Z_C, t_1, S, R, and ξ.

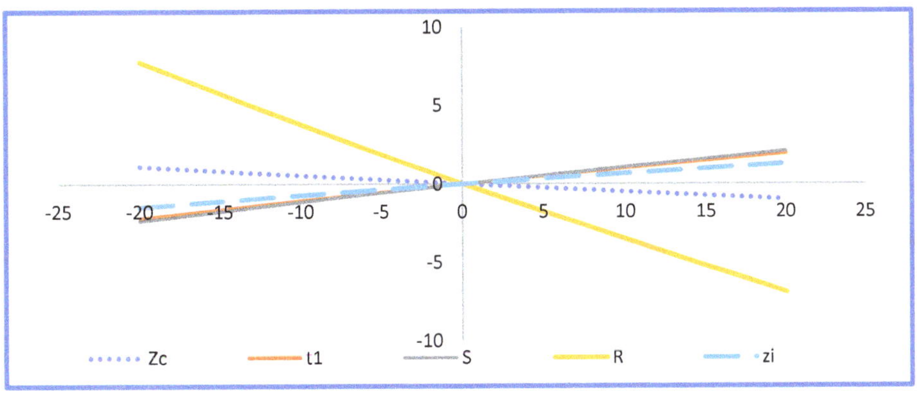

Figure 9. Impact of post optimality analysis of '$[c_{bL}, c_{bU}]$' on Z_C, t_1, S, R, and ξ.

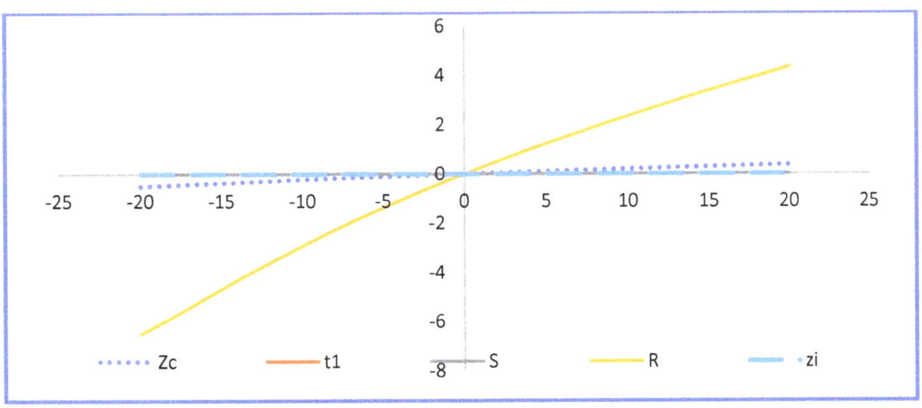

Figure 10. Impact of post optimality analysis of 'e_1' on Z_C, t_1, S, R, and ξ.

From the Figures 5–10, following implications can be observed.

(i) The center of average profit (Z_C) is highly sensitive w. r. to the selling price (p) and interval-valued purchase cost ($[C_{pL}, C_{pU}]$).Again, Z_C is less sensitive w. r. to interval-valued shortage cost ($[c_{bL}, c_{bU}]$), demand parameter, (a) and preservation parameter

(a_2) whereas it is insensitive w. r. to demand parameter (e_1). Further, $[C_{pL}, C_{pU}]$ and $[c_{bL}, c_{bU}]$ both have a reverse effect on the average profit.

(ii) Stock-in period (t_1) is less sensitive w. r. to the selling price (p), purchase cost $([C_{pL}, C_{pU}])$, shortage cost $([c_{bL}, c_{bU}])$, demand parameter (a_2) and demand parameter (e_1). Again, T is insensitive with respect to preservation parameter (a_2). Further, the parameters 'a', '$[C_{pL}, C_{pU}]$', '$[c_{bL}, c_{bU}]$', 'e_1' all have inverse effect on the business period 't_1'.

(iii) Initial inventory level (S) is less sensitive w. r. to purchase cost $[C_{pL}, C_{pU}]$ and with respect to selling price (p), demand parameter (a) and $([c_{bL}, c_{bU}])$. Again, it is insensitive with respect to preservation parameter a_2 and e_1. Further, it is observed that for the positive changes of the parameters 'a', '$[c_{bL}, c_{bU}]$', 'e_1', the initial inventory level (S) changes inversely.

(iv) The highest shortages level (R) is highly sensitive w. r. to selling price (p) and demand parameter purchase cost $([C_{pL}, C_{pU}])$ but (p) has the reverse effect on 'R'. Further, it is less impact w.r.to demand parameter (a), preservation parameter (a_2), $[c_{bL}, c_{bU}]$ and demand parameter (e_1). Further, it is noted that $(p)[c_{bL}, c_{bU}]$ and have a reverse effect on 'R'.

(v) Preservation cost (ξ) is equally sensitive w. r. to preservation parameter (a_2) and it is less sensitive w. r. to the selling price (p) and w. r. to $[c_{bL}, c_{bU}]$. Again, it is insensitive w. r. to parameters a, $[C_{pL}, C_{pU}]$ and e_1.

8. Managerial Implications

From the earlier observations, the following managerial insights may be suggested:

- The selling price of the item (p) and interval-valued purchasing cost $([C_{pL}, C_{pU}])$ have a significant impact on the retailer's profit per unit time. So, the decision-maker should think about the selling price of the item to increase the customers' demand as well as the smooth running of their business.
- To reduce the natural effect of the deterioration of products in the stock-in situation, preservation technology should be used to increase the average profit of the system.
- The proposed model is more appropriate for seasonal products e.g., fruits, vegetables, seasonal fishes, etc. At the beginning of the season, the demand for such type of the product increases then after a certain period it becomes stable. Finally, the demand for the product declined up to a certain level throughout the business period. So, the business period may be fixed. Keeping in mind this type of behavior of the demand of the sessional product, decision-maker should make the proper business plan to increase their profit.

9. Conclusions

In this study, an inventory model is developed for deteriorating items considering trapezoidal type demand and preservation technology to reduce the deterioration. Shortages are partially backlogged and inventory costs parameters are as interval-valued. Then the corresponding profit maximization problem is developed. Three different variants of QPSO techniques GQPSO, AQPSO, and WQPSO are used to solve this profit maximization problem. Finally, sensitivity analyses are studied graphically for Example 2 to study the impact of different parameters on the best-found policy. Also, from the statistical analysis, it is observed that both the techniques GQPSO and WQPSO are equally efficient to solve the optimization problems with interval-valued objectives.

The proposed trapezoidal type of demand is observed in the case of seasonal products. To reduce the natural phenomenon of deterioration, consideration of preservation technology makes more realistic in the modeling of inventory problems.

This work can be extended in various ways. One may consider the advertisement numbers/ cost, time, selling price as well as displayed stock level dependent demand, non-linear holding cost, inflation, etc. Furthermore, this work can be extended by considering trade credit policy, discount facility, advance payment policy.

Author Contributions: Conceptualization, R.M., A.A.S. and A.K.B.; methodology, R.M., A.A.S. and A.K.B.; software R.M., A.A.S. and A.K.B.; validation, R.M., A.A.S. and A.K.B.; formal analysis, R.M., A.A.S., A.K.B., R.K.C. and I.M.H.; investigation, R.M., A.A.S., I.M.H. and A.K.B.; resources, R.M., A.A.S. and A.K.B.; data curation, R.M. and A.A.S.; writing—original draft preparation, R.M., A.A.S. and A.K.B.; writing— R.M., A.A.S., A.K.B., R.K.C. and I.M.H.; review and editing, A.A.S., A.K.B., R.K.C. and I.M.H.; visualization A.A.S., A.K.B., R.K.C. and I.M.H. All authors have read and agreed to the published version of the manuscript.

Funding: CSIR (New Delhi) under CSIR-SRF Fellowship; the Department of Science and Technology, Government of India for FIST support (SR/FST/MSII/2017/10 (C)); the Research Supporting Project Number (RSP-2021/389), King Saud University, Riyadh, Saudi Arabia.

Institutional Review Board Statement: Not applicable.

Informed Consent Statement: Not applicable.

Data Availability Statement: Not applicable.

Acknowledgments: The first would like to acknowledge the financial support obtained from CSIR (New Delhi) under the CSIR-SRF Fellowship scheme (Sr.No.1061741522, Ref.No:18/06/2017(i)EU-V). Also, the second and third authors would like to acknowledge the Department of Science and Technology, Government of India, for FIST support (SR/FST/MSII/2017/10 (C)). The fourth author would like to acknowledge the Research Supporting Project Number (RSP-2021/389), King Saud University, Riyadh, Saudi Arabia. We would like to thank the editors of the journal as well as the anonymous reviewers for their valuable suggestions that make the paper stronger and more consistent.

Conflicts of Interest: The authors declare no conflict of interest.

References

1. Cheng, M.; Wang, G. A note on the inventory model for deteriorating items with trapezoidal type demand rate. *Comput. Ind. Eng.* **2009**, *56*, 1296–1300. [CrossRef]
2. Cheng, M.; Zhang, B.; Wang, G. Optimal policy for deteriorating items with trapezoidal type demand and partial backlogging. *Appl. Math. Model.* **2011**, *35*, 3552–3560. [CrossRef]
3. Lin, K.-P. An extended inventory models with trapezoidal type demands. *Appl. Math. Comput.* **2013**, *219*, 11414–11419. [CrossRef]
4. Chuang, K.-W.; Lin, C.-N.; Lan, C.-H. Order Policy Analysis for Deteriorating Inventory Model with Trapezoidal Type Demand Rate. *J. Netw.* **2013**, *8*, 1838–1844. [CrossRef]
5. Singh, T.; Pattnayak, H. An EOQ inventory model for deteriorating items with varying trapezoidal type demand rate and Weibull distribution deterioration. *J. Inf. Optim. Sci.* **2013**, *34*, 341–360. [CrossRef]
6. Lin, J.; Hung, K.-C.; Julian, P. Technical note on inventory model with trapezoidal type demand. *Appl. Math. Model.* **2014**, *38*, 4941–4948. [CrossRef]
7. Mishra, U. An inventory model for deteriorating items under trapezoidal type demand and controllable deterioration rate. *Prod. Eng.* **2015**, *9*, 351–365. [CrossRef]
8. Wu, J.; Skouri, K.; Teng, J.-T.; Hu, Y. Two inventory systems with trapezoidal-type demand rate and time-dependent deterioration and backlogging. *Expert Syst. Appl.* **2016**, *46*, 367–379. [CrossRef]
9. Vandana; Srivastava, H.M. An inventory model for ameliorating/deteriorating items with trapezoidal demand and complete backlogging under inflation and time discounting. *Math. Methods Appl. Sci.* **2016**, *40*, 2980–2993. [CrossRef]
10. Wu, J.; Teng, J.-T.; Skouri, K. Optimal inventory policies for deteriorating items with trapezoidal-type demand patterns and maximum lifetimes under upstream and downstream trade credits. *Ann. Oper. Res.* **2017**, *264*, 459–476. [CrossRef]
11. Garai, T.; Chakraborty, D.; Roy, T.K. Fully fuzzy inventory model with price-dependent demand and time varying holding cost under fuzzy decision variables. *J. Intell. Fuzzy Syst.* **2019**, *36*, 3725–3738. [CrossRef]
12. Xu, C.; Zhao, D.; Min, J.; Hao, J. An inventory model for nonperishable items with warehouse mode selection and partial backlogging under trapezoidal-type demand. *J. Oper. Res. Soc.* **2020**, *72*, 744–763. [CrossRef]
13. Kumar, P. Optimal policies for inventory model with shortages, time-varying holding and ordering costs in trapezoidal fuzzy environment. *Indep. J. Manag. Prod.* **2021**, *12*, 557–574. [CrossRef]
14. Kazemi, N.; Ehsani, E.; Jaber, M. An inventory model with backorders with fuzzy parameters and decision variables. *Int. J. Approx. Reason.* **2010**, *51*, 964–972. [CrossRef]
15. De, S.K.; Sana, S.S. Fuzzy order quantity inventory model with fuzzy shortage quantity and fuzzy promotional index. *Econ. Model.* **2013**, *31*, 351–358. [CrossRef]
16. Mondal, M.; Maity, A.K.; Maiti, M.K.; Maiti, M. A production-repairing inventory model with fuzzy rough coefficients under inflation and time value of money. *Appl. Math. Model.* **2013**, *37*, 3200–3215. [CrossRef]
17. Manna, A.K.; Dey, J.K.; Mondal, S.K. Controlling GHG emission from industrial waste perusal of production inventory model with fuzzy pollution parameters. *Int. J. Syst. Sci. Oper. Logist.* **2017**, *6*, 368–393. [CrossRef]

18. De, A.; Khatua, D.; Kar, S. Control the preservation cost of a fuzzy production inventory model of assortment items by using the granular differentiability approach. *Comput. Appl. Math.* **2020**, *39*, 1–22. [CrossRef]
19. Pulido-Rojano, A.; Andrea, A.; Padilla-Polanco, M.; Sánchez-Jiménez, M.; De la-Rosa, L. An optimization approach for inventory costs in probabilistic inventory models: A case study. *Ingeniare* **2020**, *28*, 383–395. [CrossRef]
20. Adak, S.; Mahapatra, G.S. Effect of reliability on varying demand and holding cost on inventory system incorporating probabilistic deterioration. *J. Ind. Manag. Optim.* **2020**, *18*, 173–193. [CrossRef]
21. Dutta, D.; Kumar, P. A partial backlogging inventory model for deteriorating items with time-varying demand and holding cost: An interval number approach. *Croat. Oper. Res. Rev.* **2015**, *6*, 321–334. [CrossRef]
22. Bhunia, A.K.; Shaikh, A.A. Investigation of two-warehouse inventory problems in interval environment under inflation via particle swarm optimization. *Math. Comput. Model. Dyn. Syst.* **2016**, *22*, 160–179. [CrossRef]
23. Bhunia, A.K.; Shaikh, A.A.; Cárdenas-Barrón, L.E. A partially integrated production-inventory model with interval valued inventory costs, variable demand and flexible reliability. *Appl. Soft Comput.* **2017**, *55*, 491–502. [CrossRef]
24. Gupta, R.K.; Bhunia, A.K.; Goyal, S.K. An application of genetic algorithm in a marketing oriented inventory model with interval-valued inventory costs and three-component demand rate dependent on displayed stock level. *Appl. Math. Comput.* **2007**, *192*, 466–478. [CrossRef]
25. Gupta, R.K.; Bhunia, A.; Goyal, S. An application of Genetic Algorithm in solving an inventory model with advance payment and interval valued inventory costs. *Math. Comput. Model.* **2009**, *49*, 893–905. [CrossRef]
26. Chakrabortty, S.; Madhumangal, P.A.L.; Nayak, P.K. An algorithm for solution of an interval-valued EOQ model. *Int. J. Optim. Control. Theor. Appl.* **2013**, *3*, 55–64. [CrossRef]
27. Mondal, R.; Shaikh, A.A.; Bhunia, A.K. Crisp and interval inventory models for ameliorating item with Weibull distributed amelioration and deterioration via different variants of quantum behaved particle swarm optimization-based techniques. *Math. Comput. Model. Dyn. Syst.* **2019**, *25*, 602–626. [CrossRef]
28. Shaikh, A.A.; Cárdenas-Barrón, L.E.; Tiwari, S. A two-warehouse inventory model for non-instantaneous deteriorating items with interval-valued inventory costs and stock-dependent demand under inflationary conditions. *Neural Comput. Appl.* **2019**, *31*, 1931–1948. [CrossRef]
29. Rahman, S.; Manna, A.K.; Shaikh, A.A.; Bhunia, A.K. An application of interval differential equation on a production inventory model with interval-valued demand via center-radius optimization technique and particle swarm optimization. *Int. J. Intell. Syst.* **2020**, *35*. [CrossRef]
30. Ruidas, S.; Seikh, M.R.; Nayak, P.K. A production inventory model with interval-valued carbon emission parameters under price-sensitive demand. *Comput. Ind. Eng.* **2021**, *154*, 107154. [CrossRef]
31. Ghare, P.M.; Schrader, G.F. An inventory model for exponentially deteriorating items. *J. Ind. Eng.* **1963**, *14*, 238–243.
32. Covert, R.P.; Philip, G.C. An EOQ Model for Items with Weibull Distribution Deterioration. *AIIE Trans.* **1973**, *5*, 323–326. [CrossRef]
33. Mahapatra, G.S.; Adak, S.; Mandal, T.K.; Pal, S. Inventory model for deteriorating items with time and reliability dependent demand and partial backorder. *Int. J. Oper. Res.* **2017**, *29*, 344–359. [CrossRef]
34. Shaikh, A.A.; Mashud, A.H.M.; Uddin, M.S.; Khan, M.A.A. Non-instantaneous deterioration inventory model with price and stock dependent demand for fully backlogged shortages under inflation. *Int. J. Bus. Forecast. Mark. Intell.* **2017**, *3*, 152–164. [CrossRef]
35. Shah, N.H.; Naik, M.K. Inventory model for non-instantaneous deterioration and price-sensitive trended demand with learning effects. *Int. J. Inventory Res.* **2018**, *5*, 60–77. [CrossRef]
36. Chen, L.; Chen, X.; Keblis, M.F.; Li, G. Optimal pricing and replenishment policy for deteriorating inventory under stock-level-dependent, time-varying and price-dependent demand. *Comput. Ind. Eng.* **2019**, *135*, 1294–1299. [CrossRef]
37. Mahmoodi, A. Joint pricing and inventory control of duopoly retailers with deteriorating items and linear demand. *Comput. Ind. Eng.* **2019**, *132*, 36–46. [CrossRef]
38. Saha, S.; Sen, N. An inventory model for deteriorating items with time and price dependent demand and shortages under the effect of inflation. *Int. J. Math. Oper. Res.* **2019**, *14*, 377–388. [CrossRef]
39. Khakzad, A.; Gholamian, M.R. The effect of inspection on deterioration rate: An inventory model for deteriorating items with advanced payment. *J. Clean. Prod.* **2020**, *254*, 120117. [CrossRef]
40. Khan, A.-A.; Shaikh, A.A.; Panda, G.C.; Bhunia, A.K.; Konstantaras, I. Non-instantaneous deterioration effect in ordering decisions for a two-warehouse inventory system under advance payment and backlogging. *Ann. Oper. Res.* **2020**, *289*, 243–275. [CrossRef]
41. Xu, C.; Liu, X.; Wu, C.; Yuan, B. Optimal Inventory Control Strategies for Deteriorating Items with a General Time-Varying Demand under Carbon Emission Regulations. *Energies* **2020**, *13*, 999. [CrossRef]
42. Dye, C.-Y. The effect of preservation technology investment on a non-instantaneous deteriorating inventory model. *Omega* **2012**, *41*, 872–880. [CrossRef]
43. Wahab, M.; Mamun, S.; Ongkunaruk, P. EOQ models for a coordinated two-level international supply chain considering imperfect items and environmental impact. *Int. J. Prod. Econ.* **2011**, *134*, 151–158. [CrossRef]
44. Zhao, L. An Inventory Model under Trapezoidal Type Demand, Weibull-Distributed Deterioration, and Partial Backlogging. *J. Appl. Math.* **2014**, *2014*, 1–10. [CrossRef]

45. Taleizadeh, A.A.; Moshtagh, M.S.; Moon, I. Optimal decisions of price, quality, effort level and return policy in a three-level closed-loop supply chain based on different game theory approaches. *Eur. J. Ind. Eng.* **2017**, *11*, 486. [CrossRef]
46. Rahman, M.S.; Duary, A.; Shaikh, A.A.; Bhunia, A.K. An application of parametric approach for interval differential equation in inventory model for deteriorating items with selling-price-dependent demand. *Neural Comput. Appl.* **2020**, *32*, 14069–14085. [CrossRef]
47. Dey, B.K.; Bhuniya, S.; Sarkar, B. Involvement of controllable lead time and variable demand for a smart manufacturing system under a supply chain management. *Expert Syst. Appl.* **2021**, *184*, 115464. [CrossRef]
48. Shaikh, A.A.; Das, S.C.; Bhunia, A.K.; Sarkar, B. Decision support system for customers during availability of trade credit financing with different pricing situations. *RAIRO Rech. Opérationnelle* **2021**, *55*, 1043–1061. [CrossRef]
49. Jabbarzadeh, A.; Aliabadi, L.; Yazdanparast, R. Optimal payment time and replenishment decisions for retailer's inventory system under trade credit and carbon emission constraints. *Oper. Res.* **2019**, *21*, 589–620. [CrossRef]
50. Singh, S.R.; Rathore, H. Optimal Payment Policy with Preservation Technology Investment and Shortages Under Trade Credit. *Indian J. Sci. Technol.* **2015**, *8*, 203. [CrossRef]
51. Tayal, S.; Singh, S.R.; Sharma, R. An integrated production inventory model for perishable products with trade credit period and investment in preservation technology. *Int. J. Math. Oper. Res.* **2016**, *8*, 137. [CrossRef]
52. Mishra, U.; Tijerina-Aguilera, J.; Tiwari, S.; Cárdenas-Barrón, L.E. Retailer's Joint Ordering, Pricing, and Preservation Technology Investment Policies for a Deteriorating Item under Permissible Delay in Payments. *Math. Probl. Eng.* **2018**, *2018*, 1–14. [CrossRef]
53. Mishra, U.; Cárdenas-Barrón, L.E.; Tiwari, S.; Shaikh, A.A.; Treviño-Garza, G. An inventory model under price and stock-dependent demand for controllable deterioration rate with shortages and preservation technology investment. *Ann. Oper. Res.* **2017**, *254*, 165–190. [CrossRef]
54. Bardhan, S.; Pal, H.; Giri, B.C. Optimal replenishment policy and preservation technology investment for a non-instantaneous deteriorating item with stock-dependent demand. *Oper. Res.* **2017**, *19*, 347–368. [CrossRef]
55. Shah, N.H.; Chaudhari, U.; Jani, M.Y. Optimal control analysis for service, inventory and preservation technology investment. *Int. J. Syst. Sci. Oper. Logist.* **2019**, *6*, 130–142. [CrossRef]
56. Das, S.C.; Zidan, A.; Manna, A.K.; Shaikh, A.A.; Bhunia, A.K. An application of preservation technology in inventory control system with price dependent demand and partial backlogging. *Alex. Eng. J.* **2020**, *59*, 1359–1369. [CrossRef]
57. Khanna, A.; Jaggi, C.K. An inventory model under price and stock-dependent demand for controllable deterioration rate with shortages and preservation technology investment: Revisited. *OPSEARCH* **2021**, *58*, 181–202.
58. Rahman, S.; Duary, A.; Khan, A.-A.; Shaikh, A.A.; Bhunia, A.K. Interval valued demand related inventory model under all units discount facility and deterioration via parametric approach. *Artif. Intell. Rev.* **2021**, 1–40. [CrossRef]
59. Hsu, P.H.; Wee, H.M.; Teng, H.M. Preservation technology investment for deteriorating inventory. *Int. J. Prod. Econ.* **2010**, *124*, 388–394. [CrossRef]
60. Hasan, R.; Mashud, A.H.M.; Daryanto, Y.; Wee, H.M. A non-instantaneous inventory model of agricultural products considering deteriorating impacts and pricing policies. *Kybernetes* **2020**, *50*, 2264–2288. [CrossRef]
61. Mashud, A.; Khan, M.; Uddin, M.; Islam, M. A non-instantaneous inventory model having different deterioration rates with stock and price dependent demand under partially backlogged shortages. *Uncertain Supply Chain Manag.* **2018**, *6*, 49–64. [CrossRef]
62. Yang, C.-T.; Dye, C.-Y.; Ding, J.-F. Optimal dynamic trade credit and preservation technology allocation for a deteriorating inventory model. *Comput. Ind. Eng.* **2015**, *87*, 356–369. [CrossRef]
63. Das, S.C.; Manna, A.K.; Rahman, S.; Shaikh, A.A.; Bhunia, A.K. An inventory model for non-instantaneous deteriorating items with preservation technology and multiple credit periods-based trade credit financing via particle swarm optimization. *Soft Comput.* **2021**, *25*, 5365–5384. [CrossRef]

mathematics

MDPI

Article

New Algorithm to Solve Mixed Integer Quadratically Constrained Quadratic Programming Problems Using Piecewise Linear Approximation

Loay Alkhalifa [1,*] and Hans Mittelmann [2]

1 Department of Mathematics, College of Sciences and Arts, Qassim University, Ar Rass 51921, Saudi Arabia
2 School of Math & Stat Sciences, Arizona State University, Tempe, AZ 85287, USA; mittelmann@asu.edu
* Correspondence: loay.alkhalifa@qu.edu.sa

Abstract: Techniques and methods of linear optimization underwent a significant improvement in the 20th century which led to the development of reliable mixed integer linear programming (MILP) solvers. It would be useful if these solvers could handle mixed integer nonlinear programming (MINLP) problems. Piecewise linear approximation (PLA) is one of most popular methods used to transform nonlinear problems into linear ones. This paper will introduce PLA with brief a background and literature review, followed by describing our contribution before presenting the results of computational experiments and our findings. The goals of this paper are (a) improving PLA models by using nonuniform domain partitioning, and (b) proposing an idea of applying PLA partially on MINLP problems, making them easier to handle. The computational experiments were done using quadratically constrained quadratic programming (QCQP) and MIQCQP and they showed that problems under PLA with nonuniform partition resulted in more accurate solutions and required less time compared to PLA with uniform partition.

Keywords: mixed integer nonlinear programming; piecewise linear approximation; branch and bound

Citation: Alkhalifa, L.; Mittelmann, H. New Algorithm to Solve Mixed Integer Quadratically Constrained Quadratic Programming Problems Using Piecewise Linear Approximation. *Mathematics* **2022**, *10*, 198. https://doi.org/10.3390/math10020198

Academic Editors: Humberto Rocha and Ana Maria Rocha

Received: 23 November 2021
Accepted: 4 January 2022
Published: 9 January 2022

Publisher's Note: MDPI stays neutral with regard to jurisdictional claims in published maps and institutional affiliations.

1. Introduction

The rapid advances of Linear Programming (LP), Non-Linear Programming (NLP), and Mixed Integer Linear Programming (MILP) techniques and algorithms in the 20th century led to the development of robust MILP solvers that can easily handle problems with millions of variables. On the other hand, methods that deal with Mixed Integer Non-Linear Programming (MINLP), which is the most difficult class of optimization, started to improve recently. Even with the current improvements in the MINLP solvers, it would be a great step forward if MINLP problems could be solved globally by MILP solvers. In principle, this can be done by approximating the MINLP problem by an MILP one, but solving this approximation by an MILP solver will probably be harder than solving the original problem by an MINLP solver. Recent discussions about software solvers and their development can be found in [1,2]. More detail about MINLP and their algorithms can be found in [3–6], and a recent general survey was introduced by [7]. For some real life optimization applications, we refer to [8–11].

One of the methods to remodel MINLP as MILP is the piecewise linear approximation (PLA) ([7]). This method takes an advantage of the fact that any continuous function can be approximated by a piecewise linear one. Replacing every nonlinear function in the MINLP model by their PLAs will yield an MILP model. Because of the size of the new approximated model, the PLA was not introduced to do full approximations to the MINLP models. Instead, it was used mostly to find linear under/over estimators to some of the functions involved in the models.

The procedures of approximating a nonlinear function $f(x)$, where $x \in [x_l, x_u] \subseteq \mathbb{R}$, by a piecewise Linear (PL) function are simple. First, the domain of the variables x is divided into n intervals by introducing the breakpoints $x_0 = x_l < x_1 < x_2 < \cdots < x_n = x_u$. Then the function f is evaluated at each breakpoint, and the lines that connect the points $(x_i, f(x_i))$ and $(x_{i+1}, f(x_{i+1}))$ form the desired PL function, denoted by $\overline{f}$. The PLA idea was also extended to higher dimensional functions.

Increasing the number of breakpoints will increase the accuracy of the approximation, but it will result in problem size growth. This major drawback may restrict the PLA benefits to functions of only few dimensions. Ref. [12] discuss minimizing the number of breakpoints needed to approximate a nonlinear function up to a given tolerance. This subject will not be discussed here. When doing PLA, introducing breakpoints requires adding both binary and continuous variables to the optimization problem in addition to new constraints. This is done by setting the binary variables to be SOS1 or SOS2 (please see [13]). The number of the new variables and constraints may increase exponentially with the number of dimensions of the function and it varies depending on the model used to do the PLA.

The following section will present a brief literature review on the PLA modeles. The remainder of the paper is organized as follows: in Section 2, few approaches on how to improve the PLA models are introduced. One approach shows how to take an advantage of a local solution to choose the breakpoints by nonuniform partitioning, and our contribution will be using this partitioning to produce better PLA. The other approach is to apply the PLA on only a part of the optimization problem. The computational results are given in Section 3, where the tests are applied to continuous and discrete problems.

Literature Review

PLA dates back to the 1950s (see, for example, [14–16]), and since then, many PLA models were introduced. The convex combination model (CC) (also called the λ-model in some sources) is one of the most common PLA models and it was introduced by [15]. Another model that was introduced by [14] is the incremental model. The incremental model requires one less binary variable and one less continuous variable compared to the CC model, but it needs nearly twice the number of new constraints. Modifications were done to the CC model and the incremental model resulting in the disaggregated convex combination model ([16,17]) and the multiple choice model ([18]), respectively. Piecewise linear relaxation techniques are widely used by many algorithms, and these techniques are different from PLA, even though they share some steps. In this paper, we are interested only on PLA, and we refer to [19–21], for further reading about piecewise linear relaxations.

The efficiency of MILP solvers will be affected if the PLA is done using many breakpoints, especially for higher dimensional functions. More accurate PLA can be obtained by increasing the number of breakpoints, but in all models mentioned above, the introduced binary variables will be almost as many as the breakpoints. Therefore the MILP solvers might not be able to deal with the size of the resulting MILP problem. Introducing less breakpoints will result in small approximated problems but it might lead to a bad approximation.

Many attempts have been made to deal with the size issues, but these attempts do not resolve the problem completely. A major improvement in this area took a place when [22,23] introduced a technique that allows PLA models to use dramatically fewer binary variables. They applied the technique to both versions of the convex combination model and denoted it by logarithmic model, and it was later applied to other PLA models. To do the PLA with $n+1$ breakpoints using the logarithmic model, only $\lceil \log_2 n \rceil$ binary variables are required, instead of approximately n required by other PLA models. In this paper, we will use the CC and logarithmic disaggregated convex combination (LOG) models to do the computational experiments on quadratically constrained quadratic programming (QCQP) and Mixed Integer QCQP (MIQCQP).

A good overview of the models mentioned in this section is presented by [24], and it was shown that they are equivalent in terms of the feasibility of their solutions. In term of tightness, it is desired for a PLA model to be locally ideal (the vertices of its LP relaxation satisfy the integrality constraints of the original MINLP problem). Ref. [25] shows that the incremental model is superior to the CC model since the incremental model is locally ideal (this is supported by the computational comparison that will be discussed below). Ref. [23] proves that all PLA models mentioned earlier are locally ideal except the CC model. However, the CC model has the sharpness property, i.e,. the projection of the vertices of the model onto the original set of the variables is exactly the convex hull of the set. Moreover, it can be shown that any locally ideal model is sharp. More recent theoretical comparison between the models was given by [26].

In the computational experiments made by [23], they approximated functions with one variable in 100 test instances, and used CPLEX to solve the PLAs. It was observed that for less than 10 breakpoints, all models performed well with the multiple choice model being slightly better. As the number of breakpoints increases, the logarithmic models started to gain the upper hand. When 33 breakpoints were used, the logarithmic models are almost 20 times faster than the incremental model, which is more than two times faster than the CC and multiple choice models. The disaggregated convex combination is much slower than the rest. Similar outcomes resulted from testing 100 problems where the approximated functions have two variables. A less extensive experiment was done by [27], and it was concluded that in some cases, it is better to use the incremental model rather than the logarithmic one, even though the latter has smaller size.

2. Improving PLA Models

In this section, a few approaches to improve the PLA models will be presented. Since PLA requires introducing many binary and continuous variables and constraints, it was not studied as a stand-alone method to solve MINLP problems until recently. It is usually used as a tool in optimization algorithms to find local solutions or to under/over estimate some of the nonlinear functions. For PLA models to completely transform an already hard MINLP problem and produce an MILP problem that is easier than the original to be handled, more improvements on these models are needed. Note that even if the targeted problems in this paper have many variables, the PLA models will be applied separately only to functions of two variables or less.

One approach to improve the models is to choose the breakpoints such that the variable domain is nonuniformly partitioned. This approach was motivated by the desire to produce an accurate approximation with reasonable problem size. Unfortunately, this is rarely possible since an accurate approximation requires many breakpoints. One of the methods to overcome this issue is to study how to partition a domain. Most existing PLA models choose the locations of breakpoints based on a uniform partitioning of the domain. The approach proposed in this section, namely a nonuniform partitioning, leads to a better PLA model than the one produced by uniform partitioning with the same number of breakpoints. The details are given in Section 2.1.

Another approach that will be presented in this section is applying the PLA partially to problems with many nonlinear functions. Given a complicated MINLP problem, applying the PLA to only some of its nonlinear constraints will not make the problem solvable by an MILP solver, but it might make the problem less complicated for MINLP solvers. The idea is to identify complicated nonlinear constraints and approximate some of them by linear ones, then the modified problem is solved using MINLP solvers. An algorithm that identifies these constraints and approximates them will be provided in Section 2.2.

2.1. Choosing Breakpoints by Nonuniform Partitioning

Most PLA models, that are used within the context of optimization, rely on uniform partitioning of variable domains. A few methods to do nonuniform partitioning were introduced within other contexts. For example, [28] used nonuniform partitioning to approximate a one dimensional curve. The idea is to add more breakpoints in the part of the domain where the function has higher curvature. This method is done by solving a shortest path problem, and it was later used within the PLA approach for one and two dimension by [29]. Other methods were introduced to use nonuniform partitioning to get piecewise convex/linear relaxations, such as the one proposed by [30].

The partitioning method suggested in this section is to build the partitioning around a local solution. The density of breakpoints increases as they get closer from both sides to the local solution, which itself is a breakpoint. Assume the local solution of a variable $l \leq x \leq u$ is x^*, then a possible partitioning is to have a breakpoint in the halfway in the interval from the upper/lower bound to x^*, then another point halfway toward x^* and so on. This partitioning can be given by using the formulas

$$x^* + \frac{u - x^*}{k^i} \quad \text{and} \quad x^* - \frac{x^* - l}{k^i}, \tag{1}$$

for $k = 2$ and $i = 1, 2, \ldots$, to get the breakpoint values on both sides of x^*, as shown in Figure 1A.

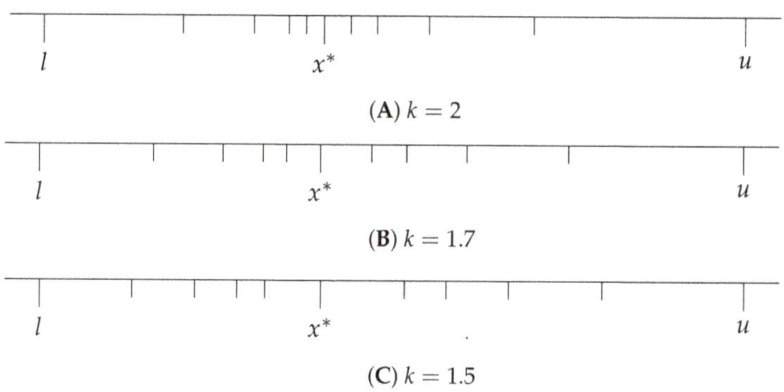

Figure 1. Partitioning Using the Formulas $x^* + \frac{u-x^*}{k^i}$ and $x^* - \frac{x^*-l}{k^i}$.

The problem with the case $k = 2$ is that it leaves half of the interval between the upper/lower bound and x^* without any breakpoints, which may affect the accuracy of the PLA. Thus, giving k different values between one and two, as in Figure 1B,C, may yield better partitioning. It can be noticed that as k gets closer to one, the density of breakpoints shifts away from x^*. If x^* happened to be at or very close to one of the bounds, the partitioning will be on one side only. In the case that the approximated function has two variables, the same logic is applied to both domains.

The PLA using this partitioning was tested for many values of $1 < k < 2$ against PLA with uniform partitioning. The results show that models with nonuniform partitioning were solved faster and yielded better solutions to most of the tested instances. Details about the targeted optimization problems and the test results will be given in Section 3.

2.2. Partial PLA

In this section, PLA will be performed only on parts of a given MINLP problem, instead of approximating all nonlinear functions. This is done by targeting complicated nonlinear functions be handled by PLA, leaving the remaining functions unchanged. The goal of this approach is to avoid introducing unnecessary variables that result from approximating simple functions. To test this approach, an algorithm is introduced to deal with problems having many nonlinear constraints by picking one constraint at every iteration and approximating it, until enough constraints are replaced.

Assume the algorithm is applied to an MINLP problem with x^* and f^* as its global optimal solution and objective function value, respectively. The algorithm starts by solving the problem (before doing any PLA) using an MINLP solver for a few nodes before it stops the solving process when a specified number of nodes, N, is reached. Then all nonlinear constraints are identified, and since the solving process was interrupted, using the current node solution will probably result in some of these constraints being violated. Now all nonviolated constraint are considered to be easy since the solver was able to satisfy them within few nodes, so the PLA will not be applied to these constraints.

Now the algorithm picks a constraint from the violated ones to be approximated using a PLA model, and this constraint is suggested to be the most violated one. As a result, the problem now has one less nonlinear constraint. Then the process is repeated until the problem has no violated constraints. Note that some of the nonlinear constraint will not be replaced, so the problem is still an MINLP one. At this point, the problem is solved regularly to produce a solution $\overline{x}^*$ to the modified problem with a function value $\overline{f}^*$. $|f^* - \overline{f}^*|$ is evaluated and the solving times of the original and modified problems are compared. Algorithm 1 shows the steps of the partial PLA approach on MIQCP problems, for some $\epsilon > 0$ and $k \in \mathbb{N}$.

Algorithm 1: An algorithm that chooses some constraints to be approximated.

```
Input:MIQCP problem;
Output:MIQCP problem with up to k constraints being approximated;
while |f̄* − f*| ≥ ε and iteration ≤ k do
    start the solving process;
    if solving is done before number of nodes reaches N then
        set the current objective value = f̄*;
        if |f̄* − f*| ≥ ε then
            add more breakpoints;
        end
    else
        stop the solving process when the tree has N nodes;
        identify the quadratic constraints;
        if number of violated constraints ≥ 1 then
            replace the most violated one with its PLA;
            set f̄* = ∞;
        else
            solve the PL problem and set the current objective value = f̄*;
        end
    end
end
```

With each iteration, one constraint is replaced by its PLA. Obviously, if k is large enough, then all violated constraints will be replaced and the resulting problem will probably be harder to solve because of the size. The computational experiments with large k came as expected and produced extremely large problems. Then k was set to one, i.e., PLA was applied to only the most violated constraint in the first iteration. For some test instances, both PLA models, that were introduced in the previous section, produced modified problems that were solved faster than the original ones, with similar solution and objective value. As the number of replaced constraints increases, the modified problem gets more complicated. Most of the tested instances appeared to be better off without PLA if $k \geq 3$. In Section 3, a summary of the test instances will be presented in addition to the main findings on applying PLA to only parts of these instances.

3. Computational Experiments

This section presents detailed reports on computational experiments that target QCQP and MIQCQP problems. The approaches that were introduced in the previous section can be applied to MINLP problems, but they were implemented to test the quadratic problems only. The same procedures can be applied to the MINLP problems by modifying the code that implements the procedures to allow it to identify other non linear functions beside the quadratic ones. Most of the chosen problems are not trivial and are based on different real world applications. The QCQP problems were taken from the kall instances in the MINLPLib library (http://www.minlplib.org/index.html, accessed on 15 May 2021), whereas the MIQCQP problems were taken from both MINLPLib and QPLIB (http://qplib.zib.de/accessed on 15 May 2021). Two dimensional functions are involved in all of the instances, so 2d PLA is required.

The nonlinear functions in the instances are approximated by linear ones using the CC and LOG models. If the breakpoints needed for CC and LOG are determined using nonuniform partitioning, then the models will be called NCC and NLOG. These four models are also used to do partial PLA to the test instances by approximating all the nonlinear functions in one constraint.

The code that implements the PLA models were written using PySCIPOpt (http://scip-interfaces.github.io/PySCIPOpt/docs/html/index.html, accessed on 15 May 2021), which is a Python interface for the global solver SCIP. The tests were applied to the instances several times before taking the average results. After the instance is read by SCIP, all quadratic constraints (or general nonlinear constraints, in case the instance is an MINLP one) are identified before the domain of every variable involved in a quadratic function is partitioned (uniformly or nonuniformly). Then SCIP adds all necessary variables and constraints needed to replace the targeted function. Finally, any quadratic constraint is deleted from the instance and linear constraints, that approximate the deleted ones, are added. Now the instance is written in a format that can be read by MILP solvers.

The local instance solutions, that are needed for the nonuniform partitioning, are generated using the solver Knitro. The comparison between the uniform and nonuniform partitioning is determined through solving the MILP problems resulting from applying CC, LOG, NCC, and NLOG to the test instances. The MILP problems are solved using CPLEX 12.10 with its default settings, and the computations are done in a Linux machine with Intel Xeon E5-2620 2GHz processor. The time limit for each problem is set to one hour.

The partial PLA tests are done by solving the original instances and the same instances with one constraint replaced. The instances are solved using SCIP Optimization Suite 6.0.0 on a Linux machine with Intel Core i5-7y54 1.2GHz processor. The default solver settings were used with a time limit of two hours. The results of the computational experiments will be described in detail in the following sections.

3.1. Continuous Variables

The computational results that are reported in this section are for the kall instances, where the objective is linear and the constraints are linear and quadratic. The quadratic constraints contain up to two functions with two variables of the forms $(x+y)^2$ or xy. Moreover, one variable functions of the form x^2 appear in the constraints of some of the problems.

Statistics of the instances are given in Table 1, where seq# is the sequential number that will refer to the corresponding instance throughout the section. The number of variables is shown under #v, and #cons (q) represents the total number of constraints including (q) quadratic ones. The best objective values found so far (according to the MINLPLib library) are entered under the obj val column header. All objective values shown in the table are proven to be the global optimum by at least 3 global solvers except instances 2, 3, 5, and 12.

The instances were solved regularly by SCIP, with time limit of 2 h. The solving times are recorded and presented in seconds in the last column of Table 1. For some instances, the limits, machine memory (ML) or time (TL), were reached before the instance is solved to optimality. Moreover, it can be observed from the solving times that some instances are trivial and solved quickly but others took time to be solved. Solving the instances here was not intended to find their global solutions, since the best solutions are already found and listed in the library and no further improvement to the solutions can be done. Instead, they were solved to give an idea about their difficulty levels, and to have a reference when it is needed to compare their results to the modified problems results.

Table 1. Statistics of continuous test instances, and the solving time by SCIP.

Instance	seq#	#v	#cons (q)	obj val	Time
kall_circles_c6a	1	18	54 (22)	2.1117	1927
kall_circles_c6b	2	18	54 (22)	1.9736	2964
kall_circles_c7a	3	20	69 (29)	2.6628	2612
kall_circles_c8a	4	22	86 (37)	2.5409	TL
kall_congruentcircles_c31	5	10	16 (4)	0.6438	1
kall_congruentcircles_c32	6	10	16 (4)	1.3759	1
kall_congruentcircles_c41	7	12	24 (6)	0.8584	1
kall_congruentcircles_c42	8	12	24 (7)	0.8584	1
kall_congruentcircles_c51	9	14	34 (11)	1.073	12
kall_congruentcircles_c52	10	14	34 (11)	1.5371	4
kall_congruentcircles_c62	11	16	46 (16)	1.2876	16
kall_congruentcircles_c63	12	16	46 (16)	1.2876	11
kall_congruentcircles_c72	13	18	60 (22)	1.9663	225
kall_diffcircles_10	14	24	71 (45)	11.9355	6356 (ML)
kall_diffcircles_5a	15	14	24 (11)	5.1162	63
kall_diffcircles_5b	16	14	24 (11)	5.1162	44
kall_diffcircles_6	17	16	31 (16)	7.7879	102
kall_diffcircles_7	18	18	40 (22)	7.1531	177
kall_diffcircles_8	19	20	49 (28)	14.4813	3350
kall_diffcircles_9	20	22	60 (36)	13.3503	5118 (ML)

Before discussing the results of instance PLAs, one important factor about the instances has to be taken into account: the sizes of variable domains. When the variable involved in PLA of a function has a large domain, the domain will need more breakpoints to get a good approximation. However, introducing many breakpoints will affect the size, so the number of breakpoints for the computational tests is set to 10 for all domains. The domain size for each variable in the tested instances ranges between 1 and 18, with average size of 7.25 per domain. Therefore, 10 points should be enough to get good models for test purposes.

To compare between the PLA models with respect to the sizes of the produced problems, the instances were transformed into MILP problems by the models CC and LOG (using nonuniform partitioning would give the same size). Each instance was approximated by these models using 10, 20, and 30 breakpoints. Then the average numbers of constraints and binary/continuous variables per problem is recorded for each set of breakpoints, as shown in Table 2. It is apparent from the table that there is a significant advantage for the model LOG in terms of needed number of binary variables and constraints. Moreover, it will be shown later that this advantage is not outweighed by the CC model's advantage of having many fewer continuous variables.

Table 2. Average sizes per problem produced by CC and LOG models using 10, 20, and 30 breakpoints.

		10	20	30
BV	CC	6135	27,195	63,355
	LOG	303	377	414
CV	CC	3808	15,083	33,916
	LOG	22,738	90,416	203,417
Cons	CC	3983	15,259	34,092
	LOG	764	891	983

While Table 2 gives enough insight on the problem sizes, detailed statistics for every problem approximated using 10 breakpoints is given in Table 3. This table makes it easier to track the detail of every tested problem and compare between its size and computational results. It should be expected that most problems will not be solved easily by CPLEX due to the large number of binaries and constraints.

Table 3. The sizes of the instances produced by CC and LOG models with 10 breakpoints.

p#	CC			LOG		
	BV	CV	Cons	BV	CV	Cons
1	6966	4318	4526	344	25,818	871
2	6966	4318	4526	344	25,818	871
3	9234	5720	5991	456	34,220	1152
4	11,826	7322	7678	584	43,822	1473
5	1134	710	744	56	4210	149
6	1134	710	744	56	4210	149
7	2106	1312	1376	104	7812	271
8	2106	1312	1376	104	7812	271
9	3402	2114	2218	168	12,614	433
10	3402	2114	2218	168	12,614	433
11	5022	3116	3270	248	18,616	635
12	5022	3166	3270	248	18,616	635
13	6966	4318	4532	344	25,818	877
14	14,742	9124	9535	728	54,624	1800
15	3402	2114	2208	168	12,614	423
16	3402	2114	2208	168	12,614	423
17	5022	3118	3255	248	18,616	620
18	6966	4318	4512	344	25,818	857
19	9234	5720	5977	456	34,220	1132
20	11,826	7322	7652	584	43,822	1447

3.1.1. Uniform vs. Nonuniform Partitioning

The computational experiments on solving the MILP problems approximating the selected instances were made by CPLEX with a time limit of one hour per problem. The objective of these experiments is to compare between uniform and nonuniform partitioning. The nonuniform partitioning is done using different values for the parameter k in the formulas in Equation (1).

Table 4 compares the results obtained from solving the CC and NCC problems, with $k = 1.5, 1.7$. For each model, the solving time is listed in seconds or as TL if the time limit is reached. The objective value of the best integer solution found within the time limit is listed under BI, and F means the solver failed to find an integer solution. Initially, 10 breakpoints were used by the models, but CPLEX failed to find a solution for most of the problems, so they were regenerated using 7 points. The third column measures the difference, if applicable, between the best integer value, $\overline{f^*}$ and the best value of the original instance, f^*, which can be found in Table 1. Smaller $|f^* - \overline{f^*}|$ value indicates that the approximation is acceptable. Note that for problems that reached time limit, the best integer value could keep improving if there is no time limit.

Given the fact that most of the original instances were solved by SCIP to global optimality within less than an hour (Table 1), it can be confirmed that complete transformation of MINLP problems into MILP ones will not make the problems easier, especially when approximating many functions. Nonetheless, this is not the goal of the experiments since the main purpose is to compare partitioning methods for PLA models. The aspects of the comparison between uniform and nonuniform partitioning will be the accuracy of the approximation and the time needed to solve the approximation.

Table 4. Results of solving MILP problems produced by CC and NCC models using 7 breakpoints.

p#	CC			NCC ($k = 1.5$)			NCC ($k = 1.7$)		
	Time	BI	$\lvert f^* - \overline{f^*}\rvert$	Time	BI	$\lvert f^* - \overline{f^*}\rvert$	Time	BI	$\lvert f^* - \overline{f^*}\rvert$
1	TL	F	NA	TL	1.184	0.9277	TL	2.47	0.3583
2	TL	F	NA	TL	3.324	1.3504	TL	F	NA
3	TL	F	NA	TL	2.0317	0.6311	TL	2.6628	0
4	TL	F	NA	TL	F	NA	TL	F	NA
5	18	0.5724	0.0714	29	0.6009	0.0429	81	0.5847	0.0591
6	11	1.366	0.0099	8	1.3831	0.0072	5	1.3544	0.0215
7	1	0.8084	0.05	1	0.8584	0	1	0.8584	0
8	TL	0.8084	0.05	133	0.8269	0.0315	110	0.8052	0.0532
9	TL	1.5373	0.4643	TL	0.7367	0.3363	TL	0.5087	0.5643
10	1200	1.473	0.0641	TL	1.5371	0	TL	1.588	0.0509
11	TL	1.221	0.0666	TL	1.2876	0	TL	1.2876	0
12	856	1.118	0.1696	TL	1.2876	0	TL	1.2876	0
13	TL	1.6689	0.2974	TL	1.9663	0	TL	1.817	0.1493
14	TL	F	NA	TL	F	NA	TL	F	NA
15	TL	5.2619	0.1457	TL	5.1162	0	TL	6.479	1.3628
16	1400	2.838	2.2782	527	3.172	1.9442	TL	3.4476	1.6686
17	3400	7.5885	0.1994	TL	7.2298	0.558	TL	F	NA
18	TL	F	NA	TL	F	NA	TL	F	NA
19	TL	F	NA	TL	15.431	0.9497	TL	17.611	3.1297
20	TL	F	NA	TL	F	NA	TL	F	NA

The data listed under the time columns indicate that all models are close in terms of the number of problems that were solved within the time limit. However, for problems that reached the time limit, the model with uniform partitioning failed to find an integer solution in eight problems, compared to four problems for the NCC model with $k = 1.5$. This implies that problems generated by CC models with nonuniform partitioning usually find feasible solutions faster than with uniform partitioning. The same outcome resulted when other nonuniform cases were tested for $k = 1.4, 1.6, 1.8$. Even when both uniform and nonuniform cases resulted in a problem that produced an integer solution within the time limit, the gap in the nonuniform case is smaller most of the times. Therefore, it can be concluded that problems produced by NCC models are usually solved faster than the ones produced by CC models.

The quality of the approximation is better tested without the time limit, where the solver runs until the global solution is found, and then the solutions of the original and approximated problems are compared. In spite of that, the table provides enough data to compare the quality of CC and NCC approximations. It can be observed that the integer solution is the same as the solution of the original instance in many problem for both cases of nonuniform partitioning, while the uniform partitioning never produced an identical integer solution to the original one. Moreover, for the cases that the integer solution is not the same as the original one, the difference between the two solutions is mostly less in the NCC cases.

The computational results for solving problems produced by the models LOG and NLOG are summarized in Table 5. The test setting and comparison aspects for these models are the same as the CC and NCC models, except for the number of breakpoints where 11 is used for this test. The outcomes also turned out to be similar: the nonuniform partitioning improved the models in both speed and accuracy. Both NLOG models resulted in more problems that were solved within time limit, compared to LOG models; and when a problem is solved within time limit for all models, almost always the NLOG problem needs less solving time. Moreover, the difference $\lvert f^* - \overline{f^*}\rvert$ is mostly smaller in the NLOG cases. It should be mentioned that in nonuniform partitioning, having a local solution as one of the breakpoints helped improving the model's quality.

Table 5. Results of solving MILP problems produced with LOG and NLOG models using 11 breakpoints.

p#	LOG			NLOG ($k = 1.5$)			NLOG ($k = 1.7$)		
	Time	**BI**	$\lvert f^* - \overline{f^*}\rvert$	**Time**	**BI**	$\lvert f^* - \overline{f^*}\rvert$	**Time**	**BI**	$\lvert f^* - \overline{f^*}\rvert$
1	TL	F	NA	TL	F	NA	TL	F	NA
2	TL	F	NA	TL	1.9488	0.0248	TL	F	NA
3	TL	F	NA	TL	2.6628	0	TL	2.6628	0
4	TL	F	NA	TL	F	NA	TL	F	NA
5	8	0.5435	0.1003	6	0.6009	0.0429	73	0.5847	0.0591
6	10	1.368	0.0079	2	1.3831	0.0072	157	1.3544	0.0215
7	3	0.8324	0.026	1	0.8584	0	47	0.8584	0
8	37	0.8348	0.0236	16	0.8269	0.0315	133	0.8052	0.0532
9	TL	2.944	1.871	TL	1.036	0.037	TL	0.378	0.695
10	1764	1.51	0.0271	194	1.5371	0	915	1.5371	0
11	TL	1.651	0.3634	1042	1.2876	0	821	1.2876	0
12	346	1.038	0.2496	231	1.2876	0	478	1.2876	0
13	TL	2.097	0.1307	TL	1.9663	0	TL	1.9663	0
14	TL	F	NA	TL	F	NA	TL	F	NA
15	2980	5.038	0.0782	TL	5.1162	0	161	4.6298	0.4864
16	TL	4.201	0.9152	191	3.172	1.9442	1625	3.4476	1.6686
17	TL	7.642	0.1459	450	7.2298	0.5581	2516	7.572	0.2159
18	TL	F	NA	TL	6.411	0.7421	TL	F	NA
19	TL	F	NA	935	7.182	7.2993	TL	2.288	12.1933
20	TL	F	NA	TL	7.217	6.1333	TL	F	NA

Comparing between the data in Tables 4 and 5 makes it clear that the logarithmic models are significantly better than the non logarithmic ones. This confirms the fact that for a PLA model, increasing the number of binary variables has more negative impact than increasing the continuous variables. Even though the logarithmic models used 4 points more (11 compared to 7), which considerably affects the size, the solving times turned out to be much less than the times for the non logarithmic case. Therefore, using logarithmic models allows introducing more breakpoints and, consequently, producing more accurate approximation. For example, the average difference between optimal value of the original instance and approximation, $\lvert f^* - \overline{f^*}\rvert$, per LOG problem is 0.328, while it is 0.667 for CC problems.

The computational experiments in this section have shown that full PLA of MINLP problems will not make the new problems any better, yet they prove that the improvement of PLA models is promising. In addition to the major development of the PLA model caused by logarithmic formulation, different components of the model can also be improved. For example, it was demonstrated that nonuniform partitioning based on local solutions can add more improvement to the models. Finally, the uses of PLA are not limited to the full PLA; it can be useful if applied to only parts of an MINLP problem, as will be shown in the next section.

3.1.2. Partial PLA

PLA was applied partially to different QCQP and MIQCQP using Algorithm 1. If the number of approximated constraints is large, then the modified problem gets more complicated and the original problem is better off without this approximation. However, in this section, the algorithm is applied with one iteration to many instances belonging to different groups from the MINLPLib library. Interestingly, this approach worked on some instances, and produced problems that needed shorter solving times. Table 6 summarizes the statistics of some of these instances, in addition to computational results of solving them using SCIP with 2 h time limit.

Table 6. Statistics of different test instances, and the solving results by SCIP.

Instance	seq#	#v	#cons (q)	obj val	Time	Gap %
kall_circlespolygons_c1p12	1	43	48 (21)	0.3396	TL	∞
kall_circlespolygons_c1p13	2	43	48 (21)	0.3396	TL	∞
kall_circlesrectangles_c1r12	3	49	41 (23)	0.3396	TL	∞
kall_diffcircles_5a	4	14	24 (11)	5.1162	63	0
kall_diffcircles_6	5	16	31 (16)	7.7879	102	0
pooling_foulds3stp	6	832	1089 (1024)	−8	5730	0
pooling_foulds4stp	7	832	1089 (1024)	−8	6235	0

As can be seen in the table, SCIP failed to close the gap when solving instances 1, 2, and 3; but it succeeded for the other problems. The goal of testing these instances is to show that approximating one constraint per instance aids the solver to handle the instance better. For each modified instance Table 7 presents the solving time, the optimal objective value $\overline{f^*}$, and the difference between the optimal values of the original and modified instance.

Table 7. Computational Results of Problems Produced by Partial PLA Using 10 Breakpoints.

p#	CC			LOG		
	Time	$\overline{f^*}$	$\lvert f^* - \overline{f^*}\rvert$	Time	$\overline{f^*}$	$\lvert f^* - \overline{f^*}\rvert$
1	92	0.2949	0.0447	241	0.2848	0.0548
2	258	0.2914	0.0482	358	0.3227	0.0169
3	TL	0.3396	0	TL	0.3396	∞
4	38	4.96	0.1562	29	4.96	0.1562
5	68	7.7879	0	47	7.7879	0
6	215	−8	0	2631	−8	0
7	3143	−8	0	622	−8	0

The results table shows that for the first three instances, SCIP was able to solve them and close the gap, except for the third instance, the gap was closed shortly after 2 h (when time limit was disabled) for the CC case, and the gap was ∞% in the LOG case. The solving time of the modified instance is mostly shorter than the original problem.

From instances with $|f^* - \overline{f^*}| > 0$, it can be implied that even if it is only one out of many constraints that was approximated, there will probably be an approximation error. These instances were approximated again with 15, 20, and 25 breakpoints, and that led to better optimal values but longer solving time. For example, when instance 1 was generated by CC with 15 points, SCIP needed 250 s to solve it and the objective value was 0.311; and with 25 points, it needed 440 s, yielding an objective value of 0.334.

An interesting finding is that the solving times of CC problems are shorter than those of LOG problems. This indicates that the models have similar performance when approximating problems with few variables using few breakpoints.

The experimental results suggest that partial PLA can produce less complicated problems with small or zero approximation error. Moreover, it was shown that LOG models have no advantage here in contrary to the case in the previous section. In the following section, similar computational experiments are performed on MIQCP problems.

3.2. Mixed Integer Variables

The PLA models, that were used in the previous section, were applied to problems where the variables involved in the quadratic terms can be integer. When introducing the breakpoints to an integer variable's interval, every integer value in the interval will be a breakpoint, so the number of breakpoints depends on the length of the interval. If this results in a large number of breakpoints, some integer values can be skipped so the number stays reasonable. Moreover, there is no need to have a convex combination of two breakpoints since the variable cannot take noninteger values. For all noninteger variables involved in the quadratic functions, the PLA procedures are still the same.

The objective of the computational tests is the same as in the case of continuous variables: to test the uniform against the nonuniform partitioning, and to assess the performance of the partial PLA models. A summary of the test instances' statistics is given in Table 8, where they were solved by SCIP with a time limit of two hours. Instances 1, 2, and 3 were taken from the QPLIB library and the remainder are from MINLPLib instances.

Table 8. Statistics of MIQCP test instances, and the solving results by SCIP.

Instance	seq#	#v	#cons (q)	obj val	Time	Gap %
3562	1	63	42 (7)	15	TL	305
3780	2	168	72 (12)	90.6	TL	1138.3
3816	3	187	387 (24)	7.3936	TL	27.37
blend029	4	102	213 (12)	13.359	8	0
blend146	5	222	624 (24)	45.297	TL	1.87
ex1236	6	92	55 (4)	19.6	1	0
gabriel02	7	261	597 (96)	39.6	TL	22.3
sep1	8	29	31 (6)	−510.81	1	0
st_e31	9	112	135 (5)	−2	4	0
tln4	10	24	24 (4)	8.3	6	0

Based on the solving times in Table 8, it can be concluded that half of the tested instances are challenging and the other half are not. It should be mentioned that in instances 1, 2, and 10, only integer variables are involved in the quadratic functions, so no nonuniform partitioning is done for these instances (a breakpoint is introduced at every integer value in the interval). Tables 9 and 10 show the results of solving the problems produced by the convex combination and the logarithmic models with 11 breakpoints. The computations were done using CPLEX with 2 h time limit.

Table 9. Results of Solving MILP Problems Obtained from MIQCP Instances, Produced by CC and NCC Models Using 11 Breakpoints.

p#	CC			NCC ($k = 1.5$)			NCC ($k = 1.7$)		
	Time	BI	$\lvert f^* - \overline{f^*}\rvert$	Time	BI	$\lvert f^* - \overline{f^*}\rvert$	Time	BI	$\lvert f^* - \overline{f^*}\rvert$
1	TL	19.6	4.6	similar					
2	TL	F	NA	similar					
3	TL	F	NA	TL	F	NA	TL	F	NA
4	TL	12.73	0.629	TL	12.73	0.629	TL	13.359	0
5	TL	F	NA	TL	F	NA	TL	F	NA
6	360	19.6	0	68	19.6	0	237	19.6	0
7	TL	F	NA	TL	F	NA	TL	F	NA
8	109	−510.29	0.25	88	−510.08	0.73	21	−510.08	0.73
9	331	−2.019	0.019	29	−2.087	0.087	16	−2.188	0.188
10	TL	8.3	0	similar					

The results in the tables show that CPLEX failed to find an integer solution to 3 NLOG problems and 4 problems produced by the other models. For problems for which CPLEX found integer solutions, it can be observed that the problems with nonuniform partitioning are solved faster than the uniformly partitioned problems, for both logarithmic and nonlogarithmic models. The data under $\lvert f^* - \overline{f^*}\rvert$ columns show that no clear advantage can be confirmed for any of the models in the accuracy of the approximation.

Partial PLA was applied to the same set of instances and the most violated constraint of each challenging instance was approximated using CC and NLOG models. As Table 8 shows, instances 1 and 2 have a relative gap of 305% and 1138%, respectively, after two hours of solving time. After PLA, SCIP reached the time limit and resulted in gaps of 53% and 1862% for the CC problems, and 200% and 774% for the LOG problems. The rest of instances resulted in better performances by the solver without PLA.

Table 10. Results of solving MILP problems produced with LOG and NLOG models using 11 breakpoints.

p#	CC			NCC ($k = 1.5$)			NCC ($k = 1.7$)		
	Time	BI	$\lvert f^* - \overline{f^*} \rvert$	Time	BI	$\lvert f^* - \overline{f^*} \rvert$	Time	BI	$\lvert f^* - \overline{f^*} \rvert$
1	TL	18.4	3.4			similar			
2	TL	F	NA			similar			
3	TL	F	NA	TL	6.6	0.7936	TL	6.2	1.1936
4	5088	13.359	0	756	13.359	0	970	13.359	0
5	TL	F	NA	TL	F	NA	TL	F	NA
6	146	19.6	0	29	19.6	0	41	19.6	0
7	TL	F	NA	TL	F	NA	TL	F	NA
8	103	−509.72	1.09	34	−510.08	0.73	43	−510.08	0.73
9	198	−2.01	0.01	89	−2.087	0.087	103	−2.188	0.188
10	354	8.3	0			similar			

4. Discussion

To summarize the results of the computational experiments, two comparisons will be discussed. The first comparison is between PLA models using uniform against nonuniform partitioning. QCQP and MIQCQP problems were approximated using CC, NCC, LOG, and NLOG models. Most of the MILP problems that resulted from the nonuniformly partitioned model were solved to optimality faster than the problems resulting from the uniform partitioning. For example, out of 20 instances, the nonuniformly partitioned logarithmic model ($k = 1.5$) failed to find an integer solution within the time limit only three times compared to 8 failures in the uniform partitioning case. For the other 12 instances that an integer solution was found by both models, the nonuniformly partitioned model was faster 11 times. Moreover, the optimal solutions are more accurate (closer or identical to the optimal solutions of the original problems) in the nonuniform partitioning models. It is worth mentioning that the both logarithmic models were better than the convex combination models in the continuous variables problems. Despite introducing more breakpoints to LOG and NLOG (which lead to more accurate solutions), their resulting problems were solved faster.

The other comparison is to discuss whether applying PLA partially on an optimization problem can be useful ot not. We tested many QCQP and MIQCQP problems by solving them by SCIP, then we use Algorithm 1 to replace one quadratic constraints by its linear approximation, then we solve it again by SCIP. The results showed that partial PLA led to an easier problem form several instances. For these instances, the average solving time was around 4818 s without PLA, while it was 1573 and 1598 for PLA using CC and LOG models, respectively. Even though more QCQP problems were solved faster without the partial PLA, it was useful for SCIP in many QCQP problems. Improving some of the steps in Algorithm 1 can decrease the difficulty in more instances. For example the violated constraints selection rule can be improved by taking into account other aspects like current feasible solutions, variables bounds, optimal solution bounds, etc.

With the improvement of MINLP solvers, it can be concluded that full transformation of an MINLP problem into an MILP one will not be beneficial. However, doing PLA partially or using it as a supporting tool within MINLP algorithms can be useful, especially after showing that PLA itself can be improved.

5. Conclusions

The computational experiments performed on both continuous and mixed integer problems indicate promising results. It can be concluded that PLA models can be improved by using nonuniform partitioning depending on local solutions instead of uniform partitioning. Moreover, the tests have shown that some challenging QCQP and MIQCQP problems can be less challenging by applying partial PLA and replacing only one constraint by its linear approximation, which is only a small change to the original problem.

Some of the paper subjects can be considered for further research in the future. In this paper, we studied nonuniform against uniform partitioning, but it would be good research of our nonuniform partitioning was tested against other nonuniform partitioning methods. Moreover, it might be useful to do the partitioning depending on the function, so every nonlinear function could have its own domain partitioning. On the other hand, the partial PLA algorithm can be used as a tool of an existing MINLP algorithm. It can be improved by, for example, selecting a violated constraint based on all information that can be brought from the already explored nodes in the branch and bound tree.

Author Contributions: Writing—original draft, L.A.; H.M.; Writing—review & editing, L.A.; H.M. All authors have read and agreed to the published version of the manuscript.

Funding: The Deanship of Scientific Research, Qassim University.

Institutional Review Board Statement: Not applicable.

Informed Consent Statement: Not applicable.

Data Availability Statement: Not applicable.

Acknowledgments: The researchers would like to thank the Deanship of Scientific Research, Qassim University for funding publication of this project.

Conflicts of Interest: The authors declare no conflict of interest.

References

1. Trespalacios, F.; Grossmann, I.E. Review of mixed-integer nonlinear and generalized disjunctive programming methods. *Chem. Ing. Tech.* **2014**, *86*, 991–1012. [CrossRef]
2. Bussieck, M.R.; Vigerske, S. MINLP solver software. In *Wiley Encyclopedia of Operations Research and Management Science*; John Wiley & Sons, Inc.: Hoboken, NJ, USA, 2010; pp. 1–12.
3. Smith, E.M.B.; Pantelides, C.C. Global optimisation of nonconvex MINLPs. *Comput. Chem. Eng.* **1997**, *21*, S791–S796. [CrossRef]
4. Ryoo, H.S.; Sahinidis, N.V. A branch-and-reduce approach to global optimization. *J. Glob. Optim.* **1996**, *8*, 107–138 [CrossRef]
5. Adjiman, C.S.; Androulakis, I.P.; Floudas, C.A. Global optimization of mixed-integer nonlinear problems. *AIChE J.* **2000**, *46*, 1769–1797. [CrossRef]
6. Burlacu, R.; Geißler, B.; Schewe, L. Solving mixed-integer nonlinear programmes using adaptively refined mixed-integer linear programmes. *Optim. Methods Softw.* **2020**, *35*, 7–64. [CrossRef]
7. Belotti, P.; Kirches, C.; Leyffer, S.; Linderoth, J.; Luedtke, J.; Mahajan, A. Mixed-integer nonlinear optimization. *Acta Numer.* **2013**, *22*, 1–131. [CrossRef]
8. Bragalli, C.; D'Ambrosio, C.; Lee, J.; Lodi, A.; Toth, P. On the optimal design of water distribution networks: A practical MINLP approach. *Optim. Eng.* **2012**, *13*, 219–246. [CrossRef]
9. Pei, M.; Lin, P.; Du, J.; Li, X; Chen, Z. Vehicle dispatching in modular transit networks: A mixed-integer nonlinear programming model. *Transp. Res. Part Logist. Transp. Rev.* **2021**, *147*, 102240. [CrossRef]
10. Stopková, M.; Stopka, O; Klapita, V. Modeling the Distribution Network Applying the Principles of Linear Programming. 2017. Available online: https://www.semanticscholar.org/paper/Modeling-the-Distribution-Network-Applying-the-Stopkov%C3%A1-Stopka/dd47e3dd531f91c555e9ce14f732ccf232c06546 (accessed on 1 November 2021).
11. Toydas, M; Saraç, T. A mixed integer nonlinear model for air refueling optimization to save fuel in military deployment operations. *Int. J. Ind. Eng.* **2020**, *27*, 627–644.
12. Rebennack, S; Kallrath, J. Continuous piecewise linear delta-approximations for univariate functions: Computing minimal breakpoint systems. *J. Optim. Theory Appl.* **2015**, *167*, 617–643. [CrossRef]
13. Beale, E.M.L; Tomlin, J.A. Special facilities in a general mathematical programming system for non-convex problems using ordered sets of variables. *OR* **1970**, *69*, 447–454.
14. Markowitz, H.M; Manne, A.S. On the solution of discrete programming problems. *Econom. J. Econom. Soc.* **1957**, *25*, 84–110. [CrossRef]

15. Dantzig, G.B. On the significance of solving linear programming problems with some integer variables. *Econom. J. Econom. Soc.* **1960**, *28*, 30–44. [CrossRef]
16. Jeroslow, R.G.; Lowe, J.K. Modelling with integer variables. *Math. Program. Oberwolfach II* **1984**, 167–184. [CrossRef]
17. Meyer, R.R. Mixed integer minimization models for piecewise-linear functions of a single variable. *Discret. Math.* **1976**, *16*, 163–171. [CrossRef]
18. Balakrishnan, A; Graves, S.C. A composite algorithm for a concave-cost network flow problem. *Networks* **1989**, *19*, 175–202. [CrossRef]
19. Misener, R.; Floudas, C.A. Global optimization of mixed-integer quadratically-constrained quadratic programs (MIQCQP) through piecewise-linear and edge-concave relaxations. *Math. Program.* **2012**, *136*, 155–182. [CrossRef]
20. Castillo, P.A.C.; Castro, P.M.; Mahalec, V. Global optimization of MIQCPs with dynamic piecewise relaxations. *J. Glob. Optim.* **2018**, *71*, 691–716. [CrossRef]
21. Sundar, K.; Nagarajan, H.; Linderoth, J.; Wang, S.; Bent, R. Piecewise polyhedral formulations for a multilinear term. *Oper. Res. Lett.* **2021**, *49*, 144–149. [CrossRef]
22. Vielma, J.P.; Nemhauser, G.L. Modeling disjunctive constraints with a logarithmic number of binary variables and constraints. *Math. Program.* **2011**, *128*, 49–72. [CrossRef]
23. Vielma, J.P.; Ahmed, S.; Nemhauser, G. Mixed-integer models for nonseparable piecewise-linear optimization: Unifying framework and extensions. *Oper. Res.*, **2010**, *58*, 303–315. [CrossRef]
24. Croxton, K.L.; Gendron, B.; Magnanti, T.L. A comparison of mixed-integer programming models for nonconvex piecewise linear cost minimization problems. *Manag. Sci.* **2003**, *49*, 1268–1273. [CrossRef]
25. Padberg, M. Approximating separable nonlinear functions via mixed zero-one programs. *Oper. Res. Lett.* **2000**, *27*, 1–5. [CrossRef]
26. Sridhar, S.; Linderoth, J.; Luedtke, J. Locally ideal formulations for piecewise linear functions with indicator variables. *Oper. Res. Lett.* **2013**, *41*, 627–632. [CrossRef]
27. Geißler, B.; Martin, A.; Morsi, A; Schewe, L. Using Piecewise Linear Functions for Solving MINLPs. *Mix. Integer. Nonlinear. Program.* **2012**, 287–314._10. [CrossRef]
28. Dahl, G.; Realfsen, B. Curve Approximation and Constrained Shortest Path Problems. Preprint (Universitetet i Oslo, Institutt for informatikk). 1996. Available online: https://www.duo.uio.no/bitstream/handle/10852/9204/1/GDahl-4.pdf (accessed on 1 November 2021).
29. Vasudeva, V. Global Optimization with Piecewise Linear Approximatio. Matser's Thesis, The University of Texas at Austin, Austin, TX, USA, 2015.
30. Nagarajan, H.; Lu, M.; Wang, S.; Bent, R; Sundar, K. An adaptive, multivariate partitioning algorithm for global optimization of nonconvex programs. *J. Glob. Optim.* **2019**, *74*, 639–675. [CrossRef]

mathematics

MDPI

Article

Using Pairwise Comparisons to Determine Consumer Preferences in Hotel Selection

Nikolai Krivulin [1,*,†], Alexey Prinkov [1,†] and Igor Gladkikh [2,†]

1 Faculty of Mathematics and Mechanics, St. Petersburg State University, Universitetskaya Emb. 7/9, 199034 St. Petersburg, Russia; aprinkov@yahoo.com
2 Graduate School of Management, St. Petersburg State University, Universitetskaya Emb. 7/9, 199034 St. Petersburg, Russia; gladkikh@gsom.spbu.ru
* Correspondence: nkk@math.spbu.ru
† These authors contributed equally to this work.

Abstract: We consider the problem of evaluating preferences for criteria used by university students when selecting a hotel for accommodation during a professional development program in a foreign country. Input data for analysis come from a survey of 202 respondents, who indicated their age, sex and whether they have previously visited the country. The criteria under evaluation are location, accommodation cost, typical guests, free breakfast, room amenities and courtesy of staff. The respondents assess the criteria both directly by providing estimates of absolute ratings and ranks, and indirectly by relative estimates using ratios of pairwise comparisons. To improve the accuracy of ratings derived from pairwise comparisons, we concurrently apply the principal eigenvector method, the geometric mean method and the method of log-Chebyshev approximation. Then, the results from the direct and indirect evaluation of ratings and ranks are examined together to analyze how the results from pairwise comparisons may differ from each other and from the results of direct assessment by respondents. We apply statistical techniques, such as estimation of means, standard deviations and correlations, to the vectors of ratings and ranks provided directly or indirectly by respondents, and then use the estimates to make accurate assessment of the criteria under study.

Keywords: pairwise comparison; matrix approximation; log-Chebyshev metric; tropical optimization; consumer preference; hotel selection

MSC: 90B50; 90C47; 91B06; 41A50; 90C24

Citation: Krivulin, N.; Prinkov, A.; Gladkikh, I. Using Pairwise Comparisons to Determine Consumer Preferences in Hotel Selection. *Mathematics* **2022**, *10*, 730. https://doi.org/10.3390/math10050730

Academic Editors: Humberto Rocha and Ana Maria Rocha

Received: 21 January 2022
Accepted: 22 February 2022
Published: 25 February 2022

1. Introduction

Evaluation of preferences for alternatives based on their pairwise comparisons is a widely accepted approach in decision making, when direct assessment of the preferences is infeasible or impossible [1–4]. The approach uses the results of pairwise comparisons of alternatives on an appropriate scale, given in the form of a pairwise comparison matrix. Then, various computational methods can be applied to the matrix to make judgment on the preference of each alternatives by evaluating its individual rating (score, priority, weight) and ranking the alternatives according to the ratings.

The methods used to derive ratings from pairwise comparisons may exploit different computational techniques. These methods are mainly based on aggregation (summation) of columns in the pairwise comparison matrix to obtain a vector of ratings of alternatives, or on approximation of the pairwise comparison matrix by a symmetrically reciprocal (consistent) matrix that directly determines the vector of ratings (see, e.g., [5,6]). The key issue in deriving ratings from pairwise comparisons by different techniques is that these techniques may produce rather different or even contradictory results (see, e.g., [7–10]), which can significantly complicate or even make it impossible to unambiguously assess alternatives when making decisions in practice.

Among other problems arising from the issue indicated above, one has to recognize how much the results of different methods vary in practical problems, and how to make the right decision in case of possible inconsistency of the results. The comparison of results of assessment methods is discussed in many research studies [8,10,11], including statistical analysis of extensive data given by many pairwise comparison matrices. However, in most cases, the source data used in the analysis are obtained by simulation [8,10] and thus are to be considered as an artificial input that may less adequately reflect the usual conditions of practice than authentic results of human evaluation.

In practical situations where the methods used can give different results, a natural but reasonable way to evaluate alternatives more unambiguously is based on the simultaneous application of several methods to make a decision that concurrently considers all the results. If these results differ significantly, the choice of one of them as the basis for making a decision does not seem entirely justified. On the contrary, the closeness and stability of the results can serve as additional arguments in favor of choosing one of them as a solution that can be considered as close to optimal.

The well-known solution approach is the principal eigenvector method [2,3,5], which defines the vector of ratings as a weighted sum of the columns in the pairwise comparison matrix, where the weights are taken proportional to the entries of the vector of ratings itself. This approach leads to a solution calculated as the eigenvector of the pairwise comparison matrix, which corresponds to the maximal eigenvalue. The geometric mean method is another widely used approach, which solves the problem by approximating the pairwise comparison matrix by a consistent matrix in the Euclidean norm on the logarithmic scale [7,12,13]. It provides the solution vector by calculating the geometric mean of the entries in each row of the pairwise comparison matrix.

An approach to the derivation of ratings from pairwise comparison matrices, which applies the Chebyshev approximation on the logarithmic scale (the log-Chebyshev approximation), is developed in [14–17]. The proposed solution technique is based on methods and results of tropical mathematics which is concerned with the study and application of algebraic systems with idempotent operations [18–21]. The approximation problem is formulated and solved as an optimization problem in terms of tropical mathematics (a tropical optimization problem), which yields a complete analytical solution given in a parametric form ready for numerical computation and formal analysis.

The problem of evaluating alternatives from their pairwise comparisons often arises in studying consumer preferences in various markets including the market of hotel services. Specifically, for hotel suppliers, it is crucial to identify the most and least important criteria used by consumers in hotel selection. The problem of assessment criteria for hotel selection by various customer groups is studied in many researches (see, e.g., [22–30]). These studies mainly concentrate on investigation of preferences of business and (or) leisure travelers, who often constitute the largest segment of hotel guests and thus are of prime interest [23–26,30], but sometimes more specific customer groups such as university students are under consideration [29].

As a source of information to understand how different factors (hotel attributes), such as location, accommodation cost, free breakfast or amenities, may affect the choice of a hotel, consumer surveys are widely used which can be administered online [26,28] or using a paper-and-pencil format [22,24,27,29,30]. A typical survey can ask respondents for absolute estimates that rate and rank the factors directly or for relative estimates in the form of pairwise comparison ratios for the factors.

The pairwise comparison data are then used to make a final assessment of factors by applying one of the methods of rating alternatives from pairwise comparisons. However, many studies rely on results obtained using only one method, which can lead to inaccurate or wrong conclusions because different methods may produce ambiguous results. To improve the quality of the assessment, it seems necessary to verify the results of one method against the results of other available methods, which makes the development and experimental study of a combined approach, which integrates several solution methods together with procedures of data analysis, a highly relevant issue.

In this paper, we consider the problem of evaluating preferences for criteria used by university students when selecting a hotel for accommodation during a short-term professional development program in a foreign country. As input data for analysis, we are restricted by the results of a survey of 202 respondents who indicated their age, sex and whether they have previous experience of visiting the country. The criteria evaluated by respondents include location, accommodation cost, typical guests, free breakfast, room amenities and courtesy of staff. The respondents assess the criteria both directly by providing estimates of absolute ratings and ranks, and indirectly by relative estimates that are represented as ratios of pairwise comparisons.

The purpose of this applied study is twofold. First, it is to investigate the possibilities of combining different methods of rating and ranking criteria together with statistical procedures to provide more accurate assessment of the criteria, to describe an applied technique intended to improve the quality of assessment, and to demonstrate the implementation of this technique to the problem of evaluating criteria for hotel selection. The second objective is to explore by statistical procedures the results obtained by different methods from pairwise comparisons provided by respondents to gain additional insight and understanding about the similarity and difference between these results.

As the key technique to improve the accuracy of ratings derived from the results of pairwise comparisons, we concurrently apply the methods of principal eigenvector, geometric mean and log-Chebyshev approximation. Then, the results obtained by the direct and indirect assessment of ratings and ranks are examined together to analyze how the results from pairwise comparisons may differ from each other and from the results of direct assessment. The analysis involves statistical techniques, such as estimation of means, standard deviations and correlations, applied to the vectors of ratings and ranks of criteria, which are provided directly or indirectly by respondents.

The paper is organized as follows. We start in Section 2 with a brief overview of the methods used below to derive ratings from pairwise comparisons, including the method based on the log-Chebyshev approximation. Section 3 presents the basic definitions, notation and results of tropical mathematics, which are then used to give direct formulas for calculating vectors of ratings using log-Chebyshev approximation. In Section 4, we formulate the problem of evaluating criteria for hotel selection, which motivates and illustrates the study, and describe the survey data used in the solution.

Section 5 offers results of statistical analysis of ratings and ranks derived from the survey data. First, we investigate whether characteristics of respondents (age, sex, etc.) may affect the degree of difference or similarity between vectors of ratings or ranks obtained for each respondent by different methods. Furthermore, statistical results of the analysis and comparison of the vectors of ratings for the criteria are given including estimates of means, standard deviations and correlations. We conclude the section with the results of comparison of rank vectors that order criteria according to their preferences. In Section 6, we present some discussions, and in Section 7 offer concluding remarks.

2. Evaluation of Alternatives from Pairwise Comparisons

Suppose there are n alternatives $\mathcal{A}_1, \dots, \mathcal{A}_n$ for making decisions, which are compared in pairs. The results of the comparisons are represented in the form of a pairwise comparison matrix $\boldsymbol{A} = (a_{ij})$ of dimension $n \times n$, where the element $a_{ij} > 0$ indicates by how many times the alternative $\mathcal{A}_i$ is more preferable than $\mathcal{A}_j$. The problem of pairwise comparisons is to find a vector $\boldsymbol{x} = (x_i)$ of absolute ratings (scores, priorities, weights) of alternatives on the basis of the relative estimates given by pairwise comparisons.

Consider a pairwise comparison matrix $\boldsymbol{A}$ and note that its elements satisfy the condition $a_{ji} = 1/a_{ij}$ (and hence $a_{ii} = 1$) for all $i, j = 1, \dots, n$, and therefore the matrix $\boldsymbol{A}$ is symmetrically reciprocal. Furthermore, if the transitive property in the form of equality $a_{ij} = a_{ik}a_{kj}$ is fulfilled for all $i, j, k = 1, \dots, n$, then the matrix $\boldsymbol{A}$ is called consistent.

The elements of a consistent pairwise comparison matrix $\boldsymbol{A}$ are represented as $a_{ij} = x_i/x_j$ for all $i, j = 1, \ldots, n$, where x_i are the components of some positive vector $\boldsymbol{x}$ determined up to a positive factor. Moreover, it directly follows from the equality $a_{ij} = x_i/x_j$ that $\boldsymbol{x} = (x_i)$ is a solution of the pairwise comparison problem of interest.

In practical applications, the matrices obtained by pairwise comparisons of alternatives are usually not consistent. In this case, a problem arises of finding a consistent matrix that is close to (approximates) the original pairwise comparison matrix. Any vector that determines this consistent matrix is then taken as the vector of absolute ratings. To find an appropriate consistent matrix of pairwise comparisons various heuristic procedures and approximation methods are used (see, e.g., [2,5,6]).

Heuristic methods for solving the problem usually apply techniques of aggregation (summation) of the columns in the pairwise comparison matrix. The columns are added up with coefficients (weights) that are dependent on the technique. The resulting vector, which generates some consistent matrix, serves as a solution to the problem. As examples of the heuristic approach, one can consider the weighted column sum method and the principal eigenvector method.

The most common in practice is the method of principal eigenvector proposed by T. Saaty in the 1970s [2,3,5,31]. The method assumes that the weights of columns for aggregation are set proportional to the components of the solution vector of absolute ratings. This leads to a solution of the problem that takes the form of the principal eigenvector $\boldsymbol{x}$ of the pairwise comparison matrix $\boldsymbol{A}$, which corresponds to the maximum eigenvalue λ of the matrix and satisfies the equality

$$\boldsymbol{A}\boldsymbol{x} = \lambda\boldsymbol{x}.$$

The approximation methods solve a problem of minimizing an error in the approximation of the pairwise comparison matrix $\boldsymbol{A} = (a_{ij})$ by a consistent matrix $\boldsymbol{X} = (x_i/x_j)$. The application of these methods offers a mathematically justified approach, which however can result in very hard optimization problems. The complexity of the solution essentially depends on the metric and scale used to estimate the approximation error.

To measure the error on the standard linear scale, different metrics may be used including the Euclidean, Manhattan and Chebyshev distance functions. However, minimizing the errors in linear scale usually leads to complex nonlinear multiextremal optimization problems that are difficult to solve [5,32], and hence is not common in practice.

If the approximation error is evaluated in logarithmic scale, then the solution may become less complicated and can sometimes be obtained in analytical form [7,12,13]. Note that when transforming from linear scale to logarithmic scale, the variation between entries in a pairwise comparison matrix $\boldsymbol{A}$ decreases, which reduces the impact on the overall error of large values (for example, $a_{ij} = 9$) to the detriment of small ones ($a_{ji} = 1/9$). Therefore, the use of the logarithmic scale in the approximation of the pairwise comparison matrices seems to be quite reasonable.

Consider the application of the log-Euclidean metric (the Euclidean metric on the logarithmic scale), in which the distance between the matrices $\boldsymbol{A} = (a_{ij})$ and $\boldsymbol{X} = (x_i/x_j)$ is determined using logarithm in a base greater than one by the formula

$$l_2(\boldsymbol{A}, \boldsymbol{X}) = \left(\sum_{1 \leq i,j \leq n} \left(\log a_{ij} - \log \frac{x_i}{x_j} \right)^2 \right)^{1/2}.$$

The minimum distance is immediately found by calculating the derivatives of the squared distance $l_2^2(\boldsymbol{A}, \boldsymbol{X})$ with respect to all x_i, equating the derivatives to zero, and solving the obtained equations. For a symmetrically reciprocal matrix $\boldsymbol{A}$, the result is a direct solution of the approximation problem, in which the components of the vector $\boldsymbol{x} = (x_1, \ldots, x_n)^T$ that determines the matrix $\boldsymbol{X}$ are given in the parametric form

$$x_i = \left(\prod_{j=1}^{n} a_{ij} \right)^{1/n} u, \qquad u > 0, \qquad i = 1, \ldots, n.$$

The obtained vector $\boldsymbol{x}$ (usually normalized in rectangular metric, i.e., with respect to the sum of components) is called the solution of the problem of pairwise comparisons by the method of geometric mean [7,12,13].

The distance between the matrices $\boldsymbol{A}$ and $\boldsymbol{X}$ in the Chebyshev metric in logarithmic scale (the log-Chebyshev distance) is defined as

$$l_\infty(\boldsymbol{A}, \boldsymbol{X}) = \max_{1 \leq i,j \leq n} \left| \log a_{ij} - \log \frac{x_i}{x_j} \right|.$$

The problem of approximating a pairwise comparison matrix in the log-Chebyshev metric can be reduced to the following problem of minimizing a function without logarithm (see, e.g., [17,33]):

$$\min_{\boldsymbol{x} > \boldsymbol{0}} \max_{1 \leq i,j \leq n} \frac{a_{ij} x_j}{x_i}, \tag{1}$$

which is equivalent to minimizing the maximum relative error [33,34], which is given by

$$\max_{1 \leq i,j \leq n} \left| \frac{a_{ij} - x_i / x_j}{a_{ij}} \right|.$$

In contrast to both methods of the principal eigenvalue and geometric mean, which always lead to a single solution vector (up to a positive factor), the solution based on the log-Chebyshev approximation may be nonunique.

The existence of multiple solutions instead of a single one, on the one hand, can make it difficult to choose the most preferable alternative in practice. On the other hand, the availability of a set of different solutions to the problem extends the possibility of making optimal decisions, for example by taking into account additional constraints to narrow the solutions. In addition, due to the rather approximate nature of the model of pairwise comparisons for which the inconsistency of pairwise judgments is typical, the existence of several solutions to the problem seems to be quite natural.

Suppose that the approximation procedure results in a set $\mathcal{S}$ of optimal vectors, rather than in a single optimal vector of ratings. As the "best" and "worst" solutions to the problem, we can take those two vectors from $\mathcal{S}$ that best and worst differentiate the alternatives with highest and lowest ratings [16,17].

Consider the ratio between the maximum and minimum elements of the vector $\boldsymbol{x} = (x_i)$, which is called the Hilbert (interval, range) seminorm, and given by

$$\max_{1 \leq i \leq n} x_i \Big/ \min_{1 \leq j \leq n} x_j = \max_{1 \leq i \leq n} x_i \times \max_{1 \leq j \leq n} x_j^{-1}.$$

The solution vector with the maximum Hilbert seminorm is taken as the best differentiating solution, whereas the vector with the minimum seminorm is as the worst differentiating solution. These vectors are obtained by solving the following problems:

$$\max_{\boldsymbol{x} \in \mathcal{S}} \max_{1 \leq i \leq n} x_i \times \max_{1 \leq j \leq n} x_j^{-1}, \tag{2}$$

$$\min_{\boldsymbol{x} \in \mathcal{S}} \max_{1 \leq i \leq n} x_i \times \max_{1 \leq j \leq n} x_j^{-1}. \tag{3}$$

Problems (1)–(3), which arise from the log-Chebyshev approximation in the pairwise comparison problem, can be directly solved in analytical form, using the methods and results of tropical mathematics (see, e.g., [14–17] for further details).

3. Tropical Mathematics Based Solutions

The purpose of this section is to provide a brief overview of the basic concepts, definitions and notation of tropical mathematics, which are used to describe the solution of the pairwise comparison problem. Further information on the theory, methods and applications of tropical mathematics can be found, for example, in [18–21,35].

3.1. Elements of Tropical Mathematics

Tropical (idempotent) mathematics deals with the theory and applications of algebraic systems with idempotent operations. An operation is idempotent if its application to arguments of the same value yields this value as the result. For example, taking the maximum is an idempotent operation since $\max\{x, x\} = x$, whereas the arithmetic addition is not: $x + x = 2x$.

Tropical optimization focuses on optimization problems that are formulated and solved in terms of tropical mathematics. Models and methods of tropical optimization make it possible to find new solutions for classical and newly posed problems. Many problems can be solved directly in explicit analytic form; for other problems, only algorithmic techniques that offer solutions in numerical form are known. Applications of tropical optimization include various problems in project scheduling, location analysis, decision making and other areas.

An example of an algebraic system with an idempotent operation is the max-algebra defined on the set of non-negative real numbers $\mathbb{R}_+ = \{x \in \mathbb{R} | x \geq 0\}$. It is closed under addition denoted by $\oplus$ and defined for all $x, y \in \mathbb{R}_+$ as maximum: $x \oplus y = \max\{x, y\}$, and multiplication defined as usual. The neutral elements with respect to addition and multiplication coincide with the arithmetic zero 0 and one 1.

The tropical addition $\oplus$ is not invertible (opposite numbers do not exist), and hence a subtraction operation is undefined in max-algebra. The notion and notation of inverse elements with respect to multiplication, and exponents have the usual sense.

Vector and matrix operations are performed according to standard rules, where the arithmetic addition $+$ is replaced by $\oplus$. Specifically, multiplication of a vector or matrix by a scalar has the same result as in the standard arithmetic. In what follows, all vectors are considered column vectors until otherwise indicated. The zero vector denoted by $\mathbf{0}$, positive vector, zero matrix and identity matrix denoted by $\boldsymbol{I}$ are defined as usual.

For any nonzero column vector $\boldsymbol{x} = (x_j)$, its multiplicative conjugate transpose is the row vector $\boldsymbol{x}^- = (x_j^-)$, where $x_j^- = x_j^{-1}$ if $x_j \neq 0$, and $x_j^- = 0$ otherwise. For the vector of all ones, which is denoted as $\mathbf{1}$, the conjugate transpose is given by $\mathbf{1}^- = \mathbf{1}^T$. Multiplicative conjugate transposition of a nonzero matrix $\boldsymbol{A} = (a_{ij})$ results in the transposed matrix $\boldsymbol{A}^- = (a_{ij}^-)$, where $a_{ij}^- = a_{ji}^{-1}$ if $a_{ji} \neq 0$, and $a_{ij}^- = 0$ otherwise.

For vectors $\boldsymbol{a}_1, \ldots, \boldsymbol{a}_n$, its linear combination with nonnegative coefficients $x_1, \ldots, x_n$ is given by $x_1\boldsymbol{a}_1 \oplus \cdots \oplus x_n\boldsymbol{a}_n$. A vector $\boldsymbol{b}$ linearly depends on vectors $\boldsymbol{a}_1, \ldots, \boldsymbol{a}_n$ if there exist numbers $x_1, \ldots, x_n$ such that $\boldsymbol{b} = x_1\boldsymbol{a}_1 \oplus \cdots \oplus x_n\boldsymbol{a}_n$. The collinearity of two vectors has the usual sense: vectors $\boldsymbol{b}$ is collinear with $\boldsymbol{a}$ if $\boldsymbol{b} = x\boldsymbol{a}$ for some x.

The set of all linear combinations $x_1\boldsymbol{a}_1 \oplus \cdots \oplus x_n\boldsymbol{a}_n$ of a system of vectors $\boldsymbol{a}_1, \ldots, \boldsymbol{a}_n$ forms a tropical linear space. Any vector $\boldsymbol{y}$ of this space is expressed as the (tropical) product of the matrix $\boldsymbol{A} = (\boldsymbol{a}_1, \ldots, \boldsymbol{a}_n)$, which consists of the vectors in the system taken as columns, and some vector $\boldsymbol{x} = (x_1, \ldots, x_n)^T$, in the form $\boldsymbol{y} = \boldsymbol{A}\boldsymbol{x}$.

For a square matrix $A = (a_{ij})$ of order n, the nonnegative integer powers indicate repeated tropical multiplication of the matrix with itself, and are defined as $A^0 = I$, $A^p = A^{p-1}A = AA^{p-1}$ for all integer $p > 0$. The trace of the matrix is given by

$$\operatorname{tr} A = a_{11} \oplus \cdots \oplus a_{nn} = \bigoplus_{i=1}^{n} a_{ii}.$$

The spectral radius of the matrix A is calculated by the formula

$$\lambda = \operatorname{tr} A \oplus \cdots \oplus \operatorname{tr}^{1/n}(A^n) = \bigoplus_{k=1}^{n} \operatorname{tr}^{1/k}(A^k).$$

On condition that $\lambda \leq 1$, the Kleene operator (the Kleene star) is defined to map the matrix A onto the matrix

$$A^* = I \oplus A \oplus \cdots \oplus A^{n-1} = \bigoplus_{k=0}^{n-1} A^k.$$

3.2. Preliminary Results

We start with a formal criterion of linear dependence of vectors. In order to check whether a vector b is dependent on the system of vectors $a_1, \ldots, a_n$, the following result can be used [36] (see, also [35]).

Lemma 1. *Denote by $A = (a_1, \ldots, a_n)$ a matrix that consists of vectors $a_1, \ldots, a_n$ as columns. Then, a vector b is linearly dependent on $a_1, \ldots, a_n$ if and only if the following equality holds:*

$$(A(b^- A)^-)^- b = 1.$$

We now consider some results in tropical optimization that provide the basis to solve pairwise comparison problems using log-Chebyshev approximation. Suppose that, given an $(n \times n)$-matrix A, we need to find positive n-vectors x to solve the problem

$$\min_{x>0} \quad x^- A x. \tag{4}$$

A complete solution of the problem is obtained as follows (see, e.g., [37]).

Lemma 2. *Let A be a matrix with spectral radius $\lambda > 0$. Then, the minimum in problem (4) is equal to λ, and all positive solutions are given in the parametric form*

$$x = (\lambda^{-1} A)^* u, \qquad u > 0.$$

Let $B = (b_j)$ be a positive $(n \times m)$-matrix with columns $b_j = (b_{ij})$, and the problem is to obtain positive n-vectors x that provide the maximum

$$\begin{aligned} \max_{x>0} \quad & 1^T x x^- 1, \\ & x = B u, \quad u > 0. \end{aligned} \tag{5}$$

The next result offers a direct solution to the problem [16,17,38].

Lemma 3. *Denote by B_{lk} the matrix obtained from B by setting to zero all elements except b_{lk} for some indices k and l. Then, the maximum in problem (5) is equal to $\Delta = 1^T B B^- 1$, and all positive solutions are given in the parametric form*

$$x = B(I \oplus B_{lk}^- B) u, \qquad u > 0,$$

where the indices k and l are selected by the conditions

$$k = \arg\max_j \mathbf{1}^T \boldsymbol{b}_j \boldsymbol{b}_j^- \mathbf{1}, \qquad l = \arg\max_i b_{ik}^{-1}.$$

Finally, suppose that, given an $(n \times n)$-matrix $\boldsymbol{A}$ with spectral radius $\lambda > 0$, we need to find positive n-vectors $\boldsymbol{x}$ that attain the minimum

$$\begin{aligned} &\min_{\boldsymbol{x}>\mathbf{0}} \quad \mathbf{1}^T \boldsymbol{x}\boldsymbol{x}^- \mathbf{1}, \\ &\qquad\quad \boldsymbol{x} = (\lambda^{-1}\boldsymbol{A})^* \boldsymbol{u}, \quad \boldsymbol{u} > \mathbf{0}. \end{aligned} \tag{6}$$

A solution to this problem is found as follows [16,17].

Lemma 4. *Let $\boldsymbol{A}$ be a matrix with spectral radius $\lambda > 0$. Then, the minimum in problem (6) is equal to $\delta = \mathbf{1}^T(\lambda^{-1}\boldsymbol{A})^*\mathbf{1}$, and all positive solutions are given in the parametric form*

$$\boldsymbol{x} = (\delta^{-1}\mathbf{1}\mathbf{1}^T \oplus \lambda^{-1}\boldsymbol{A})^* \boldsymbol{u}, \qquad \boldsymbol{u} > \mathbf{0}.$$

Finally, note that the parametric form of solutions offered by Lemmas 2–4 defines the set of solutions to problems (4)–(6) as a linear span of columns of corresponding matrices that generate the solutions.

3.3. Application to Pairwise Comparison Problem

Consider the problem of evaluating ratings of alternatives from pairwise comparison, which is solved using the log-Chebyshev approximation of a pairwise comparison matrix. In this case, we need to obtain a solution of problem (1), and then, if the result is not unique, solutions of problems (2) and (3). Below, we use results in [14–17] to demonstrate how these problems can be analytically solved in explicit form.

We combine the solutions of the above problems into a procedure to handle the pairwise comparison problem under consideration. The procedure includes the following main steps: (i) finding the set of all solution vectors that represent absolute ratings of alternatives; (ii) checking whether the vectors determine a unique solution of the problem; and, in the case of a nonunique solution, (iii) selecting vectors which best and worst differentiate alternatives with the highest and lowest ratings (the best and worth differentiating solutions).

3.3.1. Finding All Solution Vectors

First, we observe that the calculation of all solutions to the pairwise comparison problem with a matrix $\boldsymbol{A}$ by using log-Chebyshev approximation requires solving problem (1). In the tropical algebra setting, the objective function in this problem takes the vector form $\boldsymbol{x}^-\boldsymbol{A}\boldsymbol{x}$, whereas the problem itself is written as (4) and can be solved by Lemma 2. The application of the lemma starts with the evaluation of the spectral radius for the matrix $\boldsymbol{A}$, given by

$$\lambda = \bigoplus_{k=1}^{n} \operatorname{tr}^{1/k}(\boldsymbol{A}^k). \tag{7}$$

Furthermore, we construct the matrix $\lambda^{-1}\boldsymbol{A}$ and calculate its Kleene star matrix

$$\boldsymbol{B} = (\lambda^{-1}\boldsymbol{A})^* = \bigoplus_{k=0}^{n-1} (\lambda^{-1}\boldsymbol{A})^k. \tag{8}$$

The set of all solutions of the problem is described using a vector of parameters $\boldsymbol{u}$ as

$$\boldsymbol{x} = \boldsymbol{B}\boldsymbol{u}. \qquad \boldsymbol{u} > \mathbf{0} \tag{9}$$

This result shows that these solutions form a set of linear combinations of columns (are generated by columns) in the Kleene matrix $\boldsymbol{B} = (\lambda^{-1}\boldsymbol{A})^*$.

3.3.2. Checking Uniqueness of Solution

In the framework of the pairwise comparison problem, the solution obtained is considered unique if each column in the matrix $\boldsymbol{B}$ generates the same space of vectors, which happens when all columns are collinear. To check the uniqueness of the solution, we apply a refinement procedure based on Lemma 1. The procedure examines the columns of the Kleene matrix $\boldsymbol{B} = (\boldsymbol{b}_j)$ sequentially one by one and deletes a column if it is linearly dependent on others. For each column $\boldsymbol{b}_j$, we form a matrix $\boldsymbol{B}_{(j)}$ by taking all other columns of $\boldsymbol{B}$, which have not yet been deleted, and then verify the condition

$$(\boldsymbol{B}_{(j)}(\boldsymbol{b}_j^- \boldsymbol{B}_{(j)})^-)^- \boldsymbol{b}_j = 1.$$

If this condition holds, the column $\boldsymbol{b}_j$ is deleted from the matrix $\boldsymbol{B}$, and otherwise retained. The solution is unique if the procedure results in a matrix with a single column that is taken as the solution. In the case when a matrix with several linearly independent columns is obtained to generate a set of different solutions, we turn to finding the best and worst differentiating vectors of ratings among them.

3.3.3. Best Differentiating Solution

Suppose that a solution to the log-Chebyshev approximation problem is obtained in the form $\boldsymbol{x} = \boldsymbol{B}\boldsymbol{u}$, where the matrix $\boldsymbol{B}$ has linearly independent columns. To select the best differentiating solution, we need to solve problem (2). In the framework of tropical algebra, this problem has the objective function written in the form $\boldsymbol{1}^T\boldsymbol{x}\boldsymbol{x}^-\boldsymbol{1}$, and can be represented as problem (5), which is solved using Lemma 3.

According to Lemma 3, we need to find indices k and l that satisfy the conditions

$$k = \arg\max_j \boldsymbol{1}^T \boldsymbol{b}_j \boldsymbol{b}_j^- \boldsymbol{1}, \qquad l = \arg\max_i b_{ik}^{-1}.$$

Then, we construct the matrix $\boldsymbol{B}_{lk}$ by setting to zero all elements in the matrix $\boldsymbol{B}$ except for the element b_{lk}. The best differentiating solution is then given using a vector of parameters $\boldsymbol{u}_1$ by

$$\boldsymbol{x}_1 = \boldsymbol{B}(\boldsymbol{I} \oplus \boldsymbol{B}_{lk}^- \boldsymbol{B})\boldsymbol{u}_1, \qquad \boldsymbol{u}_1 > \boldsymbol{0}.$$

3.3.4. Worst Differentiating Solution

We obtain the worst differentiating solution by solving problem (3), which in terms of tropical algebra, takes the form of (6). The solution of the latter problem by applying Lemma 4 involves evaluating

$$\delta = \boldsymbol{1}^T(\lambda^{-1}\boldsymbol{A})^*\boldsymbol{1}.$$

The worst differentiating solution is given in parametric form as

$$\boldsymbol{x}_2 = (\delta^{-1}\boldsymbol{1}\boldsymbol{1}^T \oplus \lambda^{-1}\boldsymbol{A})^*\boldsymbol{u}_2, \qquad \boldsymbol{u}_2 > \boldsymbol{0}.$$

4. Evaluation of Criteria for Hotel Selection

In this section, we consider the application of the above-discussed techniques of ratings alternatives from pairwise comparisons to the problem of evaluating consumer preferences in hotel selection. The problem is to identify priorities among criteria used by students in choosing hotels in a foreign country. The research is based on data obtained from 202 students of both sexes aged 17 to 26 years, who are to take a short-term professional development program at an educational institution in the country. Some students have visited this country and may have some experience of living there.

The students are provided with a grant that covers the travel expenses, tuition payments and food expenses during the working day. At the same time, the students must pay for accommodation on their own and choose a hotel by applying information available in the Internet on booking portals and hotel websites as well as their previous experience on visiting the country.

There are six criteria that are used when choosing a hotel: C_1—location, C_2—accommodation cost, C_3—social environment (typical guests who usually stay at the hotel), C_4—free breakfast, C_5—amenities (room equipment), C_6—courtesy of hotel staff. The aim of the research is to evaluate ratings (scores, priorities, weights) and then obtain ranks of the criteria, based on the results of the survey of the students as respondents.

During the survey process, the respondent directly assesses the degree of importance (the rating) of each criterion, determines the ranks of the criteria, and compares the criteria in pairs. For the direct rating of criteria, the following absolute scale is used: $1/5$—not important, $2/5$—important, $3/5$—fairly important, $4/5$—important, 1—very important. For the pairwise comparison of criteria, the following comparative scale is used: 1—equal importance, 2—moderate importance, 3—strong importance, 4—very strong importance; 5—extreme importance of one criterion over another.

The survey results together with the parameters of respondents (sex, age, etc.) are summarized in a table. For each respondent i, the survey results include

$\boldsymbol{s}_{\mathbf{R}}(i)$, the vector of direct ratings (scores) of criteria by the respondent,

$\boldsymbol{r}_{\mathbf{R}}(i)$, the vector of direct ranks of criteria by the respondent,

$\boldsymbol{C}_{\mathbf{R}}(i)$, the matrix of pairwise comparisons of criteria, provided by the respondent.

Below, the survey results for three respondents with $i = 1, 2, 3$ are given as examples. Respondent 1, male, age 24, visited the country before, provides the following estimates:

$$\boldsymbol{s}_{\mathbf{R}}(1) = \begin{pmatrix} 3/5 \\ 1 \\ 1/5 \\ 2/5 \\ 1 \\ 1/5 \end{pmatrix} = \begin{pmatrix} 0.60 \\ 1.00 \\ 0.20 \\ 0.40 \\ 1.00 \\ 0.20 \end{pmatrix}, \qquad \boldsymbol{r}_{\mathbf{R}}(1) = \begin{pmatrix} 3 \\ 2 \\ 5 \\ 4 \\ 1 \\ 6 \end{pmatrix},$$

$$\boldsymbol{C}_{\mathbf{R}}(1) = \begin{pmatrix} 1 & 1/4 & 5 & 4 & 1/5 & 4 \\ 4 & 1 & 5 & 5 & 1/3 & 5 \\ 1/5 & 1/5 & 1 & 1/3 & 1/5 & 2 \\ 1/4 & 1/5 & 3 & 1 & 1/5 & 3 \\ 5 & 3 & 5 & 5 & 1 & 5 \\ 1/4 & 1/5 & 1/2 & 1/3 & 1/5 & 1 \end{pmatrix};$$

Respondent 2, male, age 19, never visited the country before,

$$\boldsymbol{s}_{\mathbf{R}}(2) = \begin{pmatrix} 1 \\ 1 \\ 3/4 \\ 1/2 \\ 3/4 \\ 1/2 \end{pmatrix} = \begin{pmatrix} 1.00 \\ 1.00 \\ 0.75 \\ 0.50 \\ 0.75 \\ 0.50 \end{pmatrix}, \qquad \boldsymbol{r}_{\mathbf{R}}(2) = \begin{pmatrix} 2 \\ 1 \\ 3 \\ 4 \\ 5 \\ 6 \end{pmatrix},$$

$$\boldsymbol{C}_{\mathbf{R}}(2) = \begin{pmatrix} 1 & 1 & 3 & 3 & 2 & 4 \\ 1 & 1 & 4 & 2 & 4 & 5 \\ 1/3 & 1/4 & 1 & 3 & 1 & 4 \\ 1/3 & 1/2 & 1/3 & 1 & 1 & 3 \\ 1/2 & 1/4 & 1 & 1 & 1 & 3 \\ 1/4 & 1/5 & 1/4 & 1/3 & 1/3 & 1 \end{pmatrix};$$

Respondent 3, female, age 21, never visited the country before,

$$s_{\mathbf{R}}(3)=\begin{pmatrix}4/5\\1\\1/5\\3/5\\4/5\\3/5\end{pmatrix}=\begin{pmatrix}0.80\\1.00\\0.20\\0.60\\0.80\\0.60\end{pmatrix},\qquad r_{\mathbf{R}}(3)=\begin{pmatrix}2\\1\\6\\4\\3\\5\end{pmatrix},$$

$$C_{\mathbf{R}}(3)=\begin{pmatrix}1&1/4&5&4&1/3&3\\4&1&5&5&3&5\\1/5&1/5&1&1/3&1/5&1/3\\1/4&1/5&3&1&1/4&1\\3&1/3&5&4&1&5\\1/3&1/5&3&1&1/5&1\end{pmatrix}.$$

Given the pairwise comparison matrix obtained from each respondent, the absolute ratings of criteria are found based on three computational methods. We obtain vectors of ratings by application of the principal eigenvector method and the method of geometric mean. Furthermore, we solve the problem by using the log-Chebyshev approximation in the framework of tropical algebra, which yields two vectors that are best and worst differentiate the criteria with the highest and lowest ratings (the best and worst log-Chebyshev approximation vectors). The vectors of ratings obtained are then used to rank criteria according to the values of ratings.

As a result, for each respondent i, the following vectors of ratings normalized by dividing by the maximum element, and rank vectors are calculated:

$s_{\mathbf{R}}(i)$, the direct ratings (scores) by the respondent (SR),

$r_{\mathbf{R}}(i)$, the direct ranks by the respondent (RR),

$r_{s_{\mathbf{R}}}(i)$, the ranks based on direct ratings by the respondent (RSR),

$s_{\mathbf{PE}}(i)$, the ratings by the method of principal eigenvector (SPE),

$r_{s_{\mathbf{PE}}}(i)$, the ranks based on the principal eigenvector ratings (RSPE),

$s_{\mathbf{GM}}(i)$, the ratings by the method of geometric mean (SGM),

$r_{s_{\mathbf{GM}}}(i)$, the ranks based on the geometric mean ratings (RSGM),

$s_{\mathbf{CB}}(i)$, the best ratings by the method of log-Chebyshev approximation (SCB),

$r_{s_{\mathbf{CB}}}(i)$, the ranks based on the best ratings from the log-Chebyshev approximation (RSCB),

$s_{\mathbf{CW}}(i)$, the worst ratings by the method of log-Chebyshev approximation (SCW),

$r_{s_{\mathbf{CW}}}(i)$, the ranks based on the worst ratings from the log-Chebyshev approximation (RSCW).

For respondents $i=1,2,3$ considered above as examples, we have the following results. For respondent 1, the vectors of ratings and ranks take the form

$$
r_{s_{\mathrm{R}}}(1) = \begin{pmatrix} 3 \\ 1 \\ 5 \\ 4 \\ 2 \\ 6 \end{pmatrix}, \quad s_{\mathrm{PE}}(1) = \begin{pmatrix} 0.3581 \\ 0.6526 \\ 0.1142 \\ 0.1832 \\ 1.0000 \\ 0.0943 \end{pmatrix}, \quad r_{s_{\mathrm{PE}}}(1) = \begin{pmatrix} 3 \\ 2 \\ 5 \\ 4 \\ 1 \\ 6 \end{pmatrix},
$$

$$
s_{\mathrm{GM}}(1) = \begin{pmatrix} 0.3588 \\ 0.6680 \\ 0.1190 \\ 0.1906 \\ 1.0000 \\ 0.0981 \end{pmatrix}, \quad s_{\mathrm{CB}}(1) = \begin{pmatrix} 0.3218 \\ 0.6551 \\ 0.1036 \\ 0.1581 \\ 1.0000 \\ 0.1018 \end{pmatrix}, \quad s_{\mathrm{CW}}(1) = \begin{pmatrix} 0.3218 \\ 0.6551 \\ 0.1036 \\ 0.1581 \\ 1.0000 \\ 0.1018 \end{pmatrix},
$$

$$
r_{s_{\mathrm{GM}}}(1) = \begin{pmatrix} 3 \\ 2 \\ 5 \\ 4 \\ 1 \\ 6 \end{pmatrix}, \quad r_{s_{\mathrm{CB}}}(1) = \begin{pmatrix} 3 \\ 2 \\ 5 \\ 4 \\ 1 \\ 6 \end{pmatrix}, \quad r_{s_{\mathrm{CW}}}(1) = \begin{pmatrix} 3 \\ 2 \\ 5 \\ 4 \\ 1 \\ 6 \end{pmatrix}.
$$

From the results obtained, it follows that the ranks of criteria provided by all methods for respondent 1 coincide and arrange the criteria in the following order of preference with the symbol $\succ$ used to indicate the preference relation:

$$
\mathcal{C}_5 \succ \mathcal{C}_2 \succ \mathcal{C}_1 \succ \mathcal{C}_4 \succ \mathcal{C}_6.
$$

Note that the best and worst differentiating solutions lead the same vector of ratings, and thus the method of log-Chebyshev approximation has a unique solution in this case.

For respondent 2, the calculation results in the following vectors of ratings and ranks:

$$
r_{s_{\mathrm{R}}}(2) = \begin{pmatrix} 1 \\ 2 \\ 3 \\ 5 \\ 4 \\ 6 \end{pmatrix}, \quad s_{\mathrm{PE}}(2) = \begin{pmatrix} 0.8485 \\ 1.0000 \\ 0.4462 \\ 0.3170 \\ 0.3467 \\ 0.1392 \end{pmatrix}, \quad r_{s_{\mathrm{PE}}}(2) = \begin{pmatrix} 2 \\ 1 \\ 3 \\ 5 \\ 4 \\ 6 \end{pmatrix},
$$

$$
s_{\mathrm{GM}}(2) = \begin{pmatrix} 0.8754 \\ 1.0000 \\ 0.4292 \\ 0.3184 \\ 0.3645 \\ 0.1434 \end{pmatrix}, \quad s_{\mathrm{CB}}(2) = \begin{pmatrix} 0.7500 \\ 1.0000 \\ 0.4543 \\ 0.2752 \\ 0.2500 \\ 0.1101 \end{pmatrix}, \quad s_{\mathrm{CW}}(2) = \begin{pmatrix} 1.0000 \\ 1.0000 \\ 0.4543 \\ 0.2752 \\ 0.4543 \\ 0.1667 \end{pmatrix},
$$

$$
r_{s_{\mathrm{GM}}}(2) = \begin{pmatrix} 2 \\ 1 \\ 3 \\ 5 \\ 4 \\ 6 \end{pmatrix}, \quad r_{s_{\mathrm{CB}}}(2) = \begin{pmatrix} 2 \\ 1 \\ 3 \\ 4 \\ 5 \\ 6 \end{pmatrix}, \quad r_{s_{\mathrm{CW}}}(2) = \begin{pmatrix} 1 \\ 2 \\ 3 \\ 5 \\ 4 \\ 6 \end{pmatrix}.
$$

We observe that the ranks of the criteria obtained by both methods of the principal vector and of geometric mean are the same and determine the order

$$
\mathcal{C}_2 \succ \mathcal{C}_1 \succ \mathcal{C}_3 \succ \mathcal{C}_5 \succ \mathcal{C}_4 \succ \mathcal{C}_6.
$$

Furthermore, we may consider that the ratings, which are provided by the worst differentiating vector of log-Chebyshev approximation, lead to the above ranking as well.

Indeed, this method gives equal ratings to the first and second criteria to allow these criteria to be ranked in any order, one of which is the same as above.

The best and worst differentiating vectors of ratings obtained as a result of the log-Chebyshev approximation differ in some elements and can be coupled as a vector with interval values in the form

$$\mathbf{s}_{\mathrm{C}}(2) = \begin{pmatrix} 0.7500 \ \dots \ 1.0000 \\ 1.0000 \\ 0.4543 \\ 0.2752 \\ 0.2500 \ \dots \ 0.4543 \\ 0.1101 \ \dots \ 0.1667 \end{pmatrix}.$$

The order provided by these vectors can be represented in combined form as

$$\mathcal{C}_2 \succeq \mathcal{C}_1 \succ \mathcal{C}_3 \succeq \mathcal{C}_5 \parallel \mathcal{C}_4 \succ \mathcal{C}_6,$$

where the symbol $\succeq$ denotes the weak preference relation (preferred or indifferent) and $\parallel$ indicates an undetermined preference.

Finally, the results obtained for respondent 3 are given by the vectors

$$\mathbf{r}_{\mathbf{s}_{\mathrm{R}}}(3) = \begin{pmatrix} 2 \\ 1 \\ 6 \\ 4 \\ 3 \\ 5 \end{pmatrix}, \quad \mathbf{r}_{\mathbf{s}_{\mathrm{PE}}}(3) = \begin{pmatrix} 0.3773 \\ 1.0000 \\ 0.0930 \\ 0.1630 \\ 0.6213 \\ 0.1631 \end{pmatrix}, \quad \mathbf{r}_{\mathbf{s}_{\mathrm{PE}}}(3) = \begin{pmatrix} 3 \\ 1 \\ 6 \\ 5 \\ 2 \\ 4 \end{pmatrix},$$

$$\mathbf{s}_{\mathbf{GM}}(3) = \begin{pmatrix} 0.3865 \\ 1.0000 \\ 0.0916 \\ 0.1710 \\ 0.6368 \\ 0.1728 \end{pmatrix}, \quad \mathbf{s}_{\mathbf{CB}}(3) = \begin{pmatrix} 0.3798 \\ 1.0000 \\ 0.1082 \\ 0.1755 \\ 0.6163 \\ 0.1755 \end{pmatrix}, \quad \mathbf{s}_{\mathbf{CW}}(3) = \begin{pmatrix} 0.3798 \\ 1.0000 \\ 0.1082 \\ 0.1755 \\ 0.6163 \\ 0.2279 \end{pmatrix},$$

$$\mathbf{r}_{\mathbf{s}_{\mathrm{GM}}}(3) = \begin{pmatrix} 3 \\ 1 \\ 6 \\ 5 \\ 2 \\ 4 \end{pmatrix}, \quad \mathbf{r}_{\mathbf{s}_{\mathrm{CB}}}(3) = \begin{pmatrix} 3 \\ 1 \\ 6 \\ 4 \\ 2 \\ 5 \end{pmatrix}, \quad \mathbf{r}_{\mathbf{s}_{\mathrm{CW}}}(3) = \begin{pmatrix} 3 \\ 1 \\ 6 \\ 5 \\ 2 \\ 4 \end{pmatrix}.$$

In this case, the solutions provided by the method of principal eigenvector and method of geometric mean as well as the worst differentiating solution by the log-Chebyshev approximation define the same ranks of criteria, which yields the order

$$\mathcal{C}_2 \succ \mathcal{C}_5 \succ \mathcal{C}_1 \succ \mathcal{C}_6 \succ \mathcal{C}_4 \succ \mathcal{C}_3.$$

The best and worst differentiating vectors of ratings differ only by the last element and can be combined as follows:

$$\mathbf{s}_{\mathrm{C}}(3) = \begin{pmatrix} 0.3798 \\ 1.0000 \\ 0.1082 \\ 0.1755 \\ 0.6163 \\ 0.1755 \ \dots \ 0.2279 \end{pmatrix}.$$

The order of criteria, which corresponds to these vectors, can be represented as

$$\mathcal{C}_2 \succ \mathcal{C}_5 \succ \mathcal{C}_1 \succ \mathcal{C}_6 \succeq \mathcal{C}_4 \succ \mathcal{C}_3.$$

It follows from the above examples that the results of determining the ranks of criteria, which are obtained from their pairwise comparisons by applying different methods may vary. In the next sections, we examine the difference between these results more closely by using appropriate techniques of data analysis.

5. Statistical Analysis of Ratings and Ranks

In this section, we compare ratings and ranks of criteria, obtained by different methods for all respondents participating in the survey. Ratings and ranks that are directly specified by respondents are compared with those which are based on three methods of evaluating criteria from pairwise comparisons. We examine the results of the principal eigenvector method, the geometric mean method, and the method of log-Chebyshev approximation. For the latter method, we consider two solution vectors, which best and worst differentiate the criteria if the solution is nonunique (both vectors coincide when the solution is unique).

To represent the obtained results in table form below, we use the following symbols:

SR, the direct ratings (scores) given by respondent (Scores by Respondent),

RR, the direct ranks by respondent (Ranks by Respondent),

RSR, the ranks from the direct ratings by respondent (Ranks from Scores by Respondent),

SPE, the ratings by the method of principal eigenvector (Scores by Principal Eigenvector),

RSPE, the ranks from the principal eigenvector ratings (Ranks from Scores by Principal Eigenvector),

SGM, the ratings by the method of geometric mean (Scores by Geometric Means),

RSGM, the ranks from the geometric mean ratings (Ranks from Scores by Geometric Means),

SCB, the best differentiating ratings by the method of log-Chebyshev approximation (Scores by log-Chebyshev, Best),

RSCB, the ranks from the best ratings from the log-Chebyshev approximation (Ranks from Scores by log-Chebyshev, Best),

SCW, the worst differentiating ratings from the log-Chebyshev approximation (Scores by log-Chebyshev, Worst),

RSCW, the ranks from the worst ratings from the log-Chebyshev approximation (Ranks from Scores by log-Chebyshev, Worst).

We start by counting the matches in the rank vectors obtained directly and determined through pairwise comparisons. The numbers of matches between rank vectors for each pair of the ranking methods are presented in Table 1.

Table 1. Number of matches of rank vectors.

Ranking	Ranking Method					
Method	**RR**	**RSR**	**RSPE**	**RSGM**	**RSCB**	**RSCW**
RR	202	28	56	56	59	56
RSR	28	202	20	20	18	23
RSPE	56	20	202	184	124	123
RSGM	56	20	184	202	124	125
RSCB	59	18	124	124	202	130
RSCW	56	23	123	125	130	202

The results given by Table 1 show a low level of matches (encountered for 28 out of 202 respondents) between the vectors of direct ranks (RR) and ranks determined from ratings that are specified by respondents (RSR). We can explain this inconsistency by greater variability in determining ratings by respondents, which allows one to assign the same ratings to several criteria. As a result, the ranks obtained from ratings may differ from the direct ranking, when the respondent specifies an individual rank for each criterion to indicate the number (position) of the criterion in the ranking order.

The degree of consistency of the direct ranking (RR) with the ranking on the basis of ratings obtained from pairwise comparisons (RSPE, RSGM, RSCB, RSCW) is better given by 56–59 matches of rank vectors, which are more than a quarter of respondents. Note that the maximum number of matches, which is equal to 59, is provided by the best differentiating solution of log-Chebyshev approximation (RSCB).

We note that the results of ranking criteria from pairwise comparisons demonstrate a higher level of consistency between the methods under consideration. Specifically, the largest number of matches encountered for 184 (more than 90% of) respondents is given by the outcome from the methods of principal eigenvector (RSPE) and of geometric mean (RSGM). The number of matches between the rank vectors produced by these methods and the vectors obtained using log-Chebyshev approximation (RSCB, RSCW) is within the range 123–125 (more than 60% of respondents). Finally, the rank vectors, which are obtained from the best and worst differentiating solutions found by log-Chebyshev approximation, coincide for 130 respondents.

Below, to analyze and compare the results more thoroughly, we use appropriate statistical techniques of parameter estimation and correlation analysis. As preliminary analysis of data, we investigate if parameters of respondents, such as age and sex, may affect the results of the evaluation of criteria. Furthermore, we compare the results of evaluating ratings and ranks obtained from pairwise comparisons by different methods with each other and with the ratings and ranks directly provided by respondents.

5.1. Preliminary Analysis of Data

In the analysis of indirect methods that produce results from pairwise comparison data, it seems reasonable to consider the ratings and ranks of criteria, which are directly specified by respondents, as some basis for comparison. Since these ratings and ranks are known for all respondents, we can try to select those respondents for whom the direct ranks and ranks given by direct ratings appear to be closest to each other. Considering that a part of respondents may demonstrate a higher level of consistency in the direct judgment, one can expect that these respondents are able to assess pairwise comparison of criteria more adequately and definitely.

To examine how the consistency between direct estimates of ratings and ranks may affect the results based on pairwise comparisons, we consider groups of respondents with different degrees of consistency. As a measure of inconsistency (difference) between rank vectors, we apply the Chebyshev metric (the maximum absolute componentwise difference), which is calculated for two n-vectors $\boldsymbol{a} = (a_j)$ and $\boldsymbol{b} = (b_j)$ as

$$d(\boldsymbol{a}, \boldsymbol{b}) = \max_{1 \leq j \leq n} |a_j - b_j|.$$

We consider a series of nested subsets (groups) $R_0 \subset R_1 \subset \cdots \subset R_5$, where R_i denotes the set of respondents for whom the Chebyshev metric between the vectors of direct ranks and ranks obtained from the direct ratings is less or equal to $i = 0, 1, \ldots, 5$. Table 2 shows how the total number of respondents in groups changes with increase in the upper bound for Chebyshev metric, which indicates the maximum degree of difference in groups. For each group, the percentage of respondents who visited the country and male/female percentage are also included.

Table 2. Groups of respondents according to difference between rank vectors.

Group	Maximum Component-Wise Difference	Number of Respondents in the Group	Percentage of Respondents Who Visited the Country	Male/ Female Percentage
R_0	0	28	35	33/67
R_1	1	95	19	32/68
R_2	2	167	18	30/70
R_3	3	194	17	28/72
R_4	4	201	18	28/72
R_5	5	202	19	28/72

It follows from the data presented in the table that the direct ranks and the ranks obtained from direct ratings completely coincide for 28 respondents from the group R_0, who may be considered as the most accurate evaluators. However, these respondents form a rather small part of all respondents involved in the survey, which makes it unreasonable to take this group as a good representative of the entire sample.

Furthermore, we note that the respondents in the groups $R_1, \ldots, R_5$ with less consistent direct judgments demonstrate almost the same percentage of those who visited the country (17–19%). This apparently indicates that for the most respondents, there is no relationship between visiting the country and accuracy of judgments. At the same time, for the group R_0, this percentage increases to 35%, which is in line with the idea that the accuracy of judgments should increase as respondents gain experience of staying in the country. A conclusion about the lack of systematic dependence of the accuracy of judgments on the gender and age of respondents can also be drawn for all groups.

To complete the analysis of the possible influence of the accuracy of direct judgments about ratings and ranks on the other survey results, we estimate correlations between the rank vectors directly provided by respondents and the vectors derived from pairwise comparisons. To measure correlation, we use the Kendall rank correlation coefficient, which is given for two n-vectors $\boldsymbol{a} = (a_j)$ and $\boldsymbol{b} = (b_j)$ by the formula

$$\tau(\boldsymbol{a}, \boldsymbol{b}) = \frac{2}{n(n-1)} \sum_{1 \leq i < j \leq n} \operatorname{sgn}(a_i - a_j) \operatorname{sgn}(b_i - b_j).$$

For each group R_0, R_1 and R_5, we concatenate the vectors of ratings obtained for all respondents in the group to form a single vector of ratings for each method. The estimates of the correlation coefficients between concatenated vectors for the groups R_0, R_1 and R_5 are given in Tables 3–5 (the estimates for groups R_2, R_3 and R_4 do not differ significantly from the results for R_1 and R_5 and hence are omitted).

Table 3. Correlation of rank vectors for the group R_0.

Ranking Method	Ranking Method: RR	RSPE	RSGM	RSCB	RSCW
RR	1.0000	0.8482	0.8482	0.8560	0.8645
RSPE	0.8482	1.0000	1.0000	0.9558	0.9557
RSGM	0.8482	1.0000	1.0000	0.9557	0.9557
RSCB	0.8560	0.9558	0.9558	1.0000	0.9717
RSCW	0.8645	0.9557	0.9557	0.9717	1.0000

Table 4. Correlation of rank vectors for the group R_1.

Ranking	Ranking Method				
Method	**RR**	**RSPE**	**RSGM**	**RSCB**	**RSCW**
RR	1.0000	0.8184	0.8234	0.8108	0.8221
RSPE	0.8184	1.0000	0.9943	0.9450	0.9183
RSGM	0.8234	0.9943	1.0000	0.9429	0.9223
RSCB	0.8108	0.9450	0.9429	1.0000	0.9234
RSCW	0.8221	0.9183	0.9223	0.9234	1.0000

Table 5. Correlation of rank vectors for the group R_5.

Ranking	Ranking Method				
Method	**RR**	**RSPE**	**RSGM**	**RSCB**	**RSCW**
RR	1.0000	0.7939	0.7940	0.7876	0.7839
RSPE	0.7939	1.0000	0.9862	0.9312	0.9186
RSGM	0.7940	0.9862	1.0000	0.9290	0.9204
RSCB	0.7876	0.9312	0.9290	1.0000	0.9144
RSCW	0.7839	0.9186	0.9204	0.9144	1.0000

The data presented in these tables show that for all groups of respondents, the correlations vary within small limits and maintain the relative order of magnitude for different methods. This allows us to conclude that the change in accuracy of direct estimates does not significantly affect the results of ranking based on pairwise comparisons.

In the subsequent analysis, we take into account the above results and examine all respondents together rather than divide them into groups according to the accuracy of direct judgments or parameters of respondents.

5.2. Comparison of Ratings

We now examine results of evaluating ratings, obtained by different methods from pairwise comparisons and specified directly by respondents. First, we calculate the means and standard deviations of vectors of ratings for each rating method. The mean vectors can be considered as the most representative (typical) vectors over all respondents for each method to give some general idea of the absolute ratings of criteria in hotel selection. The standard deviations describing the spread of the vectors can characterize both the differences in the perception of the relative importance of criteria by respondents and the ability of the methods to reinforce or smooth these differences.

The results of calculating the mean vectors presented in Table 6 show that on average, criterion 2 is always rated higher than the others, next come criterion 1 and then 5. The other criteria have lower mean ratings, which give the order $(3,4,6)$ for all methods except the direct rating with the order $(6,3,4)$. Examination of the output of individual respondents indicates that the second criterion is evaluated above the others for more than 60% respondents. Specifically, this result encounters in the direct rating (SR) for 165 respondents, in both methods of principal eigenvector (SPE) and geometric mean (SGM) for 125 respondents, and in the best and worst differentiating solutions of log-Chebyshev approximation (SCB and SCW) for 123 and 147 respondents respectively.

It follows from Table 6 that the direct ratings by respondents are sufficiently different from the other results. This can be explained by a rather primitive scale used in the survey, which includes only 5 points. At the same time, the evaluation of mean vectors for all indirect methods of rating criteria gives quite close results. We note that the mean vectors of the log-Chebyshev approximation appear as lower and upper boundaries for the mean vectors for both methods of principal eigenvector and geometric mean.

Table 6. Mean vectors of rating of criteria.

Criterion	Rating Method				
Number	SR	SPE	SGM	SCB	SCW
1	0.7480	0.6268	0.6272	0.5693	0.6313
2	0.9428	0.8877	0.8915	0.8504	0.9025
3	0.5784	0.3238	0.3231	0.2861	0.3511
4	0.5588	0.3044	0.3053	0.2735	0.3291
5	0.7331	0.5279	0.5303	0.4910	0.5581
6	0.6149	0.2543	0.2558	0.2252	0.2879

To measure the overall variability of ratings for all methods, we use the average standard deviation over criteria and a total (vector) standard deviation defined as the square root of the sum of squares of standard deviations for each criterion. The results of calculation are given in Table 7 and say, in particular, that the total standard deviations for the methods of principal eigenvector and geometric mean almost coincide, and both methods have slightly lower variability of ratings than the log-Chebyshev approximation.

Table 7. Variability estimates for vectors of ratings.

Variability	Rating Method				
Measure	SR	SPE	SGM	SCB	SCW
Average deviation	0.2037	0.2202	0.2198	0.2252	0.2339
Total deviation	0.5046	0.5477	0.5478	0.5613	0.5796

Table 8 demonstrates the standard deviation of ratings for each pair of criteria and methods. We note that criteria 2 and 6 most often have the smallest or the next smallest standard deviation for all methods, which suggests that the real assessment of these criteria differs the least from one respondent to another.

Table 8. Standard deviation of ratings of each criterion.

Criterion	Rating Method				
Number	SR	SPE	SGM	SCB	SCW
1	0.2188	0.2578	0.2611	0.2685	0.2633
2	0.1379	0.1837	0.1807	0.2175	0.1879
3	0.2255	0.2300	0.2279	0.2163	0.2426
4	0.2248	0.2052	0.2039	0.1963	0.2150
5	0.1978	0.2769	0.2798	0.2889	0.2914
6	0.2173	0.1676	0.1656	0.1639	0.2032

Next, we turn to the estimates of correlation between vectors of ratings produced by the direct and indirect assessment methods. For each method, we concatenate the vectors of all respondents into a single vector of ratings, and then estimate the Pearson correlation coefficients between the concatenated vectors. The correlation coefficients calculated for all pairs of methods are given in Table 9.

The results presented in this table show that the correlation between ratings obtained from pairwise comparisons is close to one for all pairs of methods, which means that any two concatenated vectors of ratings are close to be linearly dependent (collinear). Specifically, the vectors found by the methods of principal eigenvector and geometric mean correlate at the level of 0.9973, and thus can be considered poorly distinguishable from the correlation point of view.

Table 9. Correlation of vectors of ratings for all pairs of methods.

Rating Method	Rating Method				
	SR	SPE	SGM	SCB	SCW
SR	1.0000	0.6942	0.6918	0.6619	0.6889
SPE	0.6942	1.0000	0.9973	0.9613	0.9622
SGM	0.6918	0.9973	1.0000	0.9631	0.9576
SCB	0.6619	0.9613	0.9631	1.0000	0.9164
SCW	0.6889	0.9622	0.9576	0.9164	1.0000

We now consider the correlations of ratings produced by all methods for each individual criterion. Tables 10–12 demonstrate the estimates for criteria 1, 2 and 6 (the results for criteria 3, 4 and 5 are similar to those for criterion 1 and thus omitted).

Table 10. Correlation of ratings of criterion 1 for all pairs of methods.

Rating Method	Rating Method				
	SR	SPE	SGM	SCB	SCW
SR	1.0000	0.5815	0.5639	0.5151	0.5758
SPE	0.5815	1.0000	0.9949	0.9215	0.9408
SGM	0.5639	0.9949	1.0000	0.9250	0.9220
SCB	0.5151	0.9215	0.9250	1.0000	0.8723
SCW	0.5758	0.9408	0.9220	0.8723	1.0000

The results obtained for each criterion indicate that there is a high correlation at the level of $0.9929\ldots0.9959$ between the ratings produced by the method of principal eigenvector and the method of geometric mean.

Table 11. Correlation of ratings of criterion 2 for all pairs of methods.

Rating Method	Rating Method				
	SR	SPE	SGM	SCB	SCW
SR	1.0000	0.4243	0.4314	0.3778	0.4348
SPE	0.4243	1.0000	0.9959	0.8887	0.9217
SGM	0.4314	0.9959	1.0000	0.8899	0.9239
SCB	0.3778	0.8887	0.8899	1.0000	0.8004
SCW	0.4348	0.9217	0.9239	0.8004	1.0000

Furthermore, the ratings provided by the best differentiating solutions of log-Chebyshev approximation are usually more correlated with the ratings by the method of geometric mean ($0.8899\ldots0.9543$), whereas the ratings from the worst differentiating solutions are more correlated with those from the method of principal eigenvector ($0.8938\ldots0.9408$). Finally, the correlation between ratings from the best and worst differentiating solutions are somewhat lower and lay in the range $0.7776\ldots0.8752$.

Table 12. Correlation of ratings of criterion 6 for all pairs of methods.

Rating Method	Rating Method				
	SR	SPE	SGM	SCB	SCW
SR	1.0000	0.4926	0.4873	0.4569	0.4842
SPE	0.4926	1.0000	0.9929	0.9301	0.8938
SGM	0.4873	0.9929	1.0000	0.9189	0.8911
SCB	0.4569	0.9301	0.9189	1.0000	0.7776
SCW	0.4842	0.8938	0.8911	0.7776	1.0000

To summarize the results of the comparison of ratings, we can conclude that for the survey data under study, all indirect methods of rating criteria consistently produce highly correlated vectors of ratings, with the highest correlation very close to one between the results of the methods of principal eigenvector and geometric mean. The good agreement between the results of indirect methods can be explained by the natural origin of the initial data obtained as results of human judgment, which provide rationale for more consistent pairwise comparisons than artificial simulated data.

In conclusion, we note low correlations between direct and indirect ratings, which are within the range $0.3778\ldots0.5993$. This relative disagreement may serve as an illustration of the known fact that it may be difficult for typical respondents to make a correct judgment when more than two alternatives are simultaneously evaluated, whereas they normally assess pairwise comparisons more accurately.

5.3. *Comparison of Ranks*

We start with the ranks derived from ratings given by the mean vectors in Table 6. The mean vector of ratings that are directly provided by respondents yields the rank vector $(2,1,5,6,3,4)$, which arranges the criteria in the following order:

$$\mathcal{C}_2 \succ \mathcal{C}_1 \succ \mathcal{C}_5 \succ \mathcal{C}_6 \succ \mathcal{C}_3 \succ \mathcal{C}_4.$$

To see how frequently this order coincides with the order derived from the vectors of ratings of individual respondents, we calculate the Hamming distance between (the number of different elements in) the corresponding rank vectors. The results of counting the number of rank vectors, which are obtained from vectors of ratings for all methods, within fixed distances from the vector $(2,1,5,6,3,4)$ are given in Table 13.

Table 13. Number of vectors within fixed distance from $(2,1,5,6,3,4)$.

Maximum	Rating Method				
Hamming Distance	**RSR**	**RSPE**	**RSGM**	**RSCB**	**RSCW**
0	5	3	3	4	4
1	5	3	3	4	4
2	23	35	32	35	31
3	61	68	65	66	64
4	122	135	130	123	132
5	178	185	186	186	188
6	202	202	202	202	202

The mean vectors of ratings derived from pairwise comparisons by different methods imply a common rank vector $(2,1,5,3,4,6)$, which corresponds to the order

$$\mathcal{C}_2 \succ \mathcal{C}_1 \succ \mathcal{C}_5 \succ \mathcal{C}_3 \succ \mathcal{C}_4 \succ \mathcal{C}_6.$$

The numbers of rank vectors within fixed distances from the vector $(2,1,5,3,4,6)$ are shown in Table 14.

Both Tables 13 and 14 demonstrate low correspondence between the rank vectors derived from the mean vectors of ratings and the individual rank vectors for each respondent. In this case, we cannot consider the ranks based on the mean vectors as well justified, and need further examination of the individual rank vectors to find those vectors that occur more frequently for each method. The results of determining the five most common rank vectors for each method are presented in Table 15.

Table 14. Number of vectors within fixed distance from $(2, 1, 5, 3, 4, 6)$.

Maximum Hamming Distance	Rating Method				
	RSR	RSPE	RSGM	RSCB	RSCW
0	6	4	4	7	5
1	6	4	4	7	5
2	37	40	38	48	41
3	66	64	63	69	63
4	121	133	132	130	129
5	175	180	179	183	184
6	202	202	202	202	202

Table 15. Most frequently occurred rank vectors.

Ranking Method	Rank Vector	Order of Criteria	Number of Occurrences
RSR	$(1,2,5,3,4,6)$	$C_1 \succ C_2 \succ C_5 \succ C_3 \succ C_4 \succ C_6$	10
	$(2,1,5,3,6,4)$	$C_2 \succ C_1 \succ C_5 \succ C_3 \succ C_6 \succ C_4$	9
	$(1,2,3,5,6,4)$	$C_1 \succ C_2 \succ C_3 \succ C_5 \succ C_6 \succ C_4$	8
	$(2,5,6,1,3,4)$	$C_2 \succ C_5 \succ C_6 \succ C_1 \succ C_3 \succ C_4$	7
	$(1,2,3,4,5,6)$	$C_1 \succ C_2 \succ C_3 \succ C_4 \succ C_5 \succ C_6$	6
RSPE	$(2,1,5,3,6,4)$	$C_2 \succ C_1 \succ C_5 \succ C_3 \succ C_6 \succ C_4$	12
	$(2,1,5,4,6,3)$	$C_2 \succ C_1 \succ C_5 \succ C_4 \succ C_6 \succ C_3$	8
	$(2,1,5,4,3,6)$	$C_2 \succ C_1 \succ C_5 \succ C_4 \succ C_3 \succ C_6$	7
	$(2,5,1,3,6,4)$	$C_2 \succ C_5 \succ C_1 \succ C_3 \succ C_6 \succ C_4$	6
	$(2,1,4,5,6,3)$	$C_2 \succ C_1 \succ C_4 \succ C_5 \succ C_6 \succ C_3$	6
RSGM	$(2,1,5,3,6,4)$	$C_2 \succ C_1 \succ C_5 \succ C_3 \succ C_6 \succ C_4$	12
	$(2,1,5,4,6,3)$	$C_2 \succ C_1 \succ C_5 \succ C_4 \succ C_6 \succ C_3$	8
	$(2,1,4,5,6,3)$	$C_2 \succ C_1 \succ C_4 \succ C_5 \succ C_6 \succ C_3$	7
	$(1,2,5,4,3,6)$	$C_1 \succ C_2 \succ C_5 \succ C_4 \succ C_3 \succ C_6$	7
	$(2,1,3,5,6,4)$	$C_2 \succ C_1 \succ C_3 \succ C_5 \succ C_6 \succ C_4$	6
RSCB	$(2,1,5,3,6,4)$	$C_2 \succ C_1 \succ C_5 \succ C_3 \succ C_6 \succ C_4$	11
	$(2,5,1,3,4,6)$	$C_2 \succ C_5 \succ C_1 \succ C_3 \succ C_4 \succ C_6$	7
	$(2,1,5,3,4,6)$	$C_2 \succ C_1 \succ C_5 \succ C_3 \succ C_4 \succ C_6$	7
	$(2,1,5,4,3,6)$	$C_2 \succ C_1 \succ C_5 \succ C_4 \succ C_3 \succ C_6$	7
	$(1,2,5,4,3,6)$	$C_1 \succ C_2 \succ C_5 \succ C_4 \succ C_3 \succ C_6$	6
RSCW	$(2,1,5,3,6,4)$	$C_2 \succ C_1 \succ C_5 \succ C_3 \succ C_6 \succ C_4$	9
	$(2,5,1,3,4,6)$	$C_2 \succ C_5 \succ C_1 \succ C_3 \succ C_4 \succ C_6$	8
	$(2,1,5,4,3,6)$	$C_2 \succ C_1 \succ C_5 \succ C_4 \succ C_3 \succ C_6$	8
	$(1,2,5,4,3,6)$	$C_1 \succ C_2 \succ C_5 \succ C_4 \succ C_3 \succ C_6$	8
	$(2,5,1,4,3,6)$	$C_2 \succ C_5 \succ C_1 \succ C_4 \succ C_3 \succ C_6$	6

It follows from Table 15 that the most frequently occurred rank vector derived from ratings for all indirect methods and the second most frequent for the direct ratings by respondents is $(2, 1, 5, 3, 6, 4)$. This rank vector yields the order of criteria defined as

$$C_2 \succ C_1 \succ C_5 \succ C_3 \succ C_6 \succ C_4,$$

and seems to provide a more accurate assessment of the ranks of criteria for the set of respondents participating in the survey.

We conclude with the results in Table 16, which demonstrate the Kendall correlation between concatenated rank vectors obtained by direct and indirect methods (this table actually presents the same results as in Table 5 used in the preliminary analysis of data).

Table 16. Correlations of rank vectors for all methods.

Ranking Method	Ranking Method				
	RR	**RSPE**	**RSGM**	**RSCB**	**RSCW**
RR	1.0000	0.7939	0.7940	0.7876	0.7839
RSPE	0.7939	1.0000	0.9862	0.9312	0.9186
RSGM	0.7940	0.9862	1.0000	0.9290	0.9204
RSCB	0.7876	0.9312	0.9290	1.0000	0.9144
RSCW	0.7839	0.9186	0.9204	0.9144	1.0000

As for the correlation of vectors of ratings obtained from pairwise comparisons, the corresponding rank vectors are highly correlated for all methods. As before, the methods of principal eigenvector and geometric mean have the highest correlation of 0.9863, which is very close to one. At the same time, the correlation of rank vectors obtained by the log-Chebyshev approximation and by both methods of principal eigenvector and geometric mean increase to $0.9187 \ldots 0.9313$. This shows that in terms of rank vectors, the difference between the results of all methods remains quite small, and therefore, the results of one method can be used to validate the results of the other methods.

6. Discussion

In this study, we considered a problem of the assessment of 6 criteria used by university students for hotel selection when attending a short-term educational program in a foreign country. The data available for the assessment procedure include the results of a survey of 202 respondents who report on their age, sex and whether they have previously visited the country, and evaluate the criteria by different ways. In the course of evaluation, each respondent directly estimates ratings and indicate ranks of criteria, and then compares the criteria in pairs to provide an input for subsequent derivation of ratings from the pairwise comparisons by an appropriate computational method.

To provide more accuracy of the assessment, we concurrently applied three methods of rating alternatives from pairwise comparisons, where the results of the most common approaches, the method of principal eigenvector and method of geometric mean, were considered in line with results of log-Chebyshev approximation. Moreover, the ratings derived by these methods and corresponding ranks of criteria were compared with the ratings and ranks provided by respondents directly.

First, we investigated the possible influence of parameters and properties of respondents on the degree of difference and similarity of results provided by different methods of deriving ratings. Specifically, we examined the assumption that the consistency between direct ranks and ratings given by respondents may affect the results of calculating ranks from pairwise comparisons by these methods. As our analysis has shown, for the majority of respondents, the consistency of the direct judgments, as well as parameters of respondents do not have a significant impact on the stability of the results, which made it inappropriate to divide respondents into groups for separate analysis.

Furthermore, we considered ratings provided by respondents directly or indirectly through pairwise comparisons of criteria. Calculation of the mean vectors of ratings has indicated that on average all indirect methods produce very close results. Specifically, the mean vectors of the methods of principal eigenvector and geometric mean appear to be closest to each other, whereas the mean vectors of the best and worst differentiating results of log-Chebyshev approximation form lower and upper bounds for the other two mean vectors. The mean vector of direct ratings by respondents is rather different from the mean vectors of indirect methods, which is due to a rough 5-point scale that narrows the range of assessment values used by respondents.

On average all methods equally determine the highest ratings of criteria in the order $(2,1,5)$. The other criteria are ordered by all indirect methods as $(3,4,6)$, and by the ratings directly provided by respondents as $(6,3,4)$. The standard deviation of ratings of criterion 2 has the smallest or the next smallest value among the criteria, which speaks in favor of a unanimous assessment of this criterion by all respondents. As a result, the analysis of mean vectors of ratings suggests two possible candidates for the optimal rank vector: $(2,1,5,3,4,6)$ and $(2,1,5,6,3,4)$.

To give further insight into the consistency or inconsistency of ratings derived by different methods, we estimated Pearson correlation coefficients between vectors of ratings obtained for all respondents. We have found a high correlation close to one between the results of the methods of principal eigenvector and geometric mean. The correlation between ratings from these two methods and from the log-Chebyshev approximation is somewhat lower, but still close to one. The good agreement of the ratings obtained from pairwise comparisons by all methods can be seen as a result of the natural and meaningful character of the input data that reflect mechanisms of human judgment and hence have a certain level of immanent consistency.

As the next step of the investigation, we examined the vectors of ranks provided by direct and indirect assessments. First, we checked whether the rank vectors $(2,1,5,3,4,6)$ or $(2,1,5,6,3,4)$ derived from the mean vectors of ratings actually occur in the results based on the data provided by individual respondents. We have found that these vectors of ranks are relatively rare as results of all assessment methods, and thus cannot be considered as appropriate solutions. At the same time, calculating the number of the most frequently obtained rank vectors has shown that the most occurred vector is $(2,1,5,3,6,4)$ for all methods except for the ranking from direct ratings by respondents, where this vector is the second most frequent. We note that the last vector of ranks provides the same order for three most important criteria as the two vectors considered above, and differ from them only for the last three least important criteria.

It follows from the analysis of the results obtained that for the given survey data, one can take the rank vector $(2,1,5,3,6,4)$ as the most accurate assessment of criteria for hotel selection. Assuming the respondents are representative of the total population of university students, we can conclude that the typical student ranks the criteria of hotel selection in the following order (from the most important to the least important):

1. Accommodation cost, C_2;
2. Hotel location, C_1;
3. Room amenities, C_5;
4. Typical guests, C_3;
5. Courtesy of staff, C_6;
6. Free breakfast, C_4.

As another less definite but more realistic conclusion, we can group criteria into two levels. The first level consists of the criterion C_2 (accommodation cost), which is ranked first most frequently, and then criteria C_1 (hotel location) and C_5 (room amenities), which can sometimes change the order of each other. On the second level of lower importance are the criteria C_3 (typical guests), C_6 (courtesy of staff) and C_4 (free breakfast), which can go in almost any order.

Finally, we estimated Kendall correlation coefficients for rank vectors derived from pairwise comparisons. As in the case of ratings, the correlations between rank vectors have been found close to one, which shows once again that for the pairwise comparison data provided by respondents, the method of principal eigenvector, the method of geometric mean and the method of log-Chebyshev approximation produce quite consistent outcome. The similarity of assessment results can serve as an additional argument in support of accepting the obtained order of criteria as the optimal solution.

7. Conclusions

The problem of evaluating preferences of criteria for choice is a key component in the multicriteria decision making procedure of assessment alternatives, and it is of independent interest in many applications including marketing research and practice. However, the existing methods to handle the problem, which are based on both direct evaluation of each criterion and indirect evaluation, where the criteria are compared in pairs, are known to give different and even opposite results. In the case when different methods may lead to inconsistent outcomes, a natural approach is to solve the problem by using several most widely used or mathematically justified methods concurrently and then combine the results to provide a more consistent and justified final solution.

In the paper, we have presented an approach that is intended to improve quality of the evaluation of criteria used by students in hotel selection when attending an educational program abroad. The approach is based on implementation of both direct and indirect methods of evaluating criteria, followed by statistical procedures to examine similarity and difference of the results. In addition, we have compared the solutions obtained by three methods of evaluating criteria from pairwise comparisons, including the methods of principal eigenvector, geometric mean and log-Chebyshev approximation.

By applying this technique, we have derived a solution that ranks the criteria of hotel selection in a way, which combines the results of different methods to provide the basis for more accurate assessment of the criteria. The comparison of the solutions produced by the indirect methods under study has shown a fairly high degree of similarity of results. We consider that the combined assessment technique described in the paper may serve as a template for a useful solution approach to handle real world problems of evaluating alternatives, where the accuracy and reliability of results is of prime importance and thus require additional formal and experimental justification.

As directions of future research, one can consider an extension of the approach to incorporate additional methods of evaluating alternatives from pairwise comparisons, including interval, linguistic and fuzzy pairwise comparisons. Further development of the technique into efficient algorithms and procedures to support decision making processes in practical problems presents another promising line of research.

Author Contributions: Conceptualization, N.K., A.P. and I.G.; Data curation, N.K., A.P. and I.G.; Formal analysis, N.K., A.P. and I.G.; Funding acquisition, N.K.; Investigation, N.K. and A.P.; Methodology, N.K., A.P. and I.G.; Project administration, N.K.; Resources, I.G.; Software, N.K. and A.P.; Supervision, N.K.; Validation, N.K. and A.P.; Visualization, A.P.; Writing—original draft, N.K. and A.P.; Writing—review & editing, N.K. All authors have read and agreed to the published version of the manuscript.

Funding: This research was funded by the Russian Foundation for Basic Research grant number 20-010-00145.

Institutional Review Board Statement: Not applicable.

Informed Consent Statement: Not applicable.

Data Availability Statement: Not applicable.

Acknowledgments: The authors are very grateful to the anonymous referees for their constructive and helpful comments and suggestions.

Conflicts of Interest: The authors declare no conflict of interest.

References

1. Thurstone, L.L. A law of comparative judgment. *Psychol. Rev.* **1927**, *34*, 273–286. [CrossRef]
2. Saaty, T.L. *The Analytic Hierarchy Process*, 2nd ed.; RWS Publications: Pittsburgh, PA, USA, 1990.
3. Saaty, T.L. On the measurement of intangibles: A principal eigenvector approach to relative measurement derived from paired comparisons. *Not. Am. Math. Soc.* **2013**, *60*, 192–208. [CrossRef]
4. Gavalec, M.; Ramík, J.; Zimmermann, K. *Decision Making and Optimization*; Lecture Notes in Economics and Mathematical Systems; Springer: Cham, Switzerland, 2015; Volume 677. [CrossRef]
5. Saaty, T.L.; Vargas, L.G. Comparison of eigenvalue, logarithmic least squares and least squares methods in estimating ratios. *Math. Model.* **1984**, *5*, 309–324. [CrossRef]
6. Choo, E.U.; Wedley, W.C. A common framework for deriving preference values from pairwise comparison matrices. *Comput. Oper. Res.* **2004**, *31*, 893–908. [CrossRef]
7. Barzilai, J.; Cook, W.; Golany, B. Consistent weights for judgements matrices of the relative importance of alternatives. *Oper. Res. Lett.* **1987**, *6*, 131–134. [CrossRef]
8. Ishizaka, A.; Lusti, M. How to derive priorities in AHP: A comparative study. *Cent. Eur. J. Oper. Res.* **2006**, *14*, 387–400. [CrossRef]

9. Tran, N.M. Pairwise ranking: Choice of method can produce arbitrarily different rank order. *Linear Algebra Appl.* **2013**, *438*, 1012–1024. [CrossRef]
10. Mazurek, J.; Perzina, R.; Ramík, J.; Bartl, D. A numerical comparison of the sensitivity of the geometric mean method, eigenvalue method, and best–worst method. *Mathematics* **2021**, *9*, 554. [CrossRef]
11. Barzilai, J. Deriving weights from pairwise comparison matrices. *J. Oper. Res. Soc.* **1997**, *48*, 1226–1232. [CrossRef]
12. Narasimhan, R. A geometric averaging procedure for constructing supertransitive approximation to binary comparison matrices. *Fuzzy Sets Syst.* **1982**, *8*, 53–61. [CrossRef]
13. Crawford, G.; Williams, C. A note on the analysis of subjective judgment matrices. *J. Math. Psychol.* **1985**, *29*, 387–405. [CrossRef]
14. Krivulin, N. Rating alternatives from pairwise comparisons by solving tropical optimization problems. In Proceedings of the 2015 12th International Conference on Fuzzy Systems and Knowledge Discovery (FSKD), Zhangjiajie, China, 15–17 August 2015; Tang, Z., Du, J., Yin, S., He, L., Li, R., Eds.; IEEE: Piscataway, NJ, USA, 2015; pp. 162–167. [CrossRef]
15. Krivulin, N. Using tropical optimization techniques to evaluate alternatives via pairwise comparisons. In Proceedings of the 2016 Proceedings of the Seventh SIAM Workshop on Combinatorial Scientific Computing, Albuquerque, NM, USA, 10–12 October 2016; Gebremedhin, A.H., Boman, E.G., Ucar, B., Eds.; SIAM: Philadelphia, PA, USA, 2016; pp. 62–72. [CrossRef]
16. Krivulin, N. Methods of tropical optimization in rating alternatives based on pairwise comparisons. In *Operations Research Proceedings 2016*; Fink, A., Fügenschuh, A., Geiger, M.J., Eds.; Springer: Cham, Switzerland, 2018; pp. 85–91. [CrossRef]
17. Krivulin, N.; Sergeev, S. Tropical implementation of the Analytical Hierarchy Process decision method. *Fuzzy Sets Syst.* **2019**, *377*, 31–51. [CrossRef]
18. Baccelli, F.L.; Cohen, G.; Olsder, G.J.; Quadrat, J.P. *Synchronization and Linearity*; Wiley Series in Probability and Statistics; Wiley: Chichester, UK, 1993.
19. Kolokoltsov, V.N.; Maslov, V.P. *Idempotent Analysis and Its Applications*; Mathematics and Its Applications; Springer: Dordrecht, The Netherlands, 1997; Volume 401. [CrossRef]
20. Golan, J.S. *Semirings and Affine Equations over Them*; Mathematics and Its Applications; Kluwer Acad. Publ.: Dordrecht, The Netherlands, 2003; Volume 556. [CrossRef]
21. Heidergott, B.; Olsder, G.J.; van der Woude, J. *Max Plus at Work*; Princeton Series in Applied Mathematics; Princeton University Press: Princeton, NJ, USA, 2006.
22. Tsaur, S.H.; Tzeng, G.H. Multiattribute decision making analysis for customer preference of tourist hotels. *J. Travel Tour. Mark.* **1996**, *4*, 55–69. [CrossRef]
23. Dolnicar, S.; Otter, T. Which hotel attributes matter? A review of previous and a framework for future research. In Proceedings of the 9th Annual Conference of the Asia Pacific Tourism Association (APTA), Sydney, Australia, 6–9 July 2003; University of Technology Sydney: Sydney, Australia, 2003; Volume 1, pp. 176–188.
24. Yavas, U.; Babakus, E. Dimensions of hotel choice criteria: Congruence between business and leisure travelers. *Int. J. Hosp. Manag.* **2005**, *24*, 359–367. [CrossRef]
25. Hsieh, L.F.; Lin, L.H.; Lin, Y.Y. A service quality measurement architecture for hot spring hotels in Taiwan. *Tour. Manag.* **2008**, *29*, 429–438. [CrossRef]
26. Jones, P.; Chen, M.M. Factors determining hotel selection: Online behaviour by leisure travellers. *Tour. Hosp. Res.* **2011**, *11*, 83–95. [CrossRef]
27. Chang, W.Y. A study on the key success factors of service quality for international hotels. *Acta Oecon.* **2014**, *64*, 25–37. [CrossRef]
28. Yu, S.-M.; Wang, J.; Wang, J.-Q.; Li, L. A multi-criteria decision-making model for hotel selection with linguistic distribution assessments. *Appl. Soft Comput.* **2018**, *67*, 741–755. [CrossRef]
29. Tümer, M.; Aghaei, İ.; Lasisi, T.T. Assessment of housing choice criteria for the universities' students in North Cyprus using AHP method. *Eur. J. Manag. Res.* **2019**, *3*, 65–86.
30. Goral, R. Prioritizing the factors which affect the selection of hotels by consumers traveling for vacation with analytical hierarchy process (AHP) method. *J. Tour. Manag. Res.* **2020**, *7*, 11–31. [CrossRef]
31. Saaty, T.L. A scaling method for priorities in hierarchical structures. *J. Math. Psychol.* **1977**, *15*, 234–281. [CrossRef]
32. Chu, M.T. On the optimal consistent approximation to pairwise comparison matrices. *Linear Algebra Appl.* **1998**, *272*, 155–168. [CrossRef]
33. Krivulin, N. Using tropical optimization techniques in bi-criteria decision problems. *Comput. Manag. Sci.* **2020**, *17*, 79–104. [CrossRef]
34. Elsner, L.; van den Driessche, P. Max-algebra and pairwise comparison matrices. *Linear Algebra Appl.* **2004**, *385*, 47–62. [CrossRef]
35. Cuninghame-Green, R.A. Minimax algebra and applications. In *Advances in Imaging and Electron Physics*; Hawkes, P.W., Ed.; Academic Press: San Diego, CA, USA, 1994; Volume 90, pp. 1–121. [CrossRef]
36. Krivulin, N.K. Solution of generalized linear vector equations in idempotent algebra. *Vestnik St. Petersburg Univ. Math.* **2006**, *39*, 16–26.
37. Krivulin, N. Extremal properties of tropical eigenvalues and solutions to tropical optimization problems. *Linear Algebra Appl.* **2015**, *468*, 211–232. [CrossRef]
38. Krivulin, N. A maximization problem in tropical mathematics: A complete solution and application examples. *Informatica* **2016**, *27*, 587–606. [CrossRef]

Σ mathematics

Article

Impact of COVID-19 on Supply Chains: A Hybrid Trade Credit Policy

Ping Ruan [1], Yung-Fu Huang [2] and Ming-Wei Weng [2,*]

[1] Zhongshan Institute, Management School, University of Electronic Science and Technology of China, Zhongshan 528400, China; zhongshan_rp@126.com

[2] Department of Marketing and Logistics Management, Chaoyang University of Technology, Taichung 413310, Taiwan; huf@cyut.edu.tw

* Correspondence: mwweng@cyut.edu.tw

Abstract: The COVID-19 pandemic has affected all sectors of the world's economy and society. Firms need to have disaster recovery and business sustainability plans and to be able to generate profits in order to develop. Trade credit may be a good way for firms to free up cash flow and finance short-term growth. Extensions of payment will provide firms with low-cost loans under the COVID-19 credit guarantee scheme. Implementation of hybrid trade credit activities has been shown to improve the financial crisis of many firms, and the effects are particularly evident within two-echelon supply chains. An economic order quantity (EOQ) model is derived under conditions of deteriorating items, an upstream full trade credit or cash discount, and downstream partial trade credit in a supply chain. A computer program is developed to provide a numerical solution and a numerical example is used to show the solution's form and verify that the solution gives the minimum total cost per unit time.

Keywords: partial trade credit; cash discount; deteriorating items; EOQ; COVID-19

MSC: 49J55; 65A05; 68M07

Citation: Ruan, P.; Huang, Y.-F.; Weng, M.-W. Impact of COVID-19 on Supply Chains: A Hybrid Trade Credit Policy. *Mathematics* **2022**, *10*, 1209. https://doi.org/10.3390/math10081209

Academic Editors: Humberto Rocha and Ana Maria Rocha

Received: 1 March 2022
Accepted: 1 April 2022
Published: 7 April 2022

1. Introduction

The COVID-19 pandemic has had an unprecedented impact on health, economic, and financial systems around the world. From an economic and industry point of view, COVID-19 has brought uncertainties and disruptions to international businesses and supply chains. Thus, a supply chain may change the distribution network and route. Early theorization on the basic economic order quantity (EOQ) model that assumes instant payment, constant demand, and no shortages can be traced back to Harris [1]. Suppliers adopted a resolution for a hybrid payment strategy to sustain business during the COVID-19 crisis. COVID-19 could be the black swan event that finally forces many firms, and entire industries, to rethink and transform their global supply chain model. These shortages and supply-chain disruptions are significant and widespread. To protect their supply chain operations, firms may use digital supply networks, update inventory policies and planning parameters, and focus on cash flow. Some payment policies are commonly used among suppliers and retailers, such as prepayments, delays in payments, cash discounts, and the AC/DCF approach. A permissible delay in payments produces two benefits for the supplier: (1) it should attract new customers who consider it to be a type of price reduction; and (2) it should cause a reduction in the sales outstanding, since some established customers will pay more promptly in order to take advantage of permissible delays more frequently. Early theorization on the EOQ model can be traced back to Goyal [2] under the conditions of permissible delays in payments. Teng [3] amended Goyal's model [2] by considering the difference between the unit price and the unit cost and found that it makes economic sense for a well-established retailer to order a lower quantity and take the benefit of payment delays more frequently. Trade credit is used to motivate sales or decrease the on-hand inventory level to encourage customers. Numerous studies in the trade credit area can be found in the literature. Examples include Huang [4], Huang [5], Huang and Hsu [6], Hsieh et al. [7], Liao [8], Teng and Chang [9], Min et al. [10], Chen and Kang [11], Kreng and Tan [12], Lee and Rhee [13], Mathata [14], Soni and Patel [15], Ouyang et al. [16], Ouyang and Chang [17], Yang et al. [18], Chen et al. [19], Chen and Teng [20], Giri and Sharma [21], Lashgari et al. [22], and Sarkar et al. [23]. Furthermore, the classical EOQ model assumes that the purchasing cost is paid once an order is placed by a retailer. In the corporate world, companies often have to make advance payments to suppliers when their orders are large enough to be burdensome to the producer. An advance payment is a type of payment made ahead of its normal schedule; for instance, paying for a good or service before it is actually received. A prepayment is made when a selling firm receives payment from a buyer before the seller has shipped goods or provided services to the buyer. To produce a special product, the manufacturer may have to pay additional costs to set up a new process. This requires the manufacturer to obtain a fraction of the production or purchasing cost in advance. Various issues with advance payments are discussed in Maiti et al. [24], Gupta et al. [25], Thangam [26], Taleizadeh et al. [27], Zhang et al. [28], Tavakoli and Taleizadeh [29], Taleizadeh et al. [30], Shah et al. [31], Khan et al. [32], and Taleizadeh et al. [33]. Generally, in the real world, suppliers give different kinds of benefits to retailers due to advance payments. One of the popular benefits is an instant cash discount due to an advance payment. An example is a supplier who will provide a 2% discount on an invoice due in 30 days if the retailer pays within the first 10 days of receiving the invoice. Giving the buyer a small cash discount would benefit the seller as it would allow her to access the cash sooner. Khan et al. [34] proposed an inventory model for deteriorating items with a price- and stock-dependent demand rate under full/partial advance payment conditions. Shao and Meng [35] discussed the question of how to make decisions on whether the supplier's downstream enterprises should enjoy cash discounts. Several studies, including those of Huang and Chung [36], Ouyang et al. [37], Yang [38], Yang et al. [39], Feng et al. [40], Shah and Cárdenas-Barrón [41], Alshanbari et al. [42], Tripathi [43], and Mashud et al. [44], have provided extensive discussions on the applications of cash discounts. While trade credit is a powerful commercial tool for conquering new markets and building customer loyalty, it is well known that cash flow plays a pivotal role in determining firms' operation decisions. Zhou et al. [45] considered the structure of the retailer's optimal policies under different partial trade credit penalty rates. Laitinen [46] investigated the characteristics of the discounted cash flow (DCF) as a measure of a startup's financial success. Since then, several similar inventory EOQ models related to trade credit and discounted cash flow (DCF) have been proposed [47–56]. However, few studies have been done on the effect of COVID-19 on trade credit. Mashud et al. [44] showed the effect of advance and delayed payments on the retailer's total profit during the post-COVID-19 recovery period. De et al. [57] explored carbon emission issues with a production manufacturing system in the context of joint inventory control and sustainable trade credit financing for deteriorating items in a supplier–retailer–customer model in a volumetric fuzzy system. Demir and Javorcik [58] found that the impact of COVID-19 on trade finance matters included an increased risk of non-payment or non-delivery of pre-paid goods. Several studies (Agca et al. [59], Choi [60], Liu et al. [61], and Luo [62]) have suggested the effectiveness of COVID-19 in creating a trade credit policy. Some common practical issues are:

(1) The prepayment policy. This issue is key to expressing the real credit trade problem. These policies actually sustained business growth in a competitive market during the COVID-19 period;
(2) The cash discount policy. The Government's SME Recovery Loan Scheme is designed to support economic recovery and to provide continued assistance; otherwise, the supplier offers the retailer a discounted rate on an invoice in exchange for an early payment discount.

2. Problem Description

The global production and supply chain system has been disrupted due to the COVID-19 pandemic. The COVID-19 pandemic has broken most of the transportation links and distribution mechanisms between suppliers, production facilities, and customers. Therefore, in response to the challenges resulting from the COVID-19 pandemic, firms are looking to implement some credit trade policies to fill financing gaps left by engaging in both short-term (ST) and medium- and long-term (MLT) trade finance. While long-term partnerships are great for handling incremental changes during stable periods, disruptive environmental changes may require managers to consider disruptive changes to their businesses. In this paper, we specifically discuss these issues as hybrid credit trade problems during the COVID-19 period. In actuality, however, the COVID-19 pandemic has caused an unprecedented level of global disruption to economic systems and livelihoods. Zimon et al. [63] explored the trade credit management strategy in Polish group purchasing organizations during the COVID-19 pandemic. Table 1 presents a brief comparison of the results of the studies mentioned above. The following questions are often posed by suppliers and retailers as key points of interest:

(1) When is the best time to start prepayment for export items in the post-COVID-19 period?
(2) When is the best time to end prepayment for export items during the COVID-19 period?
(3) What is the optimal discount rate for items?
(4) What is the optimal selling price for items?
(5) What is the optimal production rate for items?

Table 1. Comparison of the financial policies in existing models with those in the proposed model.

References	Financial Policy					Other Consideration(s)
	EOQ/EPQ	PP	F/P	CD	DCF	
Goyal [2]	EOQ		F			Trade credit financing
Huang [4]	EOQ		F			Different payment rule
Huang [5]	EPQ		F			Two levels of trade credit policies
Huang and Hsu [6]	EOQ		P			A powerful decision-making right
Hsieh et al. [7]	EOQ		F	√		Demand and deterioration fluctuate with time
Liao [8]	EOQ		F			Non-instantaneous and exponentially deteriorating items
Teng and Chang [9]	EPQ		F			Relaxes the assumption of $N < M$
Min et al. [10]	EOQ		F			Stock-dependent demand
Chen and Kang [11]	EOQ/EPQ		F			Imperfect items/Varying permissible delays in payments
Kreng and Tan [12]	EOQ		F			Order quantity
Lee and Rhee [13]	EOQ		F			Newsvendor framework
Mathata [14]	EOQ/EPQ		P		√	Exponentially deteriorating items
Soni and Patel [15]	EOQ		F/P			Defective items/Variable production
Ouyang et al. [16]	EOQ		F			AM-GM mean inequality
Ouyang and Chang [17]	EPQ		F			Imperfect production/AM-GM inequality
Yang et al. [18]	EOQ		P			Order quantity/Limited storage capacity
Chen et al. [19]	EPQ		F/P			Convex fractional programming/Non-deteriorated items
Chen and Teng [20]	EOQ		F		√	Expiration dates/Time-varying deterioration of items
Giri and Sharma [21]	EOQ		P			Linear time-dependent demand/Shortage
Lashgari et al. [22]	EOQ	√	P			Non-instantaneous deterioration/Partial backordering
Sarkar et al. [23]	EOQ		F		√	Carbon emissions/Rework/Shortage
Maiti et al. [24]	EOQ	√	F		√	Genetic algorithm/Price-dependent demand
Gupta et al. [25]	EOQ	√	F			Real-coded genetic algorithm/Constant uniform demand
Thangam [26]	EOQ	√	F/P			Perishable items
Taleizadeh et al. [27]	EOQ	√			√	Fuzzy rough/Metaheuristic algorithms

Table 1. *Cont.*

References	Financial Policy					Other Consideration(s)
	EOQ/EPQ	**PP**	**F/P**	**CD**	**DCF**	
Zhang et al. [28]	EOQ	√	P/P		√	Time-weighted inventory
Tavakoli and Taleizadeh [29]	EOQ	√				Shortage/Percentage of purchasing cost
Taleizadeh et al. [30]	EOQ	√				Pricing/Shortage
Shah et al. [31]	EOQ	√				Fixed-lifetime/Quadratic demand/Preservation investment
Khan et al. [32]	EOQ	√				Advertising/Maximum lifetime/Shortage
Taleizadeh et al. [33]	EOQ	√	P/P			Inspection policy/Shortage/Fraction of demand
Khan et al. [34]	EOQ	√	F/P	√		Price- and stock-dependent demand
Shao and Meng [35]	EOQ			√	√	Decision tree diagram/Cost of capital
Huang and Chung [36]	EOQ	√	F	√		$I_p < I_e$
Ouyang et al. [37]	EOQ		F	√		Realistic in the modern business environment Transactions
Yang [38]	EOQ		F	√	√	Conditionally permissible delays in payments
Yang et al. [39]	EOQ			√		Delays in payments linked to order quantity
Feng et al. [40]	EPQ		F	√		$I_c \geq I_d,\ c(1-\delta)I_c \geq sI_d$
Shah and Cárdenas-Barrón [41]	EOQ			√		Order-linked credit period
Alshanbari et al. [42]	EOQ	√				Shortage/Two-parameter Weibull distribution decay rate/Advertisement
Tripathi [43]	EOQ		F	√		Time-sensitive demand/Shortage
Mashud et al. [44]	EOQ		P		√	Post-COVID-19 recovery/Price-sensitive demand/Preservation technology
Zhou et al. [45]	EOQ		P			Newsvendor/Stochastic demand
Laitinen [46]	EOQ				√	Payback/Internal rate of return (IRR)
Arcelus et al. [47]	EOQ		P	√		Special sales/Forward buying/Price-dependent demand
Stokes [48]	EOQ			√		Differential game/Working capital management/Terms of sale
Chung and Liao [49]	EOQ		F		√	Ordering quantity/Out-of-pocket inventory carrying cost
Guariglia and Mateut [50]	EOQ		F		√	Inventory investment/Coverage ratio/Financing constraints
Ho et al. [51]	EOQ		F	√		The market demand rate $D(p) = ap^{-\delta}$
Chang et al. [52]	EOQ		F		√	Trade credit linked to order quantity
Chung et al. [53]	EOQ		P/F	√	√	Mathematical solution procedures
Wu et al. [54]	EOQ		F		√	Expiration dates/ Deterioration rate at time t
Tripathi et al. [55]	EOQ		F		√	Stock-dependent demand
This paper	EOQ	√	F/P	√		COVID-19 period

Note: PP = prepayment; F/P = full/partial trade credit; CD = cash discount; DCF = discounted cash flow.

3. Notation and Assumptions

3.1. System Parameters

D	retailer's demand rate during the COVID-19 period.
P	manufacturer's production rate, where $P > D$
A	manufacturer's ordering cost per order.
θ	the item deteriorates at a constant rate θ $(0 < \theta < 1)$ per time unit.
h	the retailer's holding cost excluding interest charges, USD/per unit/year.
I_e	the retailer's interest earned per dollar per year.
I_k	the retailer's interest charged per dollar per year.
M	the upstream trade credit period in years offered by the supplier.
N	the downstream trade credit period in years offered by the retailer, where $N \leq M$.
L	the time period of prepayment.
r	the cash discount rate $0 < r < 1$.
α	the fraction of the delay in payments permitted by the supplier.
c	the unit purchasing cost.
p	the unit selling price, with $p > c$.
t_1	the production run time.
T	the length of the replenishment cycle in years.
$TVC_1(T)$	total cost per unit time (cash discount).
$TVC_2(T)$	total cost per unit time (full delay in payments).

3.2. Assumptions

This paper is based on the following assumptions:

- The rate of replenishment is considered to be infinite, while the lead time is zero;
- The inventory system involves only one item;
- An infinite planning horizon for the whole system is considered;
- The items deteriorate at a constant rate θ, where $\theta > 0$;
- Before the COVID-19 pandemic, the logistic efficiency and the latest digital technologies (e-commerce technology) were regarded as critical elements in stabilizing demand. As the pandemic continued, it understandably became challenging to stabilize and recover the retailer's demand absolutely. The demand rate, D, is known and constant.
- A discount is presented by the supplier (the manufacturer) to the retailer when the retailer agrees to delay a portion α of the prepayment for time period L. The discount rate (α) increases when $TVC_1(L - N)$ decreases during a lockdown period of the COVID-19 pandemic.

4. Model Formulation

This paper considers a two-echelon supply chain with an upstream supplier and a downstream retailer during the COVID-19 period. The structure is developed in a coordinated case. In the current COVID-19 pandemic situation, the supplier offers advance payments to the firm so that they will not cancel the order. The aim is to evaluate the effect of the cash discount and trade credit. In the replenishment period, $[0, T]$, the retailer offers a trade credit policy to customers. In the Phase I trade credit period, $[0, t_1]$, depletion of the inventory occurs due to the combined effects of production, demand, and deterioration on the replenishment cycle. Hence, the change in the inventory level can be illustrated by the following differential equation:

$$\frac{dI(t)}{dt} = P - D - \theta I(t), 0 \leq t \leq t_1, \tag{1}$$

with the boundary condition $I(t_1) = 0$. Solving Equation (1), one obtains

$$I(t) = \frac{P - D}{\theta}(1 - e^{-\theta t}), 0 \leq t \leq t_1, \tag{2}$$

In the Phase II trade credit period, $[t_1, T]$, the inventory level is decreased by the effects of demand and deterioration on the replenishment cycle. Hence, the change in the inventory level can be illustrated by the following differential equation

$$\frac{dI(t)}{dt} = -D - \theta I(t), t_1 \le t \le T, \tag{3}$$

with the boundary condition $I(t_1) = \frac{P-D}{\theta}(1 - e^{-\theta t_1})$.

Solving Equation (3) yields

$$I(t) = \frac{D}{\theta}(e^{\theta(T-t)} - 1), t_1 \le t \le T, \tag{4}$$

In considering the two-echelon supply chain issues, t_1 and T can be expressed as

$$t_1 = \frac{1}{\theta}\ln[1 + \frac{D}{P}(e^{\theta T} - 1)], \tag{5}$$

In this section, we construct an inventory model that consists of the following four elements:

- The ordering cost (OC). The retailer's ordering cost per replenishment cycle is $OC = A/T$;
- The holding cost (HC). The retailer's holding cost per replenishment cycle is $\frac{h}{T}[\int_0^{t_1} I(t)dt + \int_{t_1}^{T} I(t)dt] = \frac{h}{\theta T}(Pt_1 - DT)$;
- The deterioration cost (DC), which is calculated as $DC = \begin{cases} c(Pt_1 - DT)/\theta T \\ c(1-r)(Pt_1 - DT)/\theta T \end{cases}$;
- The purchasing cost (PC), which is calculated as $PC = \begin{cases} cD \\ (1-r)cD \end{cases}$.

4.1. Taking a Cash Discount

Based on the lengths of N, L, and $N + L$, three cases are possible: (1) $L \le T$; (2) $L - N \le T \le L$; and (3) $T \le L - N$. We discuss each case in detail below.

4.1.1. Case 1 $L \le T$

Here, the retailer pays off all items at time 0. The interest charged per unit time is

$$IC_{11} = \frac{c(1-r)I_k D}{2T}[\alpha(T-L)^2 + (1-\alpha)(T+N-L)^2], \tag{6}$$

The retailer starts selling the items from time 0 but receives money at time N. Therefore, the interest earned per unit time is

$$IE_{11} = \frac{pI_e D}{2T}[\alpha L^2 + (1-\alpha)(L-N)^2], \tag{7}$$

Consequently, the retailer's total cost, $TVC_{11}(T)$, per unit time for Case 1 is

$$\begin{aligned} TVC_{11}(T) &= \frac{A}{T} + \frac{h + c\theta(1-r)}{\theta T}(Pt_1 - DT) + (1-r)cD \\ &+ \frac{c(1-r)I_k D}{2T}[\alpha(T-L)^2 + (1-\alpha)(T+N-L)^2], \\ &- \frac{pI_e D}{2T}[\alpha L^2 + (1-\alpha)(L-N)^2] \end{aligned} \tag{8}$$

The graphical representation for Case 1 is shown in Figure 1.

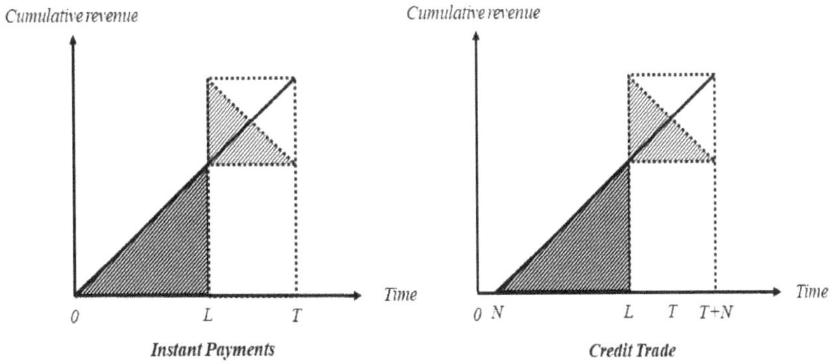

Figure 1. Graphical representation of $L \leq T$.

4.1.2. Case 2 $L - N \leq T \leq L$

In this case, the retailer receives the total revenue at time L and has to pay the supplier at $T + N$. Hence, the interest charged per unit time is

$$\mathrm{IC}_{12} = \frac{c(1-r)I_k D}{2T}(1-\alpha)(T+N-L)^2, \tag{9}$$

and the interest earned per unit time is

$$\mathrm{IE}_{12} = \frac{pI_e D}{2T}[\alpha T^2 + 2\alpha T(L-T) + (1-\alpha)(L-N)^2], \tag{10}$$

Thus, the retailer's total cost, $TVC_{12}(T)$, per unit time for Case 2 is

$$\begin{aligned} TVC_{12}(T) &= \tfrac{A}{T} + \tfrac{h+c\theta(1-r)}{\theta T}(Pt_1 - DT) + (1-r)cD \\ &+ \tfrac{c(1-r)I_k D}{2T}(1-\alpha)(T+N-L)^2 \\ &- \tfrac{pI_e D}{2T}[\alpha T^2 + 2\alpha T(L-T) + (1-\alpha)(L-N)^2] \end{aligned}, \tag{11}$$

The graphical representation for Case 2 is shown in Figure 2.

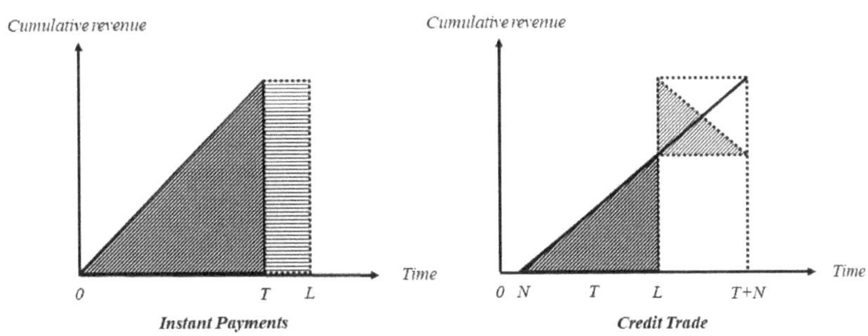

Figure 2. Graphical representation of $L - N \leq T \leq L$.

4.1.3. Case 3 $T \leq L - N$

In this case, the retailer can sell the items and receives the total revenue at time L. The annual interest earned is

$$\begin{aligned} IE_{13} &= \tfrac{pI_e D}{2T}[\alpha T^2 + 2\alpha T(L-T) + (1-\alpha)T^2 + 2(1-\alpha)T(L-T-N)] \\ &= \tfrac{pI_e D}{2}[2L - T - 2(1-\alpha)N] \end{aligned}, \tag{12}$$

From the above arguments, the retailer's annual total cost, $TVC_{13}(T)$, per unit time for Case 3 is

$$\begin{aligned} TVC_{13}(T) &= \frac{A}{T} + \frac{h+c\theta(1-r)}{\theta T}(Pt_1 - DT) + (1-r)cD \\ &- \frac{pI_eD}{2}[2L - T - 2(1-\alpha)N] \end{aligned}, \tag{13}$$

Summarizing the above cases, the retailer's total cost, $TVC_{1i}(T)$, is given by

$$TVC_{1i}(T) = \begin{cases} TVC_{11}(T), \; if \; L \leq T \\ TVC_{12}(T), \; if \; L-N \leq T \leq L \\ TVC_{13}(T), \; if \; T \leq L-N \end{cases}, \tag{14}$$

At $T = L$, we find $TVC_{11}(L) = TVC_{12}(L)$; at $T = L-N$, $TVC_{12}(L-N) = TVC_{13}(L-N)$. Hence, $TVC_{1i}(T)$ is continuous and well-defined. $TVC_{11}(T)$, $TVC_{12}(T)$, $TVC_{13}(T)$, and $TVC_1(T)$ are all defined on $T > 0$. The graphical representation for Case 3 is shown in Figure 3.

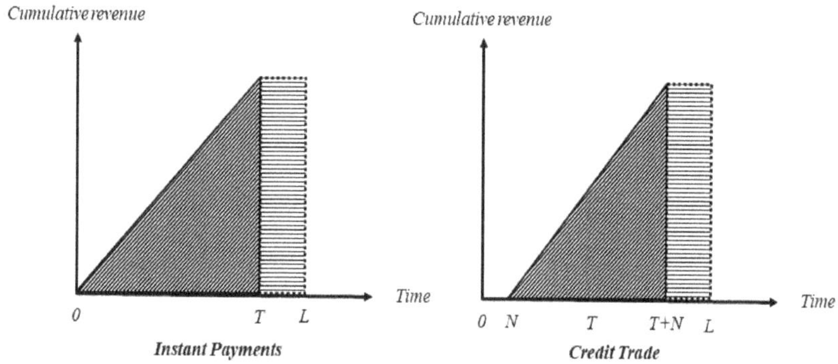

Figure 3. Graphical representation of $T \leq L - N$.

4.2. Taking a Permissible Delay

4.2.1. Case 1 $M \leq T$

In this case, the retailer receives the total revenue at time M. The interest charged per unit time is

$$IC_{21} = \frac{c(1-r)I_kD}{2T}[\alpha(T-M)^2 + (1-\alpha)(T+N-M)^2], \tag{15}$$

The interest earned per unit time is

$$IE_{21} = \frac{pI_eD}{2T}[\alpha L^2 + (1-\alpha)(L-N)^2], \tag{16}$$

From Equations (15) and (16), the annual total cost, $TVC_{21}(T)$, is given by

$$\begin{aligned} TVC_{21}(T) &= \frac{A}{T} + \frac{h+c\theta}{\theta T}(Pt_1 - DT) + cD \\ &+ \frac{cI_kD}{2T}[\alpha(T-M)^2 + (1-\alpha)(T+N-M)^2] \\ &- \frac{pI_eD}{2T}[\alpha M^2 + (1-\alpha)(M-N)^2] \end{aligned}, \tag{17}$$

4.2.2. Case 2 $M - N \leq T \leq M$

The interest charged per unit time is

$$IC_{22} = \frac{c(1-r)I_kD}{2T}(1-\alpha)(T+N-M)^2, \tag{18}$$

The interest earned per unit time is

$$IE_{22} = \frac{pI_eD}{2T}[\alpha T^2 + 2\alpha T(M-T) + (1-\alpha)(M-N)^2], \quad (19)$$

From Equations (18) and (19), the annual total cost, $TVC_{22}(T)$, is given by

$$\begin{aligned} TVC_{22}(T) &= \tfrac{A}{T} + \tfrac{h+c\theta}{\theta T}(Pt_1 - DT) + cD \\ &+ \tfrac{cI_kD}{2T}(1-\alpha)(T+N-M)^2 \\ &- \tfrac{pI_eD}{2T}[\alpha T^2 + 2\alpha T(M-T) + (1-\alpha)(M-N)^2] \end{aligned}, \quad (20)$$

4.2.3. Case 3 $T \leq M - N$

The interest earned per unit time is

$$IE_{23} = \frac{pI_eD}{2}[2M - T - 2(1-\alpha)N], \quad (21)$$

From Equation (21), the annual total cost, $TVC_{23}(T)$, is given by

$$\begin{aligned} TVC_{23}(T) &= \tfrac{A}{T} + \tfrac{h+c\theta}{\theta T}(Pt_1 - DT) + cD \\ &- \tfrac{pI_eD}{2}[2M - T - 2(1-\alpha)N] \end{aligned}, \quad (22)$$

Summarizing the above cases, the retailer's total cost, $TVC_{2i}(T)$, is given by

$$TVC_{2i}(T) = \begin{cases} TVC_{21}(T), \ if M \leq T \\ TVC_{22}(T), \ if M - N \leq T \leq M \\ TVC_{23}(T), \ if T \leq M - N \end{cases}, \quad (23)$$

At $T = M$, we find $TVC_{21}(M) = TVC_{22}(M)$; at $T = M - N$, $TVC_{22}\ (M - N) = TVC_{23}(M - N)$. Hence, $TVC_2(T)$ is continuous and well-defined. $TVC_{21}(T)$, $TVC_{22}(T)$, $TVC_{23}(T)$, and $TVC_2(T)$ are all defined on $T > 0$.

From the above argument, the annual total cost for the retailer can be expressed as:

$$TVC(T) = \begin{cases} TVC_1(T), if taking a cash discount \\ TVC_2(T), if taking a permissible delay \end{cases}, \quad (24)$$

5. Solution Procedures

The main purpose of this section is to develop a solution procedure to determine the optimal cycle time T^* to minimize the annual total cost for each case.

5.1. Taking a Cash Discount

In an attempt to minimize the annual total cost for each case, we developed solution procedures consisting of two cases, with three propositions in each case.

Proposition 1. *For $L \leq T$, if $T_{11}^* \geq L$, then $dTVC_{11}(T)/dT$ is a strictly increasing function of T and there exists a unique real solution $T_{11}^* \in [L, \infty)$ such that $TVC_{11}(T_{11}^*)$ is the minimum.*

Proof. By Theorem 3.2.10 in Cambini and Martein [64], let $q(x) = \frac{f(x)}{g(x)}$. If $f'(x)$ is a differentiable and strictly increasing function in x, $f'(x) \geq 0$ and if $g'(x)$ is a differentiable and strictly decreasing function in x, $g'(x) \geq 0$. We have shown that $q(x)$ is a concave function; therefore, there exists a unique value of x^* that minimizes $q(x^*)$. From Equation (6), we obtain $TVC_{11}(T) = \frac{f(T)}{g(T)}$, where

$$\begin{aligned} f(T) &= A + \tfrac{h+c\theta(1-r)}{\theta}(Pt_1 - DT) + (1-r)cDT \\ &+ \tfrac{c(1-r)I_kD}{2}[\alpha(T-L)^2 + (1-\alpha)(T+N-L)^2] \ , \\ &- \tfrac{pI_eD}{2}[\alpha L^2 + (1-\alpha)(L-N)^2] \end{aligned} \tag{25}$$

and

$$g(T) = T > 0, \tag{26}$$

Then, substituting into Equation (25), we take the first- and second-order derivations of $f(T)$ with respect to T and obtain

$$\begin{aligned} f'(T) &= \tfrac{h+c\theta(1-r)}{\theta}[\tfrac{PDe^{\theta T}}{P+D(e^{\theta T}-1)} - D] + (1-r)cD \\ &+ c(1-r)I_kD[T - L + (1-\alpha)N] \end{aligned} \ , \tag{27}$$

and

$$f''(T) = \frac{h+c\theta(1-r)}{\theta}[\frac{P^2D\theta e^{\theta T}\rho}{P+D(e^{\theta T}-1)^2}] + c(1-r)I_kD > 0, \tag{28}$$

Therefore, $TVC_{11}(T)$ is convex in T. The minimum value of $TVC_{11}(T)$ will occur at the point T^* that satisfies

$$\frac{dTVC_{11}(T)}{dT} = 0, \text{ if } T \geq L.$$

Next, the first-order derivatives of $TVC_{11}(T)$ with respect to T are

$$\begin{aligned} \tfrac{\mathrm{dTVC}_{11}(T)}{dT} &= \tfrac{P[h+c\theta(1-r)]}{\theta}\left\{\tfrac{DTe^{\theta T}}{P+D(e^{\theta T}-1)} - \tfrac{1}{\theta}\ln[1 + \tfrac{D}{P}(e^{\theta T}-1)]\right\} \\ &+ \tfrac{c(1-r)I_kD}{2}[T^2 - \alpha L^2 - (1-\alpha)(L-N)^2] \\ &+ \tfrac{pI_eD}{2}[\alpha L^2 - (1-\alpha)(L-N)^2] - A \\ &= 0 \end{aligned} \ , \tag{29}$$

□

Proposition 2. *If $L - N \leq T \leq L$, then $dTVC_{12}(T)/dT$ is a strictly increasing function of T and there exists a unique real solution $T^*_{12} \in [L-N, L]$ such that $TVC_{12}(T^*_{12})$ is the minimum.*

Proof. We first take the first-order derivative of $TVC_{12}(T)$ with respect to T and obtain

$$\begin{aligned} \tfrac{\mathrm{dTVC}_{12}(T)}{dT} &= \tfrac{P[h+c\theta(1-r)]}{\theta}\left\{\tfrac{DTe^{\theta T}}{P+D(e^{\theta T}-1)} - \tfrac{1}{\theta}\ln[1 + \tfrac{D}{P}(e^{\theta T}-1)]\right\} \\ &+ \tfrac{c(1-r)I_kD}{2}(1-\alpha)[T^2 - (L-N)^2] \\ &+ \tfrac{pI_eD}{2}[\alpha TL + (1-\alpha)(L-N)^2] - A \\ &= 0 \end{aligned} \ , \tag{30}$$

Since $\mathrm{dTVC}_{12}(T)/dT$ is also strictly increasing in T, the minimum value of $TVC_{12}(T)$ will occur at the point T^* that satisfies

$$\frac{\mathrm{dTVC}_{12}(T)}{dT} = 0; \text{ otherwise, } T = \left\{ \begin{array}{c} L - N if \lim_{T \to L-N^+} \frac{dTVC_{12}(T)}{dT} > 0 \\ L if \lim_{T \to L^-} \frac{dTVC_{12}(T)}{dT} < 0 \end{array} \right. \tag{31}$$

□

Proposition 3. *If $0 \leq T \leq L - N$, then $dTVC_{13}(T)/dT$ is a strictly increasing function of T and there exists a unique real solution $T^*_{13} \in (0, L-N)$ such that $TVC_{13}(T^*_{13})$ is the minimum.*

Proof. We first take the first-order derivative of $TVC_{13}(T)$ with respect to T and obtain

$$\begin{aligned}\frac{dTVC_{13}(T)}{dT} &= \frac{P[h+c\theta(1-r)]}{\theta}\left\{\frac{DTe^{\theta T}}{P+D(e^{\theta T}-1)} - \frac{1}{\theta}\ln[1+\frac{D}{P}(e^{\theta T}-1)]\right\} \\ &+\frac{pI_eD}{2}T^2 - A \\ &= 0\end{aligned} \tag{32}$$

Next, we let

$$\Delta_1 = \frac{P[h+c\theta(1-r)]}{\theta}\{\frac{D(L-N)e^{\theta(L-N)}}{P+D(e^{\theta(L-N)}-1)} - \frac{1}{\theta}\ln[1+\frac{D}{P}(e^{\theta(L-N)}-1)]\} + \frac{pI_eD}{2}(L-N)^2 - A \tag{33}$$

$$\begin{aligned}\Delta_2 &= \frac{P[h+c\theta(1-r)]}{\theta}\{\frac{DLe^{\theta L}}{P+D(e^{\theta L}-1)} - \frac{1}{\theta}\ln[1+\frac{D}{P}(e^{\theta L}-1)]\} \\ &+\frac{c(1-r)I_kD}{2}[(1-\alpha)N(2L-N)] + \frac{pI_eD}{2}[L^2-(1-\alpha)N(2L-N)] - A\end{aligned} \tag{34}$$

Since $dTVC_{13}(T)/dT$ is also strictly increasing in T, the minimum value of $TVC_{13}(T)$ will occur at the point T that satisfies $\frac{dTVC_{13}(T)}{dT} = 0$; otherwise,

$$T_{13}^* = L - N \; if \lim_{T\to L-N^-}\frac{dTVC_{13}(T)}{dT} < 0$$

□

Lemma 1. $\Delta_1 < \Delta_2$, *for* $L \geq N$.

Proof. From Proposition 2, we first take the first-order derivative of $TVC_{12}(T)$ with respect to T and obtain

$$TVC'_{12}(L-N) = \frac{\Delta_1 + \frac{pI_eD}{2}\alpha N(L-N)}{(L-N)^2} < TVC'_{12}(L) = \frac{\Delta_2}{L^2}$$

From Equation (28), since $L \geq N$, we have $\Delta_1 < \Delta_2$. □

Proposition 4.

(1) *If* $\Delta_2 < 0$, *then we obtain* $TVC_1(T^*) = TVC_1(T_{11}^*)$.
(2) *If* $\Delta_2 = 0$, *then we obtain* $TVC_1(T^*) = TVC_1(L)$.
(3) *If* $\Delta_1 < 0$ *and* $\Delta_2 > 0$, *then we obtain* $TVC_1(T^*) = TVC_1(T_{12}^*)$.
(4) *If* $\Delta_1 = 0$, *then we obtain* $TVC_1(T^*) = TVC_1(L-N)$.
(5) *If* $\Delta_1 > 0$, *then we obtain* $TVC_1(T^*) = TVC_1(T_{13}^*)$.

Proof. From (28), the first-order derivatives of $TVC_{11}(T)$ with respect to T are

$$TVC'_{11}(T) = \frac{1}{T^2}\left\{\begin{array}{c}\frac{P[h+c\theta(1-r)]}{\theta}\left\{\frac{DTe^{\theta T}}{P+D(e^{\theta T}-1)} - \frac{1}{\theta}\ln[1+\frac{D}{P}(e^{\theta T}-1)]\right\} \\ +\frac{c(1-r)I_kD}{2}[T^2-L^2+(1-\alpha)N(2L-N)] \\ +\frac{pI_eD}{2}[L^2-(1-\alpha)N(2L-N)] - A\end{array}\right\} \tag{35}$$

From Equation (35), if $\Delta_2 < 0$, since $\lim_{T\to\infty} TVC'_{11}(T) = \frac{c(1-r)I_kD}{2} > 0$ and $TVC'_{11}(L) = \frac{\Delta_2}{L^2} < 0$, the Intermediate Value Theorem implies that the root of $TVC'_{11}(T_{11}^*)$ is the unique real solution $T_{11}^* \in (L,\infty)$. From Lemma 1, since $TVC'_{12}(L-N) = \frac{\Delta_1+\frac{pI_eD}{2}\alpha N(L-N)}{(L-N)^2} < TVC'_{12}(L) = \frac{\Delta_2}{L^2} < 0$ and $\lim_{\zeta\to 0} TVC'_{13}(\zeta) < TVC'_{13}(L-N) = \frac{\Delta_1}{(L-N)^2} < 0$, and

$dTVC_{12}(T)/dT$ and $dTVC_{13}(T)/dT$ are strictly decreasing in T, the minimum value of $TVC_{12}(T)$ and $TVC_{13}(T)$ will occur at the point T^* that satisfies

$$\frac{dTVC_{12}(T)}{dT} = 0;\ \text{otherwise},\ T^* = \begin{cases} L-N, \text{ if } \lim_{T\to L-N^+} \frac{dTVC_{12}(T)}{dT} > 0 \\ N, \text{ if } \lim_{T\to N^-} \frac{dTVC_{12}(T)}{dT} < 0 \end{cases}$$

and $\frac{dTVC_{13}(T)}{dT} = 0$; otherwise, $T^* = N$, if $\lim_{T\to N^-} \frac{dTVC_{13}(T)}{dT} < 0$, respectively. In addition, it is not difficult to show that $TVC_{11}(T^*_{11}) < TVC_{11}(L) = TVC_{12}(L) < TVC_{12}(L-N) = TVC_{13}(L-N)$. Clearly, by Equations (2)–(5), $TVC_{11}(T)$, $TVC_{12}(T)$, and $TVC_{13}(T)$ are convex in T, respectively. □

5.2. Taking a Permissible Delay

In this situation, the supplier offers the retailer a trade credit. The solution procedures consist of two cases in which the business relationship is maintained during the COVID-19 period, with four propositions in each case.

Proposition 5. *If $M \le T \le \infty$, then $dTVC_{21}(T)/dT$ is a strictly increasing function of T and there exists a unique real solution $T^*_{21} \in [M, \infty]$ such that $TVC_{21}(T^*_{21})$ is the minimum.*

Proposition 6. *If $M - N \le T \le M$, then $dTVC_{22}(T)/dT$ is a strictly increasing function of T and there exists a unique real solution $T^*_{22} \in [M-N, M]$ such that $TVC_{22}(T^*_{22})$ is the minimum.*

Proposition 7. *If $0 \le T \le M - N$, then $dTVC_{23}(T)/dT$ is a strictly increasing function of T and there exists a unique real solution $T^*_{23} \in [0, M-N]$ such that $TVC_{23}(T^*_{23})$ is the minimum.*

Next, we first take the first-order derivative of $TVC_{2i}(T)$ with respect to T. Then, only one case of $TVC_{2i}(T)$ has a solution to

$$\frac{dTVC_{2i}(T)}{dT} = 0$$

We then obtain the desired results.

$$\begin{aligned} \frac{dTVC_{21}(T)}{dT} &= \frac{P(h+c\theta)}{\theta}\{\frac{DTe^{\theta T}}{P+D(e^{\theta T}-1)} - \frac{1}{\theta}\ln[1+\frac{D}{P}(e^{\theta T}-1)]\} \\ &+ \frac{(pI_e-cI_k)D}{2}[\alpha M^2+(1-\alpha)(M-N)^2] + \frac{cI_kD}{2}T^2 - A \\ &= 0 \end{aligned} \tag{36}$$

$$\begin{aligned} \frac{dTVC_{22}(T)}{dT} &= \frac{P(h+c\theta)}{\theta}\{\frac{DTe^{\theta T}}{P+D(e^{\theta T}-1)} - \frac{1}{\theta}\ln[1+\frac{D}{P}(e^{\theta T}-1)]\} \\ &+ \frac{(pI_e-cI_k)D}{2}(1-\alpha)(M-N)^2 + \frac{DT^2}{2}[\alpha pI_e+(1-\alpha)cI_k] - A \\ &= 0 \end{aligned} \tag{37}$$

$$\frac{dTVC_{23}(T)}{dT} = \frac{P(h+c\theta)}{\theta}\{\frac{DTe^{\theta T}}{P+D(e^{\theta T}-1)} - \frac{1}{\theta}\ln[1+\frac{D}{P}(e^{\theta T}-1)]\} + \frac{DT^2}{2}pI_e - A = 0 \tag{38}$$

respectively. Next, we let

$$\Delta_3 = \frac{P(h+c\theta)}{\theta}\{\frac{D(M-N)e^{\theta(M-N)}}{P+D(e^{\theta(M-N)}-1)} - \frac{1}{\theta}\ln[1+\frac{D}{P}(e^{\theta(M-N)}-1)]\} + \frac{D(M-N)^2}{2}pI_e - A \tag{39}$$

$$\begin{aligned} \Delta_4 = &\frac{P(h+c\theta)}{\theta}\left\{\frac{DMe^{\theta M}}{P+D(e^{\theta M}-1)} - \frac{1}{\theta}\ln[1+\frac{D}{P}(e^{\theta M}-1)]\right\} \\ &+ \frac{(pI_e-cI_k)D}{2}[\alpha M^2+(1-\alpha)(M-N)^2] + \frac{cI_kD}{2}M^2 - A \end{aligned} \tag{40}$$

Lemma 2. $\Delta_3 < \Delta_4$ *for* $T > 0$.

Proof. The proof is similar that of Lemma 1, we omit it here. □

Proposition 8.

(1) *If* $\Delta_4 < 0$, *then* $TVC_2^*(T^*) = TVC_2^*(T_{21}^*)$.
(2) *If* $\Delta_4 = 0$, *then* $TVC_2^*(T^*) = TVC_2^*(M)$.
(3) *If* $\Delta_3 < 0$ *and* $\Delta_4 > 0$, *then* $TVC_2^*(T^*) = TVC_2^*(T_{22}^*)$.
(4) *If* $\Delta_3 = 0$, *then* $TVC_2^*(T^*) = TVC_2^*(M-N)$.
(5) *If* $\Delta_3 > 0$, *then* $TVC_2^*(T^*) = TVC_2^*(T_{23}^*)$.

5.3. Retailer's Ordering Policies

The COVID-19 pandemic has changed retailers' payment habits, such as the share of cash transactions and average transaction values. As the economic environment has deteriorated and consumption has decreased due to the ongoing COVID-19 crisis, E-commerce has gained an advantage as a sales channel over brick-and-mortar retailers. E-commerce provides customers with access to a significant variety of products from the convenience and safety of their own home. This section describes how an effective retailer ordering policy can result in lower costs and a better understanding of sales patterns. From Equation (24), Propositions 4 and 8, we have:

(1) If $\Delta_2 < 0$ and $\Delta_4 < 0$, then $TVC^*(T^*) = \min\{TVC_1^*(T_{11}^*), TVC_2^*(T_{21}^*)\}$.
(2) If $\Delta_2 < 0$ and $\Delta_4 = 0$, then $TVC^*(T^*) = \min\{TVC_1^*(T_{11}^*), TVC_2^*(M)\}$.
(3) If $\Delta_2 < 0$, $\Delta_3 < 0$, and $\Delta_4 > 0$, then $TVC^*(T^*) = \min\{TVC_1^*(T_{11}^*), TVC_2^*(T_{22}^*)\}$.
(4) If $\Delta_2 < 0$ and $\Delta_3 = 0$, then $TVC^*(T^*) = \min\{TVC_1^*(T_{11}^*), TVC_2^*(M-N)\}$.
(5) If $\Delta_2 < 0$ and $\Delta_3 > 0$, then $TVC^*(T^*) = \min\{TVC_1^*(T_{11}^*), TVC_2^*(T_{23}^*)\}$.
(6) If $\Delta_2 = 0$ and $\Delta_4 < 0$, then $TVC^*(T^*) = \min\{TVC_1^*(L), TVC_2^*(T_{21}^*)\}$.
(7) If $\Delta_2 = 0$ and $\Delta_4 = 0$, then $TVC^*(T^*) = \min\{TVC_1^*(L), TVC_2^*(M)\}$.
(8) If $\Delta_2 = 0$, $\Delta_3 < 0$, and $\Delta_4 > 0$, then $TVC^*(T^*) = \min\{TVC_1^*(L), TVC_2^*(T_{22}^*)\}$.
(9) If $\Delta_2 = 0$ and $\Delta_3 = 0$, then $TVC^*(T^*) = \min\{TVC_1^*(L), TVC_2^*(M-N)\}$.
(10) If $\Delta_2 = 0$ and $\Delta_3 > 0$, then $TVC^*(T^*) = \min\{TVC_1^*(L), TVC_2^*(T_{23}^*)\}$.
(11) If $\Delta_1 < 0$, $\Delta_2 > 0$, and $\Delta_4 < 0$, then $TVC^*(T^*) = \min\{TVC_1^*(T_{12}^*), TVC_2^*(T_{21}^*)\}$.
(12) If $\Delta_1 < 0$, $\Delta_2 > 0$, and $\Delta_4 = 0$, then $TVC^*(T^*) = \min\{TVC_1^*(T_{12}^*), TVC_2^*(M)\}$.
(13) If $\Delta_1 < 0$, $\Delta_2 > 0$, $\Delta_3 < 0$, and $\Delta_4 > 0$, then $TVC^*(T^*) = \min\{TVC_1^*(T_{12}^*), TVC_2^*(T_{22}^*)\}$.
(14) If $\Delta_1 < 0$, $\Delta_2 > 0$, and $\Delta_3 = 0$, then $TVC^*(T^*) = \min\{TVC_1^*(T_{12}^*), TVC_2^*(M-N)\}$.
(15) If $\Delta_1 < 0$, $\Delta_2 > 0$, and $\Delta_3 > 0$, then $TVC^*(T^*) = \min\{TVC_1^*(T_{12}^*), TVC_2^*(T_{23}^*)\}$.
(16) If $\Delta_1 = 0$ and $\Delta_4 < 0$, then $TVC^*(T^*) = \min\{TVC_1^*(L-N), TVC_2^*(T_{21}^*)\}$.
(17) If $\Delta_1 = 0$ and $\Delta_4 = 0$, then $TVC^*(T^*) = \min\{TVC_1^*(L-N), TVC_2^*(M)\}$.
(18) If $\Delta_1 = 0$, $\Delta_3 < 0$, and $\Delta_4 > 0$, then $TVC^*(T^*) = \min\{TVC_1^*(L-N), TVC_2^*(T_{22}^*)\}$.
(19) If $\Delta_1 = 0$ and $\Delta_3 = 0$, then $TVC^*(T^*) = \min\{TVC_1^*(L-N), TVC_2^*(M-N)\}$.
(20) If $\Delta_1 = 0$ and $\Delta_3 > 0$, then $TVC^*(T^*) = \min\{TVC_1^*(L-N), TVC_2^*(T_{23}^*)\}$.
(21) If $\Delta_1 > 0$ and $\Delta_4 < 0$, then $TVC^*(T^*) = \min\{TVC_1^*(T_{13}^*), TVC_2^*(T_{21}^*)\}$.
(22) If $\Delta_1 > 0$ and $\Delta_4 = 0$, then $TVC^*(T^*) = \min\{TVC_1^*(T_{13}^*), TVC_2^*(M)\}$.
(23) If $\Delta_1 > 0$, $\Delta_3 < 0$, and $\Delta_4 > 0$, then $TVC^*(T^*) = \min\{TVC_1^*(T_{13}^*), TVC_2^*(T_{22}^*)\}$.
(24) If $\Delta_1 > 0$ and $\Delta_3 = 0$, then $TVC^*(T^*) = \min\{TVC_1^*(T_{13}^*), TVC_2^*(M-N)\}$.
(25) If $\Delta_1 > 0$ and $\Delta_3 > 0$, then $TVC^*(T^*) = \min\{TVC_1^*(T_{13}^*), TVC_2^*(T_{23}^*)\}$.

5.4. Algorithm

Step 1. Evaluate the solution of T according to Equations (36)–(38);
Step 2. Use Propositions 1 and 2 to determine $\min\{TVC_1(T), TVC_2(T)\}$ and the corresponding value of T;
Step 3. Let $T_{n+1} = T_n + \varepsilon$ and repeat Steps 1–2;
Step 4. If $\text{TVC}_j(T^*_{(n)}) \geq \text{TVC}_j(T^*_{(n-1)})$, then return to Step 3; otherwise, execute Step 5;
Step 5. Let $(T^*_{(n)}) = (T^*_{(n-1)})$; therefore, T^* is the optimal solution and the minimum total cost per unit time is $\text{TVC}_j(T^*)$.

6. Application Example

The practicality of the proposed model was assessed using a case study involving SMEs in Taiwan. A numerical example of this case was used to verify our analytical results, and a sensitivity analysis was used to explore trends in the optimal policies in order to obtain managerial insights for the SMEs. The COVID-19 pandemic of 2019–2021 offers a unique setting in which to examine how the supply of trade credit is impacted during a crisis that emanates from the real sector, which is radically different to a crisis that emanates from financing difficulties, such as the global financial crisis (GFC) of 2008–2009. The parallel trends of average payables during the COVID-19 period and the GFC period based on the probabilty of default of a firm are shown in Figures 4 and 5. A high probability of default is defined as 1 for the creditrisk+ of firms whose probability of default is above the median. The figures display the parallel trend of average payables for the last two years for growing firms.

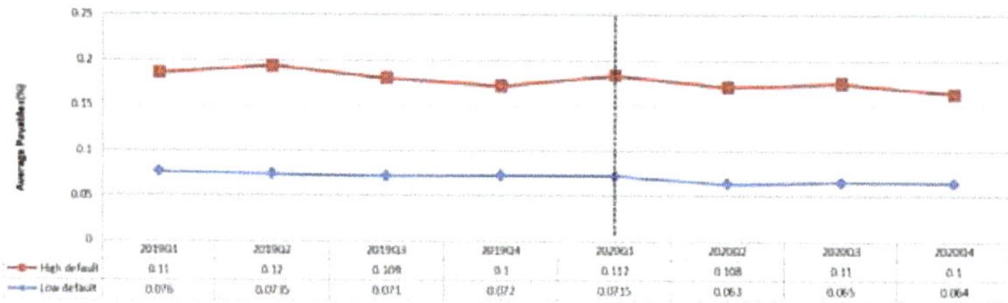

Figure 4. Parallel trend of average payables during the COVID-19 period.

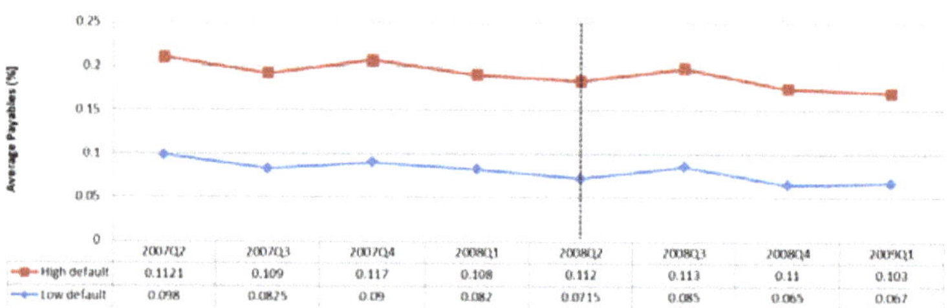

Figure 5. Parallel trend of average payables during the GFC period.

6.1. *Trade Credit and the COVID-19 Crisis*

In this section, we describe a model currently in use by Small- and Medium-Sized Enterprises (SMEs) in Taiwan. The COVID-19 pandemic outbreak forced changes in trade credit management. Managers need to answer the following essential question: will the economic uncertainty affect the speed at which the firm adjusts to the target trade credit ratio? Online retailers have also endeavoured to increase the willingness of customers to place an order by addressing the risk-adjusted return on loans (direct fiscal transfers to borrowers to help reduce their credit risk; moratoriums on loan payments). The changes in the price of a product play a vital role in customers choosing the right kinds of products, and costs are sometimes affected by the fraction of the payment delayed. Another change in the strategy for managing receivables from customers is the discount rate policy. Sales were discontinued at any price, which was connected to the offering of additional discounts or extensions to trade credit. During the COVID-19 pandemic, retailers (suppliers) expected to quickly receive payment from customers (retailers). Government assistance for firms comes in the form of loan guarantees that increase firms' access to credit as a way of loosening liquidity constraints. On the demand side, trade credit represents the firm's access to capital, especially for SMEs. The hybrid (trade credit and discount rate) policy responses to COVID-19 may entice firms provided that the trade credit is lower in periods of less-restrictive bank credit. However, given the high degree of integration of supply chains worldwide, multilateral collaboration and coordinated interventions among economies are imperative to ensure no disruptions in supply chains, help financially constrained businesses survive the pandemic, and minimize unfavorable consequences on industrial structures in the long term.

The most common terms for the use of trade credit require a retailer to make a payment within 7, 30, 60, 90, or 120 days. A percentage discount is applied if payment is made before the date agreed upon in the terms. The aim of these trade credit activities is to build long-term relationships with customers and suppliers. Figure 6 shows the trade finance in emerging markets during COVID-19. The retailer has offered a variety of trade credit agreements and the contract consists of six items: (1) a financing arrangement for the customer; (2) the customer repays the lender on the terms of the original payment (e.g., 60 days); (3) the lender pays the retailer upon approval of the invoice; (4) a financing arrangement for the manufacturer; (5) the manufacturer repays the lender on the terms of the original payment (e.g., 90 days); and (6) a commercial agreement.

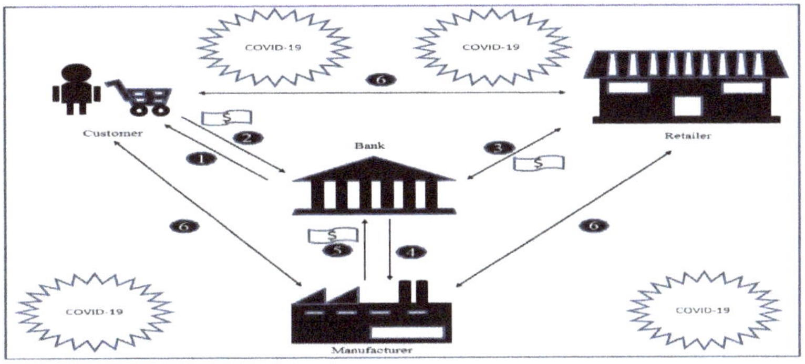

Figure 6. COVID-19 and trade finance in emerging markets.

6.2. Numerical Example

Base settings were established for the model by conducting interviews and surveys with relevant staff in the firm. In the current COVID-19 pandemic situation, the supplier offers advance payments to the firm so that they will not cancel the order. Due to the shortages in demand, the supplier offers a discount rate that is dependent on the number of installments. The firm also offers delays in payments for customers who do not have transportation and goods available. The values presented here were altered to preserve the confidentiality of the commercial information.

6.3. Sensitivity Analysis

The numerical example presented in Tables 2 and 3 were used to assess the effects of changes to system parameters (A, D, P, p, M, N, L, α, r, θ, c, and h) on the values T^*, $TVC_1(T^*_{11})$, $TVC_1(T^*_{12})$, and $TVC_1(L-N)$. Each parameter was adjusted separately (i.e., the other parameters were left unchanged) by +50%, +25%, −25%, or −50%. Our analytical results in Table 4 permit the following interesting observations and managerial insights that could be used to guide decision-making:

- The effect of decreasing the cost parameters (A, D, P, and N) would lead to a decrease in the total cost per unit time. In other words, if the costs could be reduced, then the enterprise would be able to earmark more money for the downstream trade credit period. This would also lead to a corresponding indirect increase in total profit per unit time due to decreased overall costs and/or increased sales;

- The effect of decreasing the change to θ on the value of $TVC_1(T^*_{11})$, $TVC_1(T^*_{12})$, and $TVC_1(L-N)$ is minimal; that is, a decrease of 22.472%, 22.431% and 21.61%. This indicates that attempts to increase total profits per unit time should focus on lowering the deterioration of items;
- In terms of holding cost parameters, increasing the values of the parameter h led to a corresponding decrease in T^*. This is an indication that the length of replenishment cycle times could be shortened to prevent an increase in holding costs;
- Decreasing the cost parameters (α, r) would lead to increase in the total cost per unit time. This indicates that if there were an increase in the cash discount rate, then the firm should use offers and discounts to drive customer loyalty and sales. Nonetheless, the amount spent on cash discounts could be increased to stimulate demand;
- An increase in the defect parameter (I_k or I_e) led to a corresponding increase in the total cost per unit time. This is an indication that the manufacturer can accumulate revenue by selling items and by earning interest and interest charges to reduce their finance risk.

Table 2. Let us consider an inventory system with the following data for Example 1–3.

Example 1	$A = 200$ $c = 50$ $r = 0.05$ $N = 0.05$	$D = 2000$ $h = 15$ $\alpha = 0.5$ $L = 0.08$	$P = 4000$ $I_k = 0.15$ $\theta = 0.05$	$p = 75$ $I_e = 0.1$ $M = 0.1$
Example 2	$A = 1000$ $c = 50$ $r = 0.05$ $N = 0.05$	$D = 1500$ $h = 15$ $\alpha = 0.5$ $L = 0.08$	$P = 4000$ $I_k = 0.15$ $\theta = 0.05$	$p = 75$ $I_e = 0.1$ $M = 0.1$
Example 3	$A = 1000$ $c = 50$ $r = 0.05$ $N = 0.05$	$D = 1500$ $h = 5$ $\alpha = 0.5$ $L = 0.02$	$P = 4000$ $I_k = 0.15$ $\theta = 0.05$	$p = 75$ $I_e = 0.1$ $M = 0.25$

Table 3. The optimal results of T^* and TVC^*.

Example	Conditional Expressions	Results	
		T^*	$TVC(T^*)$
1	$\Delta_1 = -185.43$, $\Delta_2 = -97.43$, $\Delta_3 = -159.38$, $\Delta_4 = -37.50$	0.35711	7661.41
2	$\Delta_1 = -85.43$, $\Delta_2 = 2.59$, $\Delta_3 = -59.38$, $\Delta_4 = 62.50$	0.33411	7691.44
3	$\Delta_1 = 0$, $\Delta_2 = 280.00$, $\Delta_3 = 285.78$, $\Delta_4 = 651.40$	0.35141	8349.57

Table 4. Results of Example 1 for three trade credit policies.

	Case 1		Case 2		Case 3	
Parameter	T	$TVC_1(T^*_{11})$	T	$TVC_1(T^*_{12})$	T	$TVC_1(L-N)$
A +50%	0.43732	8920.16	0.43358	8960.96	0.43358	9619.08
+25%	0.39924	8322.47	0.39585	8358.14	0.39585	9016.26
−25%	0.30926	6911.08	0.30671	6934.82	0.30671	7592.94
−50%	0.25248	6021.02	0.25248	6037.46	0.25047	6695.59
D +50%	0.30695	9612.78	0.30695	9647.90	0.30415	10635.1
+25%	0.32758	8684.81	0.32471	8684.81	0.32471	9540.12
−25%	0.40230	6512.89	0.39905	6539.96	0.39905	7033.55
−50%	0.48074	5181.66	0.48074	5205.02	0.47700	5534.08

Table 4. *Cont.*

	Case 1		Case 2		Case 3	
Parameter	T	$TVC_1(T_{11}^*)$	T	$TVC_1(T_{12}^*)$	T	$TVC_1(L-N)$
P +50%	0.34574	7840.23	0.34071	8718.87	0.34303	8526.90
+25%	0.35018	7769.22	0.34501	8647.86	0.34735	8456.47
−25%	0.36948	7478.04	0.36370	8356.77	0.36614	8167.84
−50%	0.39784	7096.23	0.39109	7975.42	0.39365	7789.78
p +50%	0.35527	7632.58	0.32444	8785.97	0.32631	8767.22
+25%	0.35619	7647.01	0.33727	8667.83	0.33936	8563
−25%	0.35802	7675.76	0.36820	8401.34	0.37093	8124.68
−50%	0.35893	7690.08	0.38719	8250.11	0.39036	7887.00
M +50%	0.35711	7661.40	0.35173	8540.05	0.35411	8349.57
+25%	0.35711	7661.40	0.35173	8540.05	0.35411	8349.57
−25%	0.35711	7661.40	0.35173	8540.05	0.35411	8349.57
−50%	0.35711	7661.40	0.35173	8540.05	0.35411	8349.57
N +50%	0.35713	7795.34	0.35174	8677.33	0.35405	8482.32
+25%	0.35712	7728.44	0.35174	8608.72	0.35408	8416.00
−25%	0.35708	7594.25	0.35174	8471.31	0.35412	8283.01
−50%	0.35705	7526.97	0.35174	8402.51	0.35413	8216.32
L +50%	0.35680	7229.18	0.35170	8094.87	0.35412	8240.08
+25%	0.35697	7445.61	0.35171	8317.54	0.35411	8294.83
−25%	0.35720	7876.57	0.35174	8762.40	0.35410	8404.28
−50%	0.35725	8091.11	0.35174	8984.58	0.35409	8458.98
α +50%	0.35704	7526.73	0.35257	7571.62	0.35257	8238.18
+25%	0.35707	7594.07	0.35333	7631.55	0.35333	8293.89
−25%	0.35714	7728.74	0.35488	7751.31	0.35488	8405.22
−50%	0.35717	7796.08	0.35566	7811.15	0.35566	8460.84
r +50%	0.36116	6796.87	0.35653	6842.58	0.35653	7492.26
+25%	0.35912	7229.23	0.35531	7267.04	0.35531	7920.95
−25%	0.35513	8093.41	0.35291	8115.78	0.35291	8778.12
−50%	0.35319	8525.24	0.35173	8540.05	0.35173	9206.61
c +50%	0.29967	23,976.9	0.31579	23,775.70	0.31579	24,594.2
+25%	0.32450	15,842.7	0.33329	15,743.40	0.33329	16,481.7
−25%	0.40257	−583.856	0.37941	−384.116	0.37941	193.853
−50%	0.47229	−8923.05	0.41111	−8488.82	0.41111	−7991.01
h +50%	0.34597	−6301.60	0.34325	−6273.03	0.34325	−5614.90
+25%	0.35140	680.618	0.34855	709.905	0.34855	1368.03
−25%	0.36310	14,640.7	0.35994	14,671.5	0.35994	15,329.6
−50%	0.36940	21,618.4	0.36607	21,650.1	0.36607	22,308.2
θ +50%	0.34129	17,346.3	0.33868	17,374.3	0.33868	18,032.4
+25%	0.34894	13,448.1	0.34615	13,477.0	0.34615	14,135.1
−25%	0.36585	−1899.49	0.36261	−1868.3	0.36261	−1210.17
−50%	0.36585	−1899.49	0.37174	−20,859.4	0.37174	−20,201.3
I_k +50%	0.30991	8283.38	0.32794	8069.73	0.32794	8888.17
+25%	0.33090	7987.53	0.34026	7884.91	0.34026	8623.20
−25%	0.39109	7296.58	0.36981	7488.20	0.36981	8066.16
−50%	0.43758	6879.97	0.38782	7273.79	0.38782	7771.60
I_e +50%	0.35527	7690.58	0.32631	7940.35	0.32631	8767.22
+25%	0.35619	7675.01	0.33936	7820.81	0.33936	8563.31
−25%	0.35802	7647.76	0.37093	7550.93	0.37093	8124.68
−50%	0.35893	7632.08	0.39036	7397.62	0.39036	7887.00

Figure 7 compares the full delay in payments policy with the cash discount policy. It indicates that the cash discount policy in a general supply chain model in the COVID-19 situation is determined by each member's purchase quantity and price.

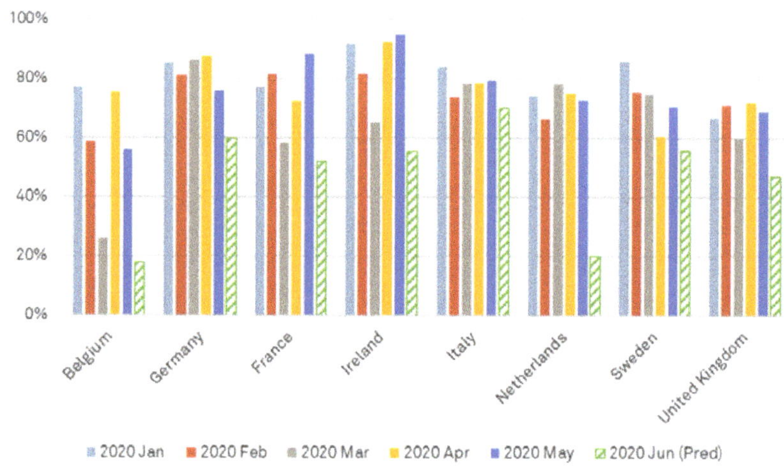

Figure 7. Minimum measures in the COVID-19 situation under the two policies.

7. Managerial Insights

In real-world business, a firm's size will affect the trade credit supply and demand side. On the demand side, the frequency of use of external financing, the proportion of credit sales, and the day sales outstanding are conditioning factors of the volume of credit purchases (for trade credit demand). On the supply side, the trade credits are conditioned by cash flow generation and by the frequency of the use of loans. Next, we describe the impact of COVID-19 on business operations, the economy, and employment at the beginning of the crisis. To help firms affected by the COVID-19 pandemic to return to normal operations, banks have implemented major policies, including cutting policy rates (the discount rate on accommodations with collateral) and providing a special accommodation facility to support bank credit for SMEs. Herein, pricing is a crucial element of business, and costs are sometimes affected by the discount rate. Are companies able to pay their trade credits on time? Figure 8 indicates the percentage of trade credit balances being paid on time in each country. The percentage of on-time payments was 32%, 23%, and 16% (lower than February's values). In this paper, we explored some important managerial insights that could help managers make decisions during the post-COVID-19 recovery period:

(i) The retailer should always examine the probability that a firm will default on its suppliers once a lockdown has been imposed, which varies depending on the degree of reliance on trade credit financing. A suggestion has been made for the retailer to be legally mandated to shut down during the first two months of the pandemic, which experienced by far the highest increase in defaults induced by trade credit payment obligations that had built up prior to the crisis.

(ii) The supplier should offer early payment discounts in order to minimize late payments, increase customer loyalty, maximize profits, and improve supplier relationships. A suggestion has been made for the manager to be able to choose to implement a discount period with a fixed percentage of savings off of an item.

(iii) As the discount is offered for the advance payment only, the retailer should often use this oppoortunity to intensify profits. A suggestion has been made for the manager to be able to choose the effective interest rate through the use of early payment discount terms.

(iv) In the COVID-19 period, suppliers should offer retailers an estimate of the payment amount and the due date through a loan service; for instance, coronavirus-related loan forgiveness options lawfully and duly declared during a COVID-19 pandemic national emergency.

(v) SMEs often have a limited number of suppliers. Firms are particularly vulnerable to the disruption of business networks and supply chains. Connections to larger operators (e.g., MNEs) and the outsourcing of business services are critical to their performance.

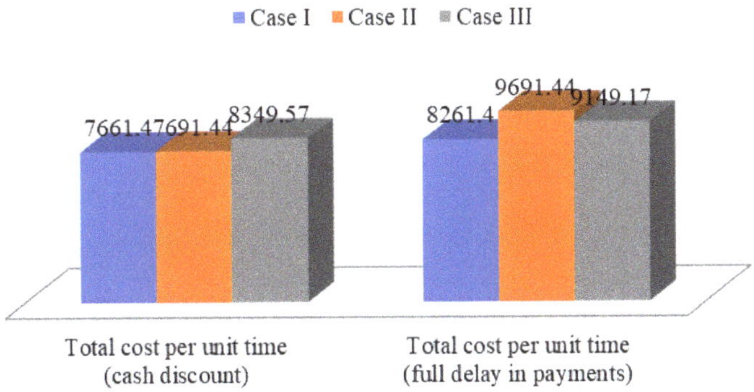

Figure 8. The percentage of trade credit balances being paid on time in each country (Source: S&P Global Market Intelligence, 5 June 2020. For illustrative purposes only).

8. Conclusions

While the production of goods and services is either reduced or paused temporarily, retailers should continue to pay at-risk suppliers to ensure cash flow and supplier survival. In this paper, we provided a hybrid trade credit policy to stimulate supplier–retailer business recovery during the COVID-19 period. Here, an alternative strategy to sustain business relationships through a hybrid payment system and discount facility considering the fraction of delayed payments was proposed. Two issues that need to be considered in this regard are: (1) due to the reduced default risk, the retailer should only provide a full trade credit policy to his/her customers with good credit; and (2) to reduce the risk of cash flow shortages and bad debt, the supplier should offer credit terms mixing a cash discount and trade credit to the retailer. Here, the supplier is likely to focus on maintaining business relationships through a hybrid payment system and discount rate policy. For example, the supplier may agree to a 2% discount off the retailer's purchasing price if payment is made within 120 days (during the COVID-19 period).

The results of this paper show that the retailer can optimize the replenishment cycle, discount rate, and time of prepayment for export items. Furthermore, we established retailer ordering policies that are given as solution procedures to determine the optimal solution under various conditions and provide a simple way to determine the optimal replenishment cycle time. The results of this paper clearly support the notion that an increase in the retailer's total cost will occur when discount rate, prepayment, and trade credit strategies are implemented wisely. The analytical formulations of the problem on the general framework described have been given. Despite the transition to cash sales in SMEs, sales with a large amount of trade credit were strongly limited, especially for new customers. In practice, suppliers allow customers a fixed period in which to settle the payment without penalty in order to increase sales and reduce on-hand inventory. The resulting nonlinear model was solved by the mathematical 12.0.0 software, and numerical examples were presented in order to illustrate the model. Demand patterns constitute an important topic to be explored in future research.

Author Contributions: Conceptualization, Y.-F.H. and P.R.; methodology, M.-W.W.; software, M.-W.W.; validation, P.R., Y.-F.H. and M.-W.W.; formal analysis, M.-W.W.; investigation, P.R.; resources, M.-W.W.; data duration, M.-W.W.; writing—original draft preparation, P.R., M.-W.W. and Y.-F.H.; writing—review and editing, P.R., M.-W.W. and Y.-F.H.; visualization, Y.-F.H.; supervision, Y.-F.H.; project administration, Y.-F.H. All authors have read and agreed to the published version of the manuscript.

Funding: Supported by Education Sciences Planning of Guangdong, China No. 420N28; National Science Foundation of China No. 72062010; Foundation for Specialty Innovation in Higher Education of Guangdong, China Philosophy and Social Sciences No. 2020WTSCX125 No. 2015WTSCX060; Education Sciences Planning of Guangdong, China No. 2021GXJK354; Guangxi Zhuang Autonomous Region Office for Philosophy and Social Sciences, Project; Philosophy and Social Sciences of Guangxi Zhuang Autonomous Region Office, China, in 2021 No. 21BGL012.

Institutional Review Board Statement: Not applicable.

Informed Consent Statement: Not applicable.

Data Availability Statement: Not applicable.

Acknowledgments: The researchers would like to thank University of Electronic Science and Technology of China for funding publication of this project.

Conflicts of Interest: The authors declare that there are no conflicts of interest.

References

1. Harris, F.W. How many parts to make at once. *Oper. Res.* **1990**, *38*, 934–1139. [CrossRef]
2. Goyal, S.K. Economic order quantity under conditions of permissible delay in payments. *J. Oper. Res. Soc.* **1985**, *36*, 335–338.
3. Teng, J.T. On the economic order quantity under conditions of permissible delay in payments. *J. Oper. Res. Soc.* **2002**, *53*, 915–918. [CrossRef]
4. Huang, Y.F. Retailer's replenishment policies under conditions of permissible delay in payments. *Yugosl. J. Oper. Res.* **2004**, *14*, 231–246. [CrossRef]
5. Huang, Y.F. Optimal retailer's replenishment decisions in the EPQ model under two levels of trade credit policy. *Eur. J. Oper. Res.* **2007**, *176*, 1577–1591. [CrossRef]
6. Huang, Y.F.; Hsu, K.H. An EOQ model under retailer partial trade credit policy in supply chain. *Int. J. Prod. Econ.* **2008**, *112*, 655–664. [CrossRef]
7. Hsieh, T.P.; Chang, H.J.; Dye, C.Y.; Weng, M.W. Optimal lot size under trade credit financing when demand and deterioration are fluctuating with time. *Int. J. Inf. Manag. Sci.* **2009**, *20*, 191–204.
8. Liao, J.J. An EOQ model with non-instantaneous receipt and exponentially deteriorating items under two-level trade credit. *Int. J. Prod. Econ.* **2008**, *113*, 852–861. [CrossRef]
9. Teng, J.T.; Chang, C.T. Optimal manufacturer's replenishment policies in the EPQ model under two levels of trade credit policy. *Eur. J. Oper. Res.* **2009**, *195*, 358–363. [CrossRef]
10. Min, J.J.; Zhou, Y.W.; Zhao, J. An inventory model for deteriorating items under stock-dependent demand and two-level trade credit. *Appl. Math. Model.* **2010**, *34*, 3273–3285. [CrossRef]
11. Chen, L.H.; Kang, F.S. Coordination between vendor and buyer considering trade credit and items of imperfect quality. *Int. J. Prod. Econ.* **2010**, *123*, 52–61. [CrossRef]
12. Kreng, V.B.; Tan, S.J. The optimal replenishment decisions under two levels of trade credit policy depending on the order quantity. *Expert. Syst. Appl.* **2010**, *37*, 5514–5522. [CrossRef]
13. Lee, C.H.; Rhee, B.D. Trade credit for supply chain coordination. *Eur. J. Oper. Res.* **2011**, *214*, 136–146. [CrossRef]
14. Mathata, G.J. An EPQ-based inventory model for exponentially deteriorating items under retailer partial trade credit policy in supply chain. *Expert. Syst. Appl.* **2012**, *39*, 3537–3550. [CrossRef]
15. Soni, H.N.; Patel, K.A. Optimal strategy for an integrated inventory system involving variable production and defective items under retailer partial trade credit policy. *Decis. Support. Syst.* **2012**, *54*, 235–247. [CrossRef]
16. Ouyang, L.Y.; Chang, C.T.; Shum, P. The EOQ with defective items and partially permissible delay in payments links to order quantity derived algebraically. *Cent. Eur. J. Oper. Res.* **2012**, *20*, 141–160. [CrossRef]
17. Ouyang, L.Y.; Chang, C.T. Optimal production lot with imperfect production process under permissible delay in payments and complete backlogging. *Int. J. Prod. Econ.* **2013**, *144*, 610–617. [CrossRef]
18. Yang, C.Y.; Ouyang, L.Y.; Hsu, C.H.; Lee, K.L. Optimal replenishment decisions under two-level trade credit with partial upstream trade credit linked to order quantity and limited storage capacity. *Math. Probl. Eng.* **2014**, *2014*, 1–14. [CrossRef]
19. Chen, S.C.; Teng, J.T.; Skouri, K. Economic production quantity models for deteriorating items with up-stream full trade credit and down-stream partial trade credit. *Int. J. Prod. Econ.* **2014**, *155*, 302–309. [CrossRef]

20. Chen, S.C.; Teng, J.T. Inventory and credit decisions for time-varying deteriorating items with up-stream and down-stream trade credit financing by discounted cash flow analysis. *Eur. J. Oper. Res.* **2015**, *243*, 566–575. [CrossRef]
21. Giri, B.C.; Sharma, S. Optimal ordering policy for an inventory system with linearly increasing demand and allowable shortages under two levels trade credit financing. *Oper. Res.* **2016**, *16*, 25–50. [CrossRef]
22. Lashgari, M.; Taleizadeh, A.A.; Sadjadi, A.A. Ordering policies for non-instantaneous deteriorating items under hybrid partial prepayment, partial trade credit and partial backordering. *J. Oper. Res. Soc.* **2018**, *69*, 1167–1196. [CrossRef]
23. Sarkar, B.; Ahmed, W.; Choi, S.B.; Tayyab, M. Sustainable inventory management for environmental impact through partial backordering and multi-trade-credit-period. *Sustainability* **2018**, *10*, 4761. [CrossRef]
24. Maiti, A.K.; Maiti, M.K.; Maiti, M. Inventory model with stochastic lead-time and price dependent demand incorporating advance payment. *Appl. Math. Model.* **2009**, *33*, 2433–2443. [CrossRef]
25. Gupta, R.K.; Bhunia, A.K.; Goyal, S.K. An application of Genetic Algorithm in solving an inventory model with advance payment and interval valued inventory costs. *Math. Comput. Model.* **2009**, *49*, 893–905. [CrossRef]
26. Thangam, A. Optimal price discounting and lot-sizing policies for perishable items in a supply chain under advance payment scheme and two-echelon trade credits. *Int. J. Prod. Econ.* **2012**, *139*, 459–472. [CrossRef]
27. Taleizadeh, A.A.; Wee, H.M.; Jolai, F. Revisiting a fuzzy rough economic order quantity model for deteriorating items considering quantity discount and prepayment. *Math. Comput. Model.* **2013**, *57*, 1466–1479. [CrossRef]
28. Zhang, Q.; Tsao, Y.C.; Chen, T.H. Economic order quantity under advance payment. *Appl. Math. Model.* **2014**, *38*, 5910–5921. [CrossRef]
29. Tavakoli, S.; Taleizadeh, A.A. An EOQ model for decaying item with full advanced payment and conditional discount. *Ann. Oper. Res.* **2017**, *259*, 415–436. [CrossRef]
30. Taleizadeh, A.A.; Tavakoli, S.; San-José, L.A. A lot sizing model with advance payment and planned backordering. *Ann. Oper. Res.* **2018**, *271*, 1001–1022. [CrossRef]
31. Shah, N.H.; Jani, M.Y.; Chaudhari, U. Optimal replenishment time for retailer under partial upstream prepayment and partial downstream overdue payment for quadratic demand. *Math. Comput. Model. Dyn. Syst.* **2018**, *24*, 1–11. [CrossRef]
32. Khan, M.A.A.; Shaikh, A.A.; Konstantaras, I.; Bhunia, A.K.; Cárdenas-Barrón, L.E. Inventory models for perishable items with advanced payment, linearly time-dependent holding cost and demand dependent on advertisement and selling price. *Int. J. Prod. Econ.* **2020**, *230*, 107804. [CrossRef]
33. Taleizadeh, A.A.; Tavassoli, S.; Bhattacharya, A. Inventory ordering policies for mixed sale of products under inspection policy, multiple prepayment, partial trade credit, payments linked to order quantity and full backordering. *Ann. Oper. Res.* **2020**, *287*, 403–437. [CrossRef]
34. Khan, M.A.A.; Shaikh, A.A.; Konstantaras, I.; Bhunia, A.K.; Cárdenas-Barrón, L.E. The effect of advance payment with discount facility on supply decisions of deteriorating products whose demand is both price and stock dependent. *Int. Trans. Oper. Res.* **2020**, *27*, 1343–1367. [CrossRef]
35. Shao, X.; Meng, H. Decision-making research about cash discounts offed by Vendors. In Proceedings of the the 4th International Conference on Operations and Supply Chain Management, Hongkong & Guangzhou, China, 25–31 July 2010; pp. 107–111.
36. Huang, Y.F.; Chung, K.J. Optimal replenishment and payment policies in the EOQ model under cash discount and trade credit. *Asia. Pac. J. Oper. Res.* **2003**, *20*, 177–190.
37. Ouyang, L.Y.; Chen, M.S.; Chuang, K.W. Economic order quantity model under cash discount and payment delay. *Int. J. Inf. Manag. Sci.* **2002**, *13*, 1–10.
38. Yang, C.T. The optimal order and payment policies for deteriorating items in discount cash flows analysis under the alternatives of conditionally permissible delay in payments and cash discount. *Top* **2010**, *18*, 429–443. [CrossRef]
39. Yang, C.T.; Pan, Q.; Ouyang, L.Y.; Teng, J.T. Retailer's optimal order and credit policies when a supplier offers either a cash discount or a delay payment linked to order quantity. *Eur. J. Ind. Eng.* **2013**, *7*, 370–392. [CrossRef]
40. Feng, H.; Li, J.; Zhao, D. Retailer's optimal replenishment and payment policies in the EPQ model under cash discount and two-level credit policy. *Appl. Math. Model.* **2013**, *37*, 3322–3339. [CrossRef]
41. Shah, N.H.; Cárdenas-Barrón, L.E. Retailer's decision for ordering and credit policies for deteriorating items when a supplier offers order-linked credit period or cash discount. *Appl. Math. Comput.* **2015**, *259*, 569–578. [CrossRef]
42. Alshanbari, H.M.; El-Bagoury, A.A.H.; Khan, M.A.A.; Mondal, S.; Shaikh, A.A.; Rashid, A. Economic order quantity model with weibull distributed deterioration under a mixed cash and prepayment scheme. *Comput. Intell. Neurosci.* **2021**, *2021*, 1–16. [CrossRef]
43. Tripathi, R.P. Innovative approach of EOQ structure for decaying items with time sensitive demand, cash-discount, shortages and permissible delay in payments. *Int. J. Appl. Comput. Math.* **2021**, *7*, 1–16. [CrossRef]
44. Mashud, A.H.M.; Hasan, M.R.; Daryanto, Y.; Wee, H.M. A resilient hybrid payment supply chain inventory model for post COVID-19 recovery. *Comput. Ind. Eng.* **2021**, *157*, 1–15. [CrossRef]
45. Zhou, Y.W.; Wen, Z.L.; Wu, X.; Cheng, M.C. A single-period inventory and payment model with partial trade credit. *Comput. Ind.Eng.* **2015**, *90*, 132–145. [CrossRef]
46. Laitinen, E.K. Discounted cash flow (DCF) as a measure of startup financial success. *Theor. Econ. Lett.* **2019**, *9*, 2997–3020. [CrossRef]

47. Arcelus, F.J.; Shah, N.H.; Srinivasan, G. Retailer's response to special sales: Price discount vs. trade credit. *Omega* **2001**, *29*, 417–428. [CrossRef]
48. Stokes, J.R. Dynamic cash discounts when sales volume is stochastic. *Q. Rev. Econ. Financ.* **2005**, *45*, 144–160. [CrossRef]
49. Chung, K.J.; Liao, J.J. The optimal ordering policy in a DCF analysis for deteriorating items when trade credit depends on the order quantity. *Int. J. Prod. Econ.* **2006**, *100*, 116–130. [CrossRef]
50. Guariglia, A.; Mateut, S. Credit channel, trade credit channel, and inventory investment: Evidence from a panel of UK firms. *J. Bank. Financ.* **2006**, *30*, 2835–2856. [CrossRef]
51. Ho, C.H.; Ouyang, L.Y.; Su, C.H. Optimal pricing, shipment and payment policy for an integrated supplier-buyer inventory model with two-part trade credit. *Eur. J. Oper. Res.* **2008**, *187*, 496–510. [CrossRef]
52. Chang, C.T.; Ouyang, L.Y.; Teng, J.T.; Cheng, M.C. Optimal ordering policies for deteriorating items using a discounted cash-flow analysis when a trade credit is linked to order quantity. *Comput. Ind. Eng.* **2010**, *59*, 770–777. [CrossRef]
53. Chung, K.J.; Lin, S.D.; Srivastava, H.M. The inventory models for deteriorating items in the discounted cash-flows approach under conditional trade credit and cash discount in a supply chain system. *Appl. Math. Inf. Sci.* **2014**, *8*, 2103–2111. [CrossRef]
54. Wu, J.; Al-khateeb, F.B.; Teng, J.T.; Cárdenas-Barrón, L.E. Inventory models for deteriorating items with maximum lifetime under downstream partial trade credits to credit-risk customers by discounted cash-flow analysis. *Int. J. Prod. Econ.* **2016**, *171*, 105–115. [CrossRef]
55. Tripathi, R.P.; Singh, D.; Aneja, S. Inventory control model using discounted cash flow approach under multiple suppliers' trade credit and stock dependent demand for deteriorating items. *Int. J. Inv. Res.* **2019**, *5*, 210–223.
56. Vayas-Ortega, G.; Soguero-Ruiz, C.; Rojo-Álvarez, J.L.; Gimeno-Blanes, F.J. On the differential analysis of enterprise valuation methods as a guideline for unlisted companies assessment (I): Empowering discounted cash flow valuation. *Appl. Sci.* **2020**, *10*, 5875. [CrossRef]
57. De, S.K.; Mahata, G.C.; Maity, S. Carbon emission sensitive deteriorating inventory model with trade credit under volumetric fuzzy system. *Int. J. Intell. Syst.* **2021**, *36*, 7563–7590. [CrossRef]
58. Demir, B.; Javorcik, B. Trade finance matters: Evidence from the COVID-19 crisis. *Oxf. Rev. Econ. Policy* **2020**, *36*, S397–S408. [CrossRef]
59. Agca, S.; Birge, J.R.; Wang, Z.; Wu, J. The Impact of COVID-19 on Supply Chain Credit Risk. 2020, pp. 1–38. Available online: https://ssrn.com/abstract=3639735 (accessed on 1 March 2022).
60. Choi, T.M. Risk analysis in logistics systems: A research agenda during and after the COVID-19 pandemic. *Transp. Res. E Logist. Transp. Rev.* **2021**, *145*, 102190. [CrossRef]
61. Liu, Y.; Zhang, Y.; Fang, H.; Chen, X. SMEs' line of credit under the COVID-19: Evidence from China. *Small. Bus. Econ.* **2022**, *58*, 807–828. [CrossRef]
62. Luo, H. COVID-19 and trade credit speed of adjustment. *Financ. Res. Lett.* **2022**, 1–8, *in press*. [CrossRef]
63. Zimon, G.; Dankiewicz, R. Trade credit management strategies in SMEs and the COVID-19 pandemic—s case of Poland. *Sustainability* **2020**, *12*, 6114. [CrossRef]
64. Cambini, A.; Martein, L. Generalized convexity and optimization. *Lect. Notes Econ. Math. Syst.* **2009**, *616*. [CrossRef]

mathematics

MDPI

Article

Roadmap Optimization: Multi-Annual Project Portfolio Selection Method

Ran Etgar [1,*] and Yuval Cohen [2]

1 Department of Industrial Engineering, Faculty of Engineering, Ruppin Academic Center, Emek Hefer 4025000, Israel
2 Department of Industrial Engineering, Afeka College for Engineering, Mivtza Kadesh 38, Tel Aviv 6910717, Israel; yuvalc@afeka.co.il
* Correspondence: ranetgar@ruppin.ac.il

Abstract: The process of project portfolio selection is crucial in many organizations, especially R&D organizations. There is a need to make informed decisions on the investment in various projects or lack thereof. As the projects may continue over more than 1 year, and as there are connections between various projects, there is a need to not only decide which project to invest in but also when to invest. Since future benefits from projects are to be depreciated in comparison with near-future ones, and due to the interdependency among projects, the question of allocating the limited resources becomes quite complex. This research provides a novel heuristic method for allocating the limited resources over multi-annual planning horizons and examines its results in comparison with an exact branch and bound solution and various heuristic ones. This paper culminates with an efficient tool that can provide both practical and academic benefits.

Keywords: metaheuristics; project selection; portfolio management; resource; R&D; roadmap; program management

MSC: 90B35

Citation: Etgar, R.; Cohen, Y. Roadmap Optimization: Multi-Annual Project Portfolio Selection Method. *Mathematics* **2022**, *10*, 1601. https://doi.org/10.3390/math10091601

Academic Editors: Humberto Rocha and Ana Maria Rocha

Received: 20 March 2022
Accepted: 27 April 2022
Published: 8 May 2022

Publisher's Note: MDPI stays neutral with regard to jurisdictional claims in published maps and institutional affiliations.

1. Introduction

Most project-based organizations are faced with a long list of proposed projects that compete for a limited set of resources such as money, manpower, and equipment [1]. Project portfolio selection (PPS) aims to find which projects an organization should take [2–4]. Needless to say, as the decision to allocate and prioritize projects today affects the organization's competitive position in the future [5], and the decisions of initiation (and termination) of projects are of a strategic nature since they involve the commitment of substantial enterprise resources [6], it is recognized worldwide that there is a need to manage projects as an overall portfolio [7] and not as separated projects [8].

Evidently, organizations wish to maximize their return on investment when selecting projects [9], and therefore, the selection process should be based on criteria that take into account this objective function [10].

The operational research problem of PPS was defined [11] as the situation where several projects are available for investment. They are different in their resource needs (both resource types and resource demand level).

The current research and method are lacking in two aspects: they do not take into consideration the time value of the projects (i.e., contribution depending on the projects' completion time), and they also do not consider the dependence between the projects (i.e., precedence and competition over resources). This research aims to present a novel formulation of the problem, namely one that incorporates theses aspects. The article also provides an exact solution algorithm (branch and bound) that can solve small to medium problems. However, due to the NP-hard nature of the problem, large-scale problems require a different approach. Thus, several metaheuristic algorithms are proposed and analyzed.

2. Literature

The research in the area of project portfolio selection has been growing intensively in the past decade, with intensive proliferation of research articles tackling a myriad of variations of this problem. Some of this proliferation was summarized in several review papers. For example, Frey and Buxmann [12] reviewed the literature on portfolio selection of IT projects. Weissenberger-Eibl and Teufel [13] provided a strategic and political review on project selection. Padhy [14] and Condé and Martens [15] reviewed six-sigma project selections. Danesh et al. [16] provided a broad review of multi-criteria portfolio management, and Mohagheghi et al. [17] reviewed models, uncertainty approaches, solution techniques, and case studies.

The simplest model of the project selection problems is single-attribute optimization under resource constraints. This problem resembles the well-known knapsack algorithm [18,19]. This problem was primarily extended to reflect synergies and uncertainty [18]. The main extensions to this type of optimization were with interactions and under uncertainty [20,21].

Another classical model of the project selection problem is financial project portfolio selection, which is based on mean profit vs. profit variance [22]. However, this model of portfolio selection is rooted in investment in the stock markets, and its use is indeed more suited to investment decisions in a portfolio of stocks and other financial assets. Thus, only a few papers on project portfolio selection adopted this modeling approach [9]. This model's assumptions and characteristics are more suited to pure investment decisions than to a company or organization decision problem.

Biobjective models were a natural evolution of single-objective optimization [20,23]. While multi-objective optimization became the research mainstay of project portfolio selection, biobjective models, due to their simplicity, still attract some research attention [19,23–25].

The multi-objective optimization models followed the aforementioned models [26–30]. The literature on multi-objective project portfolio selections proliferated [31–36] and became the main branch of the project portfolio research [16], and it is still prevalent [37]. Within this multi-objective framework, robustness became a prevalent requirement and an objective to address the uncertainties of project portfolio selection [29,32].

During the last two decades, fuzzy logic started to play an important role in decision making, and in the past decade, it made its way into the portfolio selection literature. For example, Perez and Gomez [33] and Perez et al. [38] used fuzzy constraints between projects, while others [1,35,39,40] used them in the objective function.

Another branch of research gathered both project scheduling and project selection into a single decision frame (some examples are in [19,30,41–43]).

While the above discussion dealt with project selection in technical terms, strategic project portfolio selection presents a very different approach [25,44–46]. Killen et al. [4] identify three strategic perspectives of project portfolio selection: (1) the resource-based view, (2) the dynamic capabilities view, and (3) the absorptive capacity view. Kaiser et al. [47] stressed the role of structural alignment of selected projects with the organization's values, vision, and strategy. Kopmann et al. [46] suggested fostering both deliberate and emergent strategies. Finally, Guo et al. [25] suggested balancing strategic contributions and financial returns.

While the existing reviews cover their relevant part of the literature and suggest some classifications of the project selection problems, a more formal and general classification system can contribute to them. We suggest a system that would be close to Kendall's notation in queuing theory [48]. To foster a discussion for filling this gap, we present here our initial attempt at classification of project portfolio selection's main problem types using the major characteristics of these problems. It is hoped that the suggested classification scheme will initiate a discussion (or even a debate) which will culminate in an agreed-upon standard classification method. In Figure 1, we propose a classification method for the PPS problem. The suggestion is to classify the problems by their objective type, the solution method used to make the selection, the nature of the data, and the constraints. Therefore, the proposed method has four classifiers:

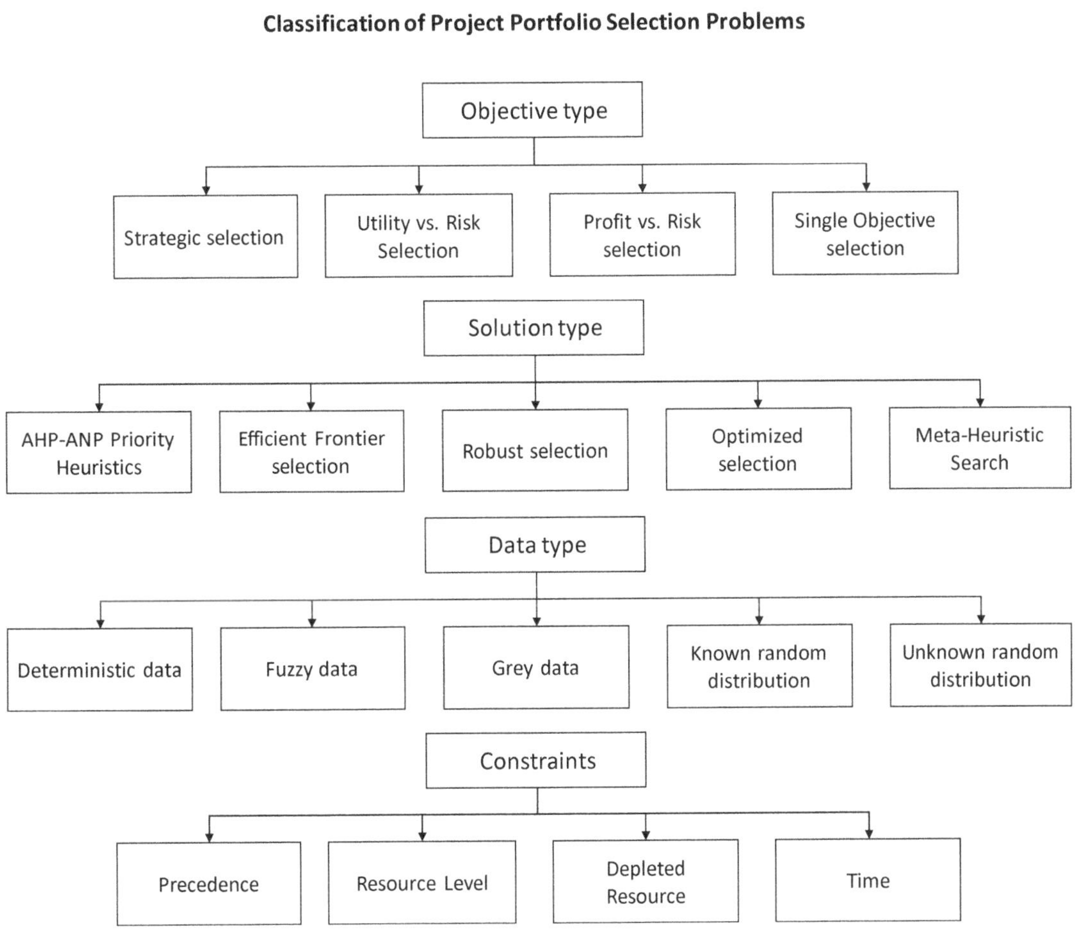

Figure 1. Project portfolio selection classification scheme.

X— Objective type: single-attribute selection, multiple-attribute selection, fitness function, profit vs. risk selection, utility vs risk selection, and strategic selection;

Y— Solution type: optimized selection, robust selection, efficient frontier, AHP/ANP, and priority-based;

Z— Data type: deterministic data, fuzzy data, and stochastic data;

W—Constraint characteristics: a combination of letters that testify for the existence of the characteristic constraint as follows: P = precedence constraints, RL = resource-level constraints, DR = depleted resource constraints, and T = time constraints.

Some examples of this classification scheme are shown in Table 1.

Table 1. Examples of the suggested classifications.

	Reference	Objective Type	Solution Type	Data Type	Constraints Combination
1	[33]	Multi-objective	Efficient frontier	Fuzzy	P, RL, DR, T
2	[49]	Strategic selection	None	Deterministic	P, RL, DR, T
3	[50]	Single-objective selection	Optimized	Deterministic	P, RL, DR, T
4	[51]	Single-objective selection	Optimized	Deterministic	P, RL, DR, T
5	[52]	Single-objective selection	Optimized	Known random distribution	P, RL, DR, T
6	[53]	Profit vs. risk selection	Optimized	Known random distribution	None
7	[32]	Multi-objective	Robust selection	Deterministic	P, RL, DR, T

As stated above, enhancements to this classification scheme and even challenges are welcome as part of future research.

The problem researched in this article is of a single objective (maximum value); the data are assumed to be deterministic, and no depleted resources are concerned. The article utilizes two solution types: optimal for small-to-medium-sized problems and metaheuristic searching for larger ones. Therefore, this problem should be classified according to the hierarchical classification of Figure 1 as multi-attribute, optimized, and deterministic.

This article does not consider random, fuzzy, or gray data.

3. Problem Description

The problem deals with an ongoing situation of R&D to develop projects. The need is to decide which projects should be scheduled for each year in the planning horizon. As a company has a limited amount of resources each year, it is impossible to perform all projects at once, and therefore, there is a need to schedule less-lucrative projects for distant years. In the case of no constraints, all projects would have been planned for the first year. However, this is not the case. Two reasons compel postponing projects to future years:

- Resource availability: Each project requires a specific level of various resources. The assumption is that there cannot be a breach of the available level of each resource.
- Precedence: As part of the R&D project, the company acquires new capabilities that can be exploited for future projects.

3.1. Problem Assumptions

- Each project has a specific value to the company. This value can be measured in the same units (typically profit, measured in dollars).
- The value of the project to the company depreciates as a function of time; that is, each year can be assigned a corresponding coefficient that expresses the depreciated value. (For example, a project assigned to year 1 has a value of 100. The same project assigned to year 2 has a value of 80. If assigned to year 3, it has an even lower value, etc.).
- All relevant resources are renewable (e.g., work hours). For each year, there are new levels of resources available. The amount of a resource that was not consumed in year n cannot be used in year n + 1.
- Each project has given levels of resources needed for its completion (e.g., programmer hours or QA hours).
- Technical precedence dependencies exist between the projects. Thus, project x can be performed based on the technology performed for project y.

3.2. Problem Notations

To assist understanding of the formulation, Table 2 depicts the notations used for the formulation.

Table 2. Problem notations.

Notation	Definition
$\overline{\mathcal{Y}}$	A vector of all the depreciation values of the years. $y_i \in \overline{\mathcal{Y}}$ is the value of year i.
$\overline{\overline{\mathcal{R}}}$	A matrix containing the available resources at each year in the planning horizon. $r_{j,i} \in \overline{\overline{\mathcal{R}}}$ is the level of resource j at year i.
$\overline{\overline{\mathcal{Q}}}$	A matrix containing the required resources for each project. $q_{k,j} \in \overline{\overline{\mathcal{Q}}}$ is the requirement of resource j of project k.
$\overline{\mathcal{P}}$	A vector of all the contributions (values) of the projects. $p_k \in \overline{\mathcal{P}}$ is the value of project k.
$\overline{\overline{\mathcal{D}}}$	A matrix containing technical dependencies. $d_{k,m} \in \overline{\overline{\mathcal{D}}}$ has the following values: $d_{k,m} = \begin{cases} 1 & if\ project\ m\ depends\ on\ project\ k \\ 0 & otherwise \end{cases}$
$\overline{\overline{\mathcal{X}}}$	Decision variable matrix. $x_{k,i} \in \overline{\overline{\mathcal{X}}}$ has the following values: $x_{k,i} = \begin{cases} 1 & if\ project\ k\ is\ to\ be\ completed\ in\ year\ i \\ 0 & otherwise \end{cases}$
$\mathcal{V}$	Total portfolio value.
$\overline{\overline{Z}}$	Auxiliary variables, denoting the years the project is in process.
$\overline{\overline{\mathcal{W}}}$	Auxiliary decision variables denoting the level of resources used by a project in any given year.
N	Number of projects in the examined portfolio.
H	Planning horizon (number of years).
R	Number of resource types.
$\overline{G}$	Genotype vector.
i,j,k,l,m,n	Indices: i,l—Year index. j—Resource index. k,m,n,p—Project index.

3.3. Problem Formulation

The objective function is for maximizing the cumulative value of the project portfolio such that

$$\max \mathcal{V} = \sum_{\forall y_i \in \overline{\mathcal{Y}}} \sum_{\forall p_k \in \overline{\mathcal{P}}} \sum_{\forall x_{k,i} \in \overline{\overline{\mathcal{D}}}} y_i p_k x_{k,i} \tag{1}$$

Each project is to end in one year only, so the relevant constraints are

$$\sum\nolimits_{\forall i} x_{k,i} = 1 \ \forall k \in \{1,2,\ldots,N\} \tag{2}$$

Since a project can stretch over more than one year (i.e., start before its final year), there is a need for an auxiliary set of variables ($\overline{\overline{Z}}$), denoting the years a project can use resources (i.e., the project cannot use resources after its ending):

$$z_{k,i} \le \sum\nolimits_{l=i}^{H} x_{k,l} \ \forall k \in \{1,2,\ldots,N\} \tag{3}$$

Thus, $z_{k,i}$ can have the maximum value of 1 for each year until its ending year and 0 onward.

To ensure that the project spreads over consecutive years, we use

$$z_{k,i} \leq z_{k,i+1} \; \forall k \in \{1,2,\ldots,N\}, \forall i \in \{1,2,\ldots,H-1\} \tag{4}$$

This notification enables the setting of the resource-consuming variables ($\overline{\overline{\mathcal{W}}}$), denoting the level of resource j consumed by project k at year i:

$$w_{k,i,j} \leq q_{k,j} z_{k,i} \; \forall k \in \{1,2,\ldots,N\}, \;\; \forall i \in \{1,2,\ldots,H\}, \;\; \forall j \in \{1,2,\ldots,P\} \tag{5}$$

Thus, a project may consume part of the resources it needs in year n and part of year n + 1 (e.g., use 2023's budget and 2024's budget). Additionally, each project must consume all the needed resources:

$$\sum_{i=1}^{H} w_{k,i,l} = q_{k,j} \; \forall k \in \{1,2,\ldots,N\}, \forall l \in \{1,2,\ldots,H\} \tag{6}$$

To prevent over-consumption of resources at any given year, the resource level constraints are

$$\sum_{\forall k} \sum_{\forall i} w_{k,j} \leq r_{j,i} \; \forall j \in \{1,2,\ldots,P\} \tag{7}$$

Finally, the technical dependencies among the projects are expressed as

$$z_{k,i} \geq z_{m,i} \; \forall d_{k,m} = 1, \forall i \in \{1,2,\ldots,H\} \tag{8}$$

A small example to help illustrate this formulation is detailed in the Appendix A.

4. Problem Complexity and Exact Solution

Although the formulation depicted in Section 3.3 is accurate and needed, it is of little contribution when trying to solve such problems. The described problem is NP-complete, yet exact solutions can be obtained via the branch and bound (B&B) method. Section 4.1 describes a B&B solution for this problem.

Since a B&B solution is practical for some of the problems, and the NP-completeness of the general problem hinders exact solutions, a practical and efficient metaheuristic solution is described in Section 5. This solution method is useful for large-sized problems and also provides an initial upper bound for the B&B algorithm.

4.1. Complexity

To prove the NP-completeness of the PPS, a reduction to a well-known problem is needed: the precedence constraint knapsack problem [54]. The reduction is performed as follows:

- Set the horizon (number of years) to 2. Thus, the projects are either assigned to the first year (i.e., inserted to the "knapsack") or to the second one (left out).
- Set the second year's value to zero ($y_2 = 0$).

As PPS is far more complicated than this, it is clear that PPS is of an NP-complete nature.

4.2. Branch and Bound Algorithm

An efficient B&B algorithm is based on the following components:

- A lower bound (LB): Any feasible solution can provide an LB. The algorithm can start with either a solution produced by the metaheuristic algorithm or a simple heuristic solution.
- An upper bound (UB): This is an efficient way to assess the maximal potential of a partial solution (i.e., branch) of the tree. The UB can serve as a convenient heuristic precedence rule (i.e., a rule to decide which branch to further develop first).
- A branching method: This is a way to create the net branches.

The following subsections describe the components of the algorithm.

4.2.1. Initial Solution

It is convenient (though not necessary) to start the run of the algorithm with a lower bound (LB). The higher the LB, the better. A reasonable algorithm should consider all important attributes of the projects, namely the resource requirements and precedence. The following algorithm can provide a decent initial solution:

1. For each resource type (j):
    1.1 Calculate for each project the ratio of the project value to its resource requirement; that is, $\theta_k = \frac{p_k}{q_{k,j}}$, where θ_k denotes the value that can be obtained from one unit of the resource by performing the project.
    1.2 For each project, calculate its total impact. The impact is calculated by aggregating the project's θ_k value and the values of all its successors (direct and indirect). For example, for project 1, the impact is $I_1 = \theta_1 + \theta_5 + \theta_7 + \theta_8$ (since projects 5, 7, and 8 are successors to project 1).
    1.3 Sort the projects in descending order by the total impact.
    1.4 Schedule the projects according to the order obtained in Step 1.3. Each project should be scheduled for the first available year.
    1.5 Calculate the total value obtained from the schedule of Step 1.4 $[\mathcal{V}(j)]$.
2. The solution is set to be $LB = \max\limits_{\forall j} \mathcal{V}(j)$.

4.2.2. Branching

The branching process is quite simple. Each "level" of the tree represents a year. Therefore, the tree depth can only be as deep as the planning horizon.

Each branch is simply a set of projects that fully utilize at least one of the resources available for that year; that is, when branching year i, each branch is a set J of all the unscheduled projects that fulfill the following requirements:

- There exists a resource type (j) for which

$$\sum_{\forall i \in J} q_{k,j} \geq r_{i,j} \tag{9}$$

- For every subset of J, denoted by J^- (i.e., $J^- \subset J$ and not $J^- = J$) and for every resource type (j), the following is true:

$$\sum_{\forall i \in J^-} q_{k,j} < r_{i,j} \tag{10}$$

The two conditions may look cumbersome, but all they mean is that the set J is a set that exploits one resource to the fullest, and any subset of J can be extended by adding project (s).

4.2.3. Bounding Rule

An efficient bounding rule should satisfy the following demands:

- Simply calculable: The rule should provide the result with a low complexity algorithm (otherwise, it provides no benefit).
- Low UB: To trim the tree as much as possible, the UB should be as low as possible.

The following algorithm provides the two demands. The algorithm is based on two relaxations of the problem: the first is ignoring the precedence constraints, and the second is ignoring the multi-resource nature of the problem (by dealing with one resource at a time).

1. For each resource type (j), the following should be performed:
    1.1 Ignore all resource requirements (other than resource j) and precedence constraints.

   1.2 The remaining problem is the multiple knapsacks problem. Solve the problem accordingly.
   1.3 As the multiple knapsacks problem is an NP-complete problem, when a project is scheduled to start in year i and finish in year $i+1$, calculate the obtained value of the project proportional to the year it was scheduled for.
   1.4 Calculate the total values of the projects (each according to its year) $UB(k)$.
2. Calculate the overall bound $UB = \min_{\forall k} UB(k)$.

The proposed algorithm is quite simple and rapid for small-scale problems.

5. Metaheuristic Search

Although the previous section describes an exact algorithm, it is impractical to apply this algorithm (or any exact algorithm) successfully to large-size problems. The algorithm provided in Section 4.2.1 may prove unsatisfactory for even medium- let alone large-scale problems.

To provide near-optimal solutions, a metaheuristic approach is suggested. The benefit of this approach is that (providing enough runtime) optimality is guaranteed. Since the first conception of metaheuristics, many approaches were suggested. The proposed solution is based on the CLONALG metaheuristics. This search method was developed by de Castro and Van Zuben [55]. This basic method can be applied to various scheduling problems, as long as the following requirements are provided [56]:

- Representation of the solution space (i.e., the "antibodies");
- An initial set of solutions.
- A procedure to create valid mutations in the antibodies (i.e., mutations creating new solutions that are included in the solution space).

Section 5.1 elaborates on the application of the first requirement to PPS. The second requirement application is provided in Section 5.2. Finally, several different mutation algorithms (third requirement) are detailed in Section 6.

5.1. Vector Representation

Although the formulation provided in Section 3 (and the notations of Table 2) is mathematically accurate, the presentation of the decision variables ($\overline{\overline{\mathcal{X}}}$) is impractical for the purposed CLONALG process. The main problem is that most possible instances of $\overline{\overline{\mathcal{X}}}$ are not feasible, either because they represent schedules that violate the dependency constraints or violate the resource constraints. An ideal representation is one in which every possible instance represents a feasible solution. Thus, the mutation process would never yield infeasible solutions. Another requirement is that all feasible solutions can be represented (i.e., the vector representation should spread the entire solution space) so the mutation process will not omit any possible solution.

To achieve this, a vector $\overline{G}$ is introduced. This vector represents the "genotype" of the solution (i.e., it is not the schedule itself), but from each possible instance of $\overline{G}$, a feasible "phenotype" ($\overline{\overline{\mathcal{X}}}$, the solution) can be derived. $\overline{G}$ is a vector of N natural numbers (from 1 to N), where g_k is the kth project to be scheduled.

The transformation from $\overline{G}$ to $\overline{\overline{\mathcal{X}}}$ ("genotype to phenotype translation") is performed as follows:

MAIN

1. For $k=1$ to N, do the following:
   1.1 If $\sum_{j=1}^{N} x_{k,j} = 0$ (i.e., g_k has not been scheduled yet): Run procedure "FindYear" with parameter k.
   1.2 If $\sum_{i=1}^{N} x_{k,j} = 1$ (i.e., g_k has already been scheduled): Continue.

FindYear(k)

1. For $m = 1$ to $k - 1$, do the following (without loss of generality, as it is assumed that if project k depends on project m, then $k > m$):
    1.1 If $d_{m,k} = 1$ and $\sum_{j=1}^{N} x_{m,j} = 0$ (project k depends on project m, and project m has not been scheduled), run procedure "SCHEDULE" with parameter m.
    1.2 Otherwlse (either project k does not depend on project m or project m has already been scheduled), continue.
2. Find the first year i in which g_k can be scheduled (has enough available resources and does not violate dependencies), and set $x_{k,i} = 1$.
3. Return

This simple algorithm provides the two requirements: (1) any genotype $\overline{G}$ can be converted to a feasible "phenotype" (feasible $\overline{\overline{\mathcal{X}}}$) through a simple process, and (2) all feasible solutions can be originated from a genotype $\overline{G}$.

5.2. Initial Solution Generation

Since the aforementioned procedure makes any vector containing the natural numbers from 1 to N to be a feasible representation of a solution, a procedure to generate an initial feasible solution is quite straightforward; any "scrambled" vector containing the numbers from 1 to N in random order will suffice. To achieve a set of these scrambled vectors, the following method was applied:

1. A matrix $\overline{\overline{\mathcal{S}}}$ with dimensions $N \times 2$ was created;
2. For i = 1 to N do the following:
    2.1 Set $s_{i,1} = i$.
    2.2 Set $s_{i,2} = U(0,1)$ (random number from the unit uniform distribution).

After this stage, the first column is filled with running numbers and the second with random numbers.

3 Sort $\overline{\overline{\mathcal{S}}}$ in ascending order according to the second column.

4 The first column of $\overline{\overline{\mathcal{S}}}$ is an initial solution vector ($\overline{G}$).

6. Mutation Generation

The presentation of the solution vector and the initial solution set generating set the ground for the central part of the CLONALG process: mutation generation and cloning. The mutations are generated by random change insertion to the genome vector ($\overline{G}$). Since the genotype vector describes the order of scheduling the projects, the mutation process will be carried out by changing this order. This section describes three approaches to the mutation process. The first and the second are "traditional" approaches, and the third one attempts to exploit clustering techniques to improve the search performance. An additional approach is also presented: a combination of the previous ones.

6.1. Minor Mutations

The most trivial and straightforward approach to changing the order of the vector members is by replacement. The simplest way is to randomly choose a project (member of the genotype vector) and replace it with its neighbor, as depicted in Figure 2. In this case, the fifth location (project 6) was chosen and replaced with its neighbor in the sixth location (project 3).

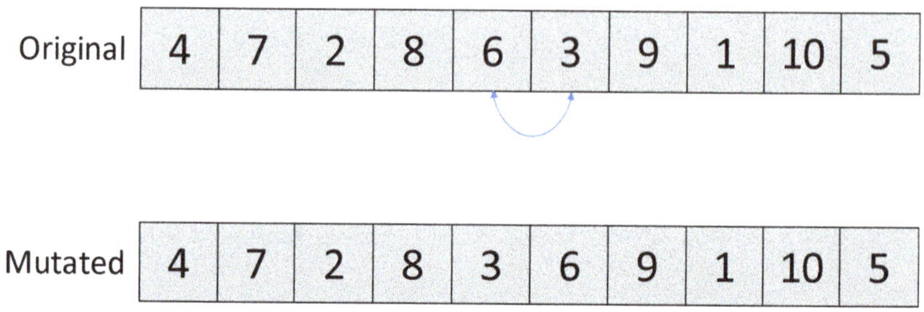

Figure 2. A simple mutation.

There is an advantage in small mutations. When in the vicinity of the optimal solution, a small mutation is less likely to cause damage and drift away from the optimal solution, as visualized in Figure 3. The small mutation, though in a wrong direction, is less likely to corrupt the solution value than the larger mutation in the correct direction.

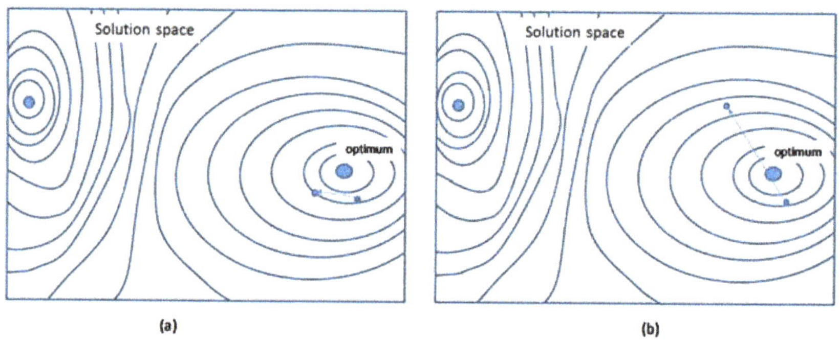

Figure 3. (**a**) A small mutation. (**b**) A large mutation.

6.2. Major Mutations

Though the minor mutations have the advantage of minimal damage when in the vicinity of the optimum, they are likely to provide only minimal improvement (i.e., many steps needed toward the optimum). To illustrate this, let us examine the case depicted in Figure 4.

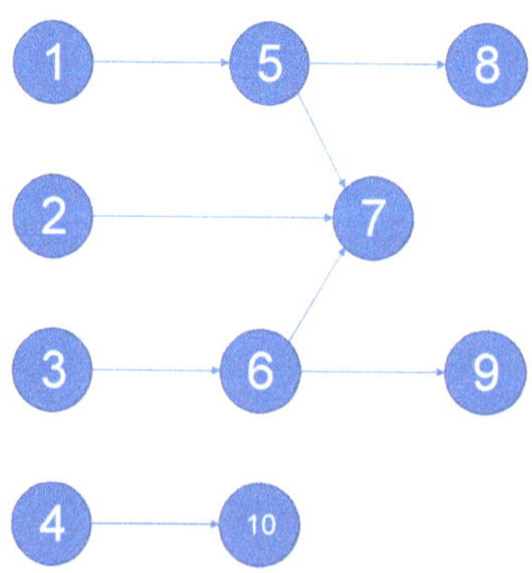

Figure 4. Projects' dependencies.

Project 7 has a high value and therefore should be scheduled ASAP. Let us assume a pre-mutated solution vector $\overline{G} = (4, 10, 1, 2, 3, 5, 6, 7, 8, 9)$. A mutation switching Projects 1 and 10 may decrease the total value (as Project 10 is postponed) without expediting the lucrative Project 7. This mutation has a high probability of being rejected (decrease in the objective function). To expedite Project 7, there is a need for several "lucky" small mutations. This sequence of small mutations will indeed appear eventually but may take quite a long time. Figure 5 visually depicts the difference between the mutation types.

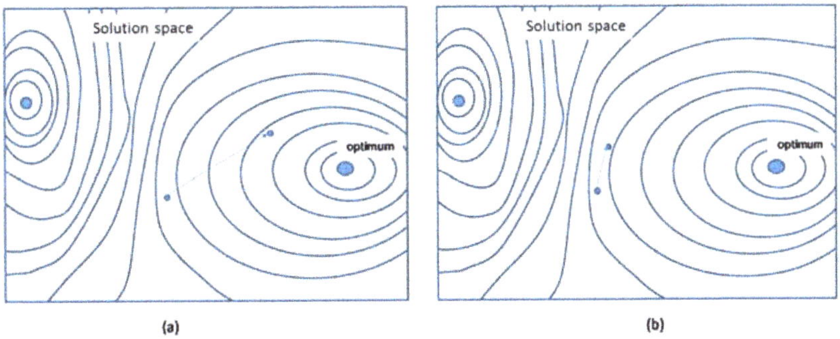

Figure 5. Comparison between the advantages of (**a**) large and (**b**) small mutations.

To improve this, another version of mutations is suggested: randomly choosing two projects in the vector and switching their locations, as depicted in Figure 6. In this case, the third location (Project 2) and the eighth location (Project 1) were switched.

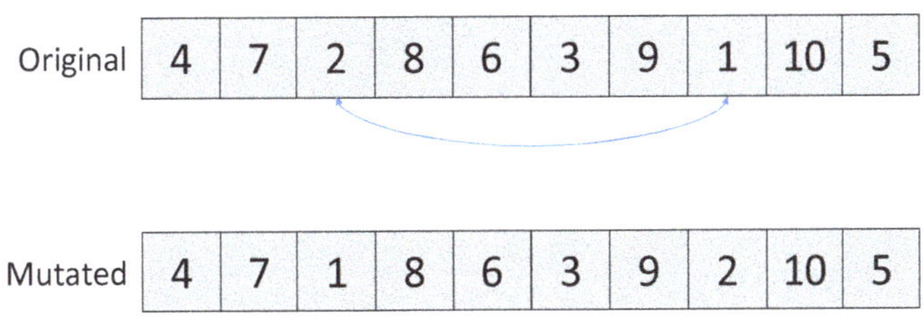

Figure 6. Switch mutation.

While this "mega-mutation" may prove lethal (i.e., significantly reduce the objective function value), it may also save the need for a lucky sequence of minor mutations.

6.3. Oriented Mutations

As claimed by Darwin, in nature, all mutations, whether large or small, are totally random and have no special direction (unlike the Lamarckism theory, which claims that they evolve toward a defined goal). The two mutation types described in Sections 6.1 and 6.2 are Darwinian; that is, they are totally random, and each project has the same probability to be selected. This full randomness has the distinct advantage of being totally unbiased but may prove inefficient. Evolutionary scientists have claimed for decades that a gene cannot be considered simply "bad", "good", or even "helpful" for the organism, but rather a set of genes operating together may prove beneficial [57–59]. For example, a set of sharp incisors and canines is of no use for an herbivore animal, nor are long intestines and complex stomachs useful for a carnivorous hunter. A gene for sharp teeth can contribute only when accompanied by other genes for a carnivorous lifestyle, yielding a cheetah for example. Not to push the natural metaphor too much, but this observation from the field of "ordinary" evolution can be adapted to the field of evolutionary metaheuristic search. If project X and project Y are both predecessors of project Z (which is very lucrative), then there is no advantage in expediting project X alone, but it is very beneficial to expedite X and Y together. In our example, expediting Project 5 may prove unbeneficial if not accompanied by Projects 6 and 7. A beneficial mutation will include several combined small mutations and thus collectively improve the total value.

The problem then is how to recognize whether project X is beneficial with project Y. The projects have dependencies, and they compete for the same resources. To predict whether two projects should be scheduled together, a technique often used by data science was exploited: the similarity coefficient method (SCM). The concept of the SCM was originally used for group technology (GT) applications [60,61], but it is now used for a wide variety of classification and optimization problems [62–64]. Therefore, the proposed method is actually a combination of three fields: SCM, oriented (Lamarckist) evolution, and PPS. Obviously, the similarity between machines and manufacturing processes (as used by GT) should be altered too.

The challenge is to keep the process of CLONALG and its random mutations while inserting "smart" mutations. The basic notion laying beneath this approach is to increase the probability of clustered-together projects to be moved simultaneously (either postponed or expedited but together) by resorting to the SCM. The first step is to calculate the similarity between the projects (i.e., likelihood of benefiting from being scheduled together). The second phase is to generate the mutation in a way that is based on this similarity.

The similarity measure will be as follows:

- Dependent projects: In the example depicted in Figure 4, Projects 5 and 6 have a common dependent in Project 7. This means that it will not be possible to gain the value of Project 7, even if Project 5 is expedited. There is a need to complete Projects 6 and 2 as well. Any small mutation expediting just one of these projects will leave the others untouched and fail to yield a major gain in value. Furthermore, any mutation that causes one of these projects to be postponed will yield a major reduction in the total value. A mutation involving all three projects may enable the expedition of the lucrative Project 7. The proposed similarity measure is (based on [65])

$$\mathbb{S}1_{k,m} = \frac{\sum_{i=1}^{N} d_{k,i} d_{m,i}}{\sum_{i=1}^{N} d_{k,i}(1 - d_{m,i}) + \sum_{i=1}^{N} d_{m,i}(1 - d_{k,i}) + \sum_{i=1}^{N} d_{k,i} d_{m,i}} \tag{11}$$

  where the latter is simply the number of projects that depend on both project k and m divided by the number of projects that depend on either of the two. For example, $\mathbb{S}1_{5,6} = \frac{1}{1+1+1} = \frac{1}{3}$, since only Project 7 depends on both projects, and there are 3 projects that depend on either 5 or 6.
- Mutual dependencies: When two projects depend on the same (or nearly the same) projects, expediting both together causes only a little more impact on the entire schedule than expediting only one (i.e., "two for almost the same price"). Therefore, the second similarity is calculated as follows:

$$\mathbb{S}2_{k,m} = \frac{\sum_{i=1}^{N} d_{i,k} d_{i,m}}{\sum_{i=1}^{N} d_{i,m}(1 - d_{i,m}) + \sum_{i=1}^{N} d_{i,m}(1 - d_{i,k}) + \sum_{i=1}^{N} d_{i,k} d_{i,m}} \tag{12}$$

- Resource requirements: Obviously, all projects compete for the same resource pool. Therefore, the third similarity is based on the measure of the level of common resources required by both projects. Two projects that require totally different resources do not compete at all. Projects that compete for the same resource, in which the resource itself is in abundance, will result in a minimal competition. If, however, both have high requirements for a low-level resource, then they are in head-to-head competition.

Therefore, the third similarity can be calculated as the ratio between the required combined resources and the availability of these resources:

$$s3_{k,m} = \sum_{j=1}^{R} \frac{q_{j,k} + q_{j,m}}{\sum_{i=1}^{H} r_{j,i}} \tag{13}$$

- This similarity measure differs from $\mathbb{S}1$ and $\mathbb{S}2$. First, it measures dissimilarity. Second, the value of $s3$ depends on the number of resources (R). The larger the value of R, the larger the value of $s3$. This poses a problem for the implementation of the CLONALG method. Therefore, the similarity measure ($\mathbb{S}3_{k,m}$) will be calculated as follows:

$$\mathbb{S}3_{k,m} = \frac{1 - s3_{k,m}}{\max\limits_{\forall n \neq p}(1 - s3_{n,p})} \tag{14}$$

Thus, the similarity measure ($\mathbb{S}3$) is set to be a 0–1 number, where 1 indicates 2 non-competing projects, and the lower the value, the higher the competition for resources.

These three similarity measures are utilized to create a similarity coefficient that incorporates all these attributes. The similarity coefficient is, therefore, the following:

$$\mathbb{SC}_{k,m} = \alpha_1 \mathbb{S}1_{k,m} + \alpha_2 \mathbb{S}2_{k,m} + \alpha_3 \mathbb{S}3_{k,m} \tag{15}$$

where $\alpha_1 + \alpha_2 + \alpha_3 = 1$ and $\alpha_1, \alpha_2, \alpha_3$ are non-negative.

The similarity coefficient is the base of the mutation generation algorithm:

1. Randomly choose a project k;
2. Create an empty set of projects Φ;
3. For each project $m = 1 \ldots N$ where $m \neq k$, do the following:
   3.1 Generate a random number $u \sim U(0,1)$;
   3.2 If $\mathbb{SC}_{k,m} > u$, then add project m to Φ.
4. Randomly choose a location l for project k;
5. Move all members of the set Φ to location l (while maintaining the inner order of Φ).

6.4. Mixed Mutations

The basic concept of the oriented mutations is that the advance toward the optimum is not limited to random mutations but also benefits from knowledge and common sense (as expressed in the similarity coefficient matrix). In Figure 7a, there is an illustration of random mutations, where the new solutions are spread randomly. Figure 7b depicts an illustration of the oriented mutations, where the new solutions are concentrated in the vicinities of the optima.

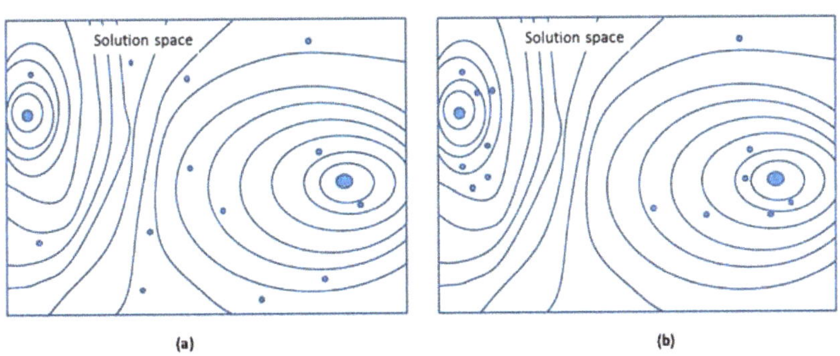

Figure 7. (**a**) Random mutations vs. (**b**) Oriented mutations.

The main risk of the oriented mutation is that the "orientation" will lower the diversification, or the ability to visit many different regions of the solution space [66]. Lower diversification will interfere with the random search and disturb the metaheuristic process that enables escape from the local optima. Though the oriented approach increases the search intensification, it is essential to find an optimal balance between intensification and diversification [67].

To overcome this risk, other approaches were examined for the mutation process. These approaches were basically to tune the level of similarity that is, in the mutation algorithm depicted in Section 6.2, Step 3 is replaced by the following:

3 For each project $m = 1 \ldots N$ where $m \neq k$, do the following:

  3.1 Generate a random number $u \sim U(0,1)$;

  3.2 If $\alpha \mathbb{SC}_{k,m} > u$, then add project m to Φ.

where α is the tuning parameter. A high value of α means that the mutation process relies more on oriented mutations. A low value of α means it is less reliant on these. When $\alpha = 0$, the mutation process is the same as depicted in Section 6.2.

7. Computational Results

7.1. Database

To examine the performance of the different approaches, a random data set of portfolios was generated. To encompass the wide variety of portfolios, the database was generated by randomly generating different cases varying in size, connectivity, and resources:

- Size: The number of projects varied between 20 and 120 projects (20, 40, 60, and 80 projects);
- Connectivity: The number of precedence connections can vary between zero (i.e., no project depends on any other one) and full connectivity (i.e., all projects can be presented as Project 1 precedes Project 2, which precedes Project 3, and so on). The problems were divided into three categories: high, low, and medium connectivity;
- Resources: For each type of resource, its scarcity can be measured by the ratio between the total demand (of all projects) and its annual availability. The resource scarcity has a strong connection to the planning horizon (the scarcer the resource, the more years are needed to complete the entire set of projects). As the number of years for the entire project was set to five, the scarcest resource was set to require seven times the annual resource level. The total number of resources varied between 1 and 3.

For each combination of size, connectivity, and resources, a set of five random portfolios was generated.

The data set was planned for 5 years (a project scheduled for year 6 is considered to be undesired and not performed).

7.2. Experiment Design

The purpose of the experiment was to assess the performance of the four approaches to CLONALG mutations (i.e., minor, major, oriented, and mixed mutations). As there was no database of optimal (or even best-known) solutions, the comparison was based on the following measures:

- Ratio to optimal solution: For smaller problems (i.e., sets of 20 projects), an optimal solution was found using B&B;
- Ratio to best-known solution: For each instance, the best value was found, and for each method, the ratio between its solution and the best solution was calculated.

To compare the various methods correctly, they were run for the exact same time. The runtime for each problem was set by running the minor mutation first until no improvement was achieved for 20 generations. The time was measured, and then all other methods were run for the same length of time, thus providing a comfortable benchmark.

The mixed method used a value of $\alpha = 0.5$.

7.3. Initial Results

The results of the experiment for small problems are depicted in Table 3. For larger problems, the results are depicted in Table 4.

Table 3. Initial results: ratio to optimal solutions.

Connectivity	Resources	Minor	Major	Oriented	Mixed
Low	1	1	1	1	1
	2	0.94	0.94	1	1
	3	0.93	0.96	1	1
Med	1	1	1	1	1
	2	0.96	1	1	1
	3	0.93	0.94	1	1
High	1	1	1	1	1
	2	0.96	0.96	1	1
	3	1	0.95	1	1

Table 4. Initial results: ratio to optimal best known.

Problem Size	Connectivity	Resources	Minor	Major	Oriented	Mixed
40	Low	1	1	1	1	1
		2	0.94	0.96	1	1
		3	0.93	0.93	0.96	1
	Med	1	1	1	1	1
		2	0.96	1	1	1
		3	0.94	0.94	1	1
	High	1	1	1	1	1
		2	0.96	0.96	1	1
		3	1	1	1	1
60	Low	1	0.95	1	1	1
		2	0.99	0.98	0.96	1
		3	0.91	0.95	0.96	1
	Med	1	0.97	1	1	1
		2	0.91	1	0.94	0.99
		3	0.99	0.97	0.95	1
	High	1	0.96	0.96	0.93	1
		2	0.95	0.97	1	1
		3	0.93	0.9	0.98	1
80	Low	1	0.91	0.92	1	1
		2	1	0.93	0.97	1
		3	0.99	0.94	0.96	0.99
	Med	1	0.9	0.96	0.97	1
		2	0.93	0.96	1	1
		3	0.94	0.96	0.96	1
	High	1	0.97	0.91	0.99	1
		2	0.91	0.99	0.97	1
		3	0.97	0.94	0.96	0.98

As can be seen from Tables 3 and 4, the mixed method outperformed all other methods (also relying only on minor mutations proven to be underperforming). A comparison between the methods is described in Figure 8.

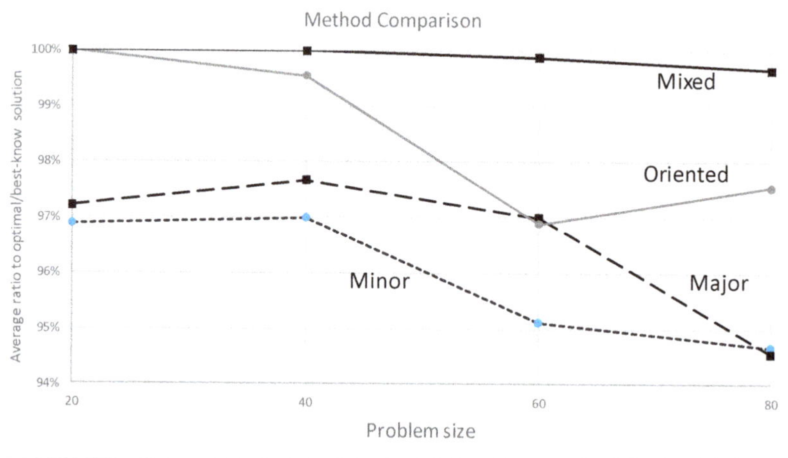

Figure 8. Comparison.

The graph of Figure 8 reveals that the "oriented" (i.e., Lamarckian) mutations indeed improved the performance of the metaheuristic search. However, mixing them with "classic" mutations provided better solutions. The question that arises is the composition of this mix. The experiment was based on the arbitrary setting of $\alpha = 0.5$ (i.e., halfway between oriented mutations and totally non-oriented ones). As the advantage of oriented mutations was established, it is interesting to further explore and find the optimal ratio of this "mix".

To test this point, a second experiment was performed. The same sets of 80 problems were used again for different values of α. The results are depicted in Figure 9. From the results, it is evident that there was almost no significance to the value of α as long as there was a presence of oriented mutations and as long as the oriented mutations monopolized the process.

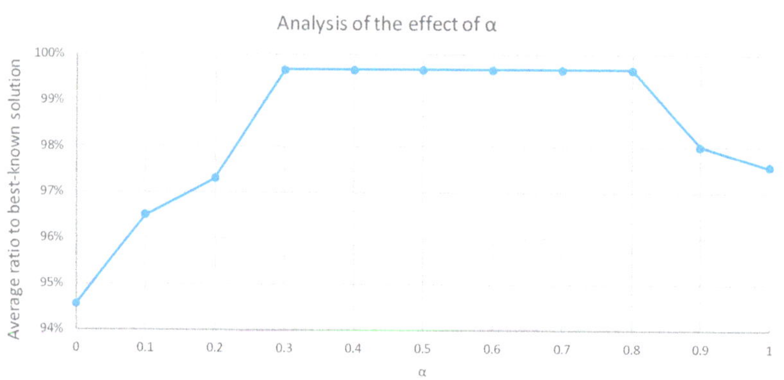

Figure 9. Effect of α.

8. Summary and Conclusions

This paper aimed to tackle the problem of project selection subject to resource constraints and technical precedence. To test this novel problem, the research developed a benchmark database of portfolios varying in size, precedence complexity, and resources. As far as we can ascertain, this is a one-of-a-kind database, and one of the outcomes of this research is to set benchmark results. This paper provides an exact formulation and an example for the new problem.

To solve the problem, a practical search approach for reaching a solution was developed. This enhancement approach would be applicable for most metaheuristic search techniques by using clustering methods that portray the attractive search zones and act as "intensificators". The proposed search was able to generate feasible, meaningful, and highly satisfactory solutions to the planning of long-horizon problems.

The proposed algorithm has both theoretical and practical implications. The practical one is its ability to upgrade the PPS problem decision making and base it on solid exact foundations. The decision-making process should be based less on "gut feelings" and more on exact and well-presented data. Furthermore, the process may enable the decision makers to be aware of the impact of various constraints and lead to improved decisions (e.g., the economic benefit of recruiting more of a specific type of engineers). The theoretical implications, on the other hand, can be derived from the metaheuristic approach; the suggested oriented search need not be limited to PPS and can be implemented in various scheduling (and perhaps other) problems.

An obvious weakness of the article is its limitation to a specific problem, where the data is deterministic and the objective function is limited to a single one (maximum gain value). In reality, the data are often fuzzy or stochastic, and the proposed model does not take this into account. It is worth mentioning that there is nothing fundamental that prevents the proposed meta-heuristic search techniques from tackling fuzzy objective functions, and this may be a suitable direction for further research.

Another direction for future research could use the presented insights to develop better algorithms that will smartly manipulate the mutation type in the different phases of the search and develop a technique or a method for optimizing the various factors of the search to better its performance.

Author Contributions: Conceptualization, R.E. and Y.C.; methodology, R.E. and Y.C.; software, R.E.; analysis, R.E.; writing, R.E. and Y.C. All authors have read and agreed to the published version of the manuscript.

Funding: This research received no external funding.

Institutional Review Board Statement: Not applicable.

Informed Consent Statement: Not Applicable.

Data Availability Statement: Not applicable.

Conflicts of Interest: The authors declare no conflict of interest.

Appendix A. Problem Example

To help visualize the problem formulation, a miniature PPS is portrayed. Whereas typical PPS may include hundreds of projects, this one includes only 10.

Appendix A.1. Projects

The set includes 10 projects with the dependencies included in Figure 4.

The dependency matrix is, therefore, the following:

$$\overline{\overline{\mathcal{D}}} = \begin{pmatrix} 0 & 0 & 0 & 0 & 1 & 0 & 0 & 0 & 0 & 0 \\ 0 & 0 & 0 & 0 & 0 & 0 & 1 & 0 & 0 & 0 \\ 0 & 0 & 0 & 0 & 0 & 1 & 0 & 0 & 0 & 0 \\ 0 & 0 & 0 & 0 & 0 & 0 & 0 & 0 & 0 & 1 \\ 0 & 0 & 0 & 0 & 0 & 0 & 0 & 1 & 0 & 0 \\ 0 & 0 & 0 & 0 & 0 & 0 & 1 & 0 & 1 & 0 \\ 0 & 0 & 0 & 0 & 0 & 0 & 0 & 0 & 0 & 0 \\ 0 & 0 & 0 & 0 & 0 & 0 & 0 & 0 & 0 & 0 \\ 0 & 0 & 0 & 0 & 0 & 0 & 0 & 0 & 0 & 0 \\ 0 & 0 & 0 & 0 & 0 & 0 & 0 & 0 & 0 & 0 \end{pmatrix}$$

Appendix A.2. Resources

The planning horizon is limited to 3 years. In this example, the number of resources is limited to one. Therefore, the matrix $\overline{\overline{\mathcal{R}}}$ is of the dimensions 3×1. Let us assume a constant level of 5 units (e.g., 5 worker years for each of the years of the entire horizon): $\overline{\overline{\mathcal{R}}}^T = (5, 5, 5, \infty)$

As can be seen, the last year has an infinite level of resources, since scheduling a project to year 4 means not performing it at all.

The demand for resources is depicted in Table A1. From this table, the matrix $\overline{\overline{\mathcal{Q}}}$ can be derived: $\overline{\overline{\mathcal{Q}}}^T = (2, 3, 1, 3, 2, 3, 1, 2, 3, 1)$

Table A1. Resource demand.

Project	Resource Requirement	Project	Resource Requirement
1	2	6	3
2	3	7	2
3	1	8	2
4	3	9	3
5	2	10	1

Appendix A.3. Planning Horizon

As mentioned, the planning horizon spreads over 3 years. The first year has a value (depreciation) of 1 (no depreciation), the second has a value of 0.8, the third has a value of 0.5, and everything that follows has a value of 0 (i.e., not planned to be developed at all). Therefore, we set $H = 4$ (i.e., the 3 years of the planning horizon plus 1 year for the projects that would not be realized). We also set the following: $\overline{\mathcal{Y}}^T = (1, 0.8, 0.5, 0)$

Appendix A.4. Projects' Values

The projects' values are depicted in Table A2. Therefore, the vector $\overline{\mathcal{P}}$ is set to $\overline{\mathcal{P}}^T = (1, 1, 1, 1, 1, 2, 8, 2, 2, 3)$.

Table A2. Projects' values.

Project	Value	Project	Value
1	1	6	2
2	1	7	8
3	1	8	2
4	1	9	2
5	1	10	3

Appendix A.5. Objective Function

From the previous notations, the objective function can be derived:

$$\max \mathcal{V} = (X_{1,1} + X_{2,1} + X_{3,1} + X_{4,1} + X_{4,1} + 3X_{6,1} + 8X_{7,1} + 2X_{8,1} + 2X_{9,1} + 3X_{10,1}) + 0.8(X_{1,2} + X_{2,2} + X_{3,2} + X_{4,2} + X_{4,2} + 3X_{6,2} + 8X_{7,2} + 2X_{8,2} + 2X_{9,2} + 3X_{10,2}) + 0.5(X_{1,3} + X_{2,3} + X_{3,3} + X_{4,3} + X_{4,3} + 3X_{6,3} + 8X_{7,3} + 2X_{8,3} + 2X_{9,3} + 3X_{10,3})$$

Appendix A.6. Single Completion Year Constraints

To ensure that a project is completed in 1 year only, the following constraints are added:

$$\begin{aligned}
X_{1,1} + X_{1,2} + X_{1,3} + X_{1,4} &= 1\\
X_{2,1} + X_{2,2} + X_{2,3} + X_{2,4} &= 1\\
X_{3,1} + X_{3,2} + X_{3,3} + X_{3,4} &= 1\\
X_{4,1} + X_{4,2} + X_{4,3} + X_{4,4} &= 1\\
X_{5,1} + X_{5,2} + X_{5,3} + X_{5,4} &= 1\\
X_{6,1} + X_{6,2} + X_{6,3} + X_{6,4} &= 1\\
X_{7,1} + X_{7,2} + X_{7,3} + X_{7,4} &= 1\\
X_{8,1} + X_{8,2} + X_{8,3} + X_{8,4} &= 1\\
X_{9,1} + X_{9,2} + X_{9,3} + X_{9,4} &= 1\\
X_{10,1} + X_{10,2} + X_{10,3} + X_{10,4} &= 1
\end{aligned}$$

Appendix A.7. Precedence Constraints

The auxiliary variables $z_{i,j}$ are set as follows:

$$\begin{gathered}
z_{1,1} \le x_{1,1} + x_{1,2} + x_{1,3} + x_{1,4},\ z_{1,2} \le x_{1,2} + x_{1,3} + x_{1,4}, z_{1,3} \le x_{1,3} + x_{1,4}, z_{1,4} \le x_{1,4}\\
z_{2,1} \le x_{2,1} + x_{2,2} + x_{2,3} + x_{2,4},\ z_{2,2} \le x_{2,2} + x_{2,3} + x_{2,4}, z_{2,3} \le x_{2,3} + x_{2,4}, z_{2,4} \le x_{2,4}\\
z_{3,1} \le x_{3,1} + x_{3,2} + x_{3,3} + x_{3,4},\ z_{3,2} \le x_{3,2} + x_{3,3} + x_{3,4}, z_{3,3} \le x_{3,3} + x_{3,4}, z_{3,4} \le x_{3,4}\\
z_{4,1} \le x_{4,1} + x_{4,2} + x_{4,3} + x_{4,4},\ z_{4,2} \le x_{4,2} + x_{4,3} + x_{4,4}, z_{4,3} \le x_{4,3} + x_{4,4}, z_{4,4} \le x_{4,4}\\
z_{5,1} \le x_{5,1} + x_{5,2} + x_{5,3} + x_{5,4},\ z_{5,2} \le x_{5,2} + x_{5,3} + x_{5,4}, z_{5,3} \le x_{5,3} + x_{5,4}, z_{5,4} \le x_{5,4}\\
z_{6,1} \le x_{6,1} + x_{6,2} + x_{6,3} + x_{6,4},\ z_{6,2} \le x_{6,2} + x_{6,3} + x_{6,4}, z_{6,3} \le x_{6,3} + x_{6,4}, z_{6,4} \le x_{6,4}\\
z_{7,1} \le x_{7,1} + x_{7,2} + x_{7,3} + x_{7,4},\ z_{7,2} \le x_{7,2} + x_{7,3} + x_{7,4}, z_{7,3} \le x_{7,3} + x_{7,4}, z_{7,4} \le x_{7,4}\\
z_{8,1} \le x_{8,1} + x_{8,2} + x_{8,3} + x_{8,4},\ z_{8,2} \le x_{8,2} + x_{8,3} + x_{8,4}, z_{8,3} \le x_{8,3} + x_{8,4}, z_{8,4} \le x_{8,4}\\
z_{9,1} \le x_{9,1} + x_{9,2} + x_{9,3} + x_{9,4},\ z_{9,2} \le x_{9,2} + x_{9,3} + x_{9,4}, z_{9,3} \le x_{9,3} + x_{9,4}, z_{9,4} \le x_{9,4}\\
z_{10,1} \le x_{10,1} + x_{10,2} + x_{10,3} + x_{10,4},\ z_{10,2} \le x_{10,2} + x_{10,3} + x_{10,4}, z_{10,3} \le x_{10,3} + x_{10,4},\\
z_{10,4} \le x_{10,4}\\
z_{1,1} \le z_{1,2} \le z_{1,3}\\
z_{2,1} \le z_{2,2} \le z_{2,3}\\
z_{3,1} \le z_{3,2} \le z_{3,3}\\
z_{4,1} \le z_{4,2} \le z_{4,3}\\
z_{5,1} \le z_{5,2} \le z_{5,3}\\
z_{6,1} \le z_{6,2} \le z_{6,3}\\
z_{7,1} \le z_{7,2} \le z_{7,3}\\
z_{8,1} \le z_{8,2} \le z_{8,3}\\
z_{9,1} \le z_{9,2} \le z_{9,3}\\
z_{10,1} \le z_{10,2} \le z_{10,3}
\end{gathered}$$

Appendix A.8. Resource Requirements

$$
\begin{array}{c}
w_{1,1} \le z_{1,1}, w_{1,2} \le z_{1,2}, w_{1,3} \le z_{1,3}, w_{1,4} \le z_{1,4} \\
w_{2,1} \le z_{2,1}, w_{2,2} \le z_{2,2}, w_{2,3} \le z_{2,3}, w_{2,4} \le z_{2,4} \\
w_{3,1} \le z_{3,1}, w_{3,2} \le z_{3,2}, w_{3,3} \le z_{3,3}, w_{3,4} \le z_{3,4} \\
w_{4,1} \le z_{4,1}, w_{4,2} \le z_{4,2}, w_{4,3} \le z_{4,3}, w_{4,4} \le z_{4,4} \\
w_{5,1} \le z_{5,1}, w_{5,2} \le z_{5,2}, w_{5,3} \le z_{5,3}, w_{5,4} \le z_{5,4} \\
w_{6,1} \le 2z_{6,1}, w_{6,2} \le 2z_{6,2}, w_{6,3} \le 2z_{6,3}, w_{6,4} \le 2z_{6,4} \\
w_{7,1} \le 8z_{7,1}, w_{7,2} \le 8z_{7,2}, w_{7,3} \le 8z_{7,3}, w_{7,4} \le 8z_{7,4} \\
w_{8,1} \le 2z_{8,1}, w_{8,2} \le 2z_{8,2}, w_{8,3} \le 2z_{8,3}, w_{8,4} \le 2z_{8,4} \\
w_{9,1} \le 2z_{9,1}, w_{9,2} \le 2z_{9,2}, w_{9,3} \le 2z_{9,3}, w_{9,4} \le 2z_{9,4}
\end{array}
$$

$$
w_{10,1} \le 3z_{10,1}, w_{10,2} \le 3z_{10,2}, w_{10,3} \le 3z_{10,3}, w_{10,4} \le 3z_{10,4}
$$

Appendix A.9. Resource Consumption Constraints

$$
\begin{array}{c}
w_{1,1} + w_{1,2} + w_{1,3} + w_{1,4} = 1 \\
w_{2,1} + w_{2,2} + w_{2,3} + w_{2,4} = 1 \\
w_{3,1} + w_{3,2} + w_{3,3} + w_{3,4} = 1 \\
w_{4,1} + w_{4,2} + w_{4,3} + w_{4,4} = 1 \\
w_{5,1} + w_{5,2} + w_{5,3} + w_{5,4} = 1 \\
w_{6,1} + w_{6,2} + w_{6,3} + w_{6,4} = 2 \\
w_{7,1} + w_{7,2} + w_{7,3} + w_{7,4} = 8 \\
w_{8,1} + w_{8,2} + w_{8,3} + w_{8,4} = 2 \\
w_{9,1} + w_{9,2} + w_{9,3} + w_{9,4} = 2 \\
w_{10,1} + w_{10,2} + w_{10,3} + w_{10,4} = 3
\end{array}
$$

Appendix A.10. Resource Limitations

The resource limitation constraints for each of the years of the planning horizon are as follows, where year 4 does not have a constraint as it is limitless:

$$
\begin{array}{c}
w_{1,1} + w_{2,1} + w_{3,1} + w_{4,1} + w_{5,1} + w_{6,1} + w_{7,1} + w_{8,1} + w_{9,1} + w_{10,1} \le 5 \\
w_{1,2} + w_{2,2} + w_{3,2} + w_{4,2} + w_{5,2} + w_{6,2} + w_{7,2} + w_{8,2} + w_{9,2} + w_{10,2} \le 5 \\
w_{1,3} + w_{2,3} + w_{3,3} + w_{4,3} + w_{5,3} + w_{6,3} + w_{7,3} + w_{8,3} + w_{9,3} + w_{10,3} \le 5
\end{array}
$$

Appendix A.11. Precedence Constraints

$z_{1,1} \le z_{5,1}$	$z_{1,2} \le z_{5,2}$	$z_{1,3} \le z_{5,3}$
$z_{5,1} \le z_{8,1}$	$z_{5,2} \le z_{8,2}$	$z_{5,3} \le z_{8,3}$
$z_{5,1} \le z_{7,1}$	$z_{5,2} \le z_{7,2}$	$z_{5,3} \le z_{7,3}$
$z_{2,1} \le z_{7,1}$	$z_{2,2} \le z_{7,2}$	$z_{2,3} \le z_{7,3}$
$z_{6,1} \le z_{7,1}$	$z_{6,2} \le z_{7,2}$	$z_{6,3} \le z_{7,3}$
$z_{3,1} \le z_{6,1}$	$z_{3,2} \le z_{6,2}$	$z_{3,3} \le z_{6,3}$
$z_{6,1} \le z_{9,1}$	$z_{6,2} \le z_{9,2}$	$z_{6,3} \le z_{9,3}$

As can be seen, even for such a small problem, the formulation is quite complicated and non-linear.

References

1. Tavana, M.; Keramatpour, M.; Santos-Arteaga, F.J.; Ghorbaniane, E. A fuzzy hybrid project portfolio selection method using Data Envelopment Analysis, TOPSIS and Integer Programming. *Expert Syst. Appl.* **2015**, *42*, 8432–8444. [CrossRef]
2. Gutiérrez, E.; Magnusson, M. Dealing with legitimacy: A key challenge for Project Portfolio Management decision makers. *Int. J. Proj. Manag.* **2014**, *32*, 30–39. [CrossRef]
3. Jonas, D. Empowering project portfolio managers: How management involvement impacts project portfolio management performance. *Int. J. Proj. Manag.* **2010**, *28*, 818–831. [CrossRef]

4. Killen, C.; Hunt, R. Dynamic capability through project portfolio management in service and manufacturing industries. *Int. J. Manag. Proj. Bus.* **2010**, *3*, 157–169. [CrossRef]
5. Meskendahl, S. The influence of business strategy on project portfolio management and its success—a conceptual framework. *Int. J. Proj. Manag.* **2010**, *28*, 807–817. [CrossRef]
6. Lechler, T.G.; Thomas, J.L. Examining new product development project termination decision quality at the portfolio level: Consequences of dysfunctional executive advocacy. *Int. J. Proj. Manag.* **2015**, *33*, 1452–1463. [CrossRef]
7. Martinsuo, M. Project portfolio management in practice and in context. *Int. J. Proj. Manag.* **2013**, *31*, 794–803. [CrossRef]
8. Pendharkar, P.C. A decision-making framework for justifying a portfolio of IT projects. *Int. J. Proj. Manag.* **2014**, *32*, 625–639. [CrossRef]
9. Dutra, C.C.; Ribeiro, J.L.D.; de Carvalho, M.M. An economic–probabilistic model for project selection and prioritization. *Int. J. Proj. Manag.* **2014**, *32*, 1042–1055. [CrossRef]
10. Meade, L.M.; Presley, A. R&D project selection using the analytic network process. *IEEE Trans. Eng. Manag.* **2002**, *49*, 59–66.
11. Jafarzadeh, M.; Tareghian, H.R.; Rahbarnia, F.; Ghanbari, R. Optima lselection of project portfolios using reinvestment strategy within a flexible time horizon. *Eur. J. Oper. Res.* **2015**, *243*, 658–664. [CrossRef]
12. Frey, T.; Buxmann, P. IT Project Portfolio Management—A Structured Literature Review. In Proceedings of the 20th European Conference on Information Systems, Barcelona, Spain, 10–13 June 2012.
13. Weissenberger-Eibl, M.A.; Teufel, B. Organizational politics in new product development project selection. *Eur. J. Innov. Manag.* **2011**, *14*, 51–73. [CrossRef]
14. Padhy, R. Six Sigma project selections: A critical review. *Int. J. Lean Six Sigma* **2017**, *8*, 244–258. [CrossRef]
15. Condé, G.C.P.; Mauro, L.M. Six sigma project generation and selection: Literature review and feature based method proposition. *Prod. Plan. Control.* **2020**, *31*, 1303–1312. [CrossRef]
16. Danesh, D.; Ryan, M.J.; Abbasi, A. Multi-criteria decision-making methods for project portfolio management: A literature review. *Int. J. Manag. Decis. Mak.* **2018**, *17*, 75–94. [CrossRef]
17. Mohagheghi, V.; Mousavi, S.M.; Antuchevičienė, J.; Mojtahedi, M. Project portfolio selection problems: A review of models, uncertainty approaches, solution techniques, and case studies. *Technol. Econ. Dev. Econ.* **2019**, *25*, 1380–1412. [CrossRef]
18. Hall, N.G.; Long, D.Z.; Qi, J.; Sim, M. Managing underperformance risk in project portfolio selection. *Oper. Res.* **2015**, *63*, 660–675. [CrossRef]
19. Tofighian, A.A.; Naderi, B. Modeling and solving the project selection and scheduling. *Comput. Ind. Eng.* **2015**, *83*, 30–38. [CrossRef]
20. Gutjahr, W.J.; Reiter, P. Bi-objective project portfolio selection and staff assignment under uncertainty. *Optimization* **2010**, *59*, 417–445. [CrossRef]
21. Ghapanchi, A.H.; Tavana, M.; Khakbaz, M.H.; Low, G. A methodology for selecting portfolios of projects with interactions and under uncertainty. *Int. J. Proj. Manag.* **2012**, *30*, 791–803. [CrossRef]
22. Mutavdzic, M.; Maybee, B. An extension of portfolio theory in selecting projects to construct a preferred portfolio of petroleum assets. *J. Pet. Sci. Eng.* **2015**, *133*, 518–528. [CrossRef]
23. Arratia, N.M.; López, F.; Schaeffer, S.E.; Cruz-Reyes, L. Static R&D project portfolio selection in public organizations. *Decis. Support Syst.* **2016**, *84*, 53–63.
24. Kalashnikov, V.; Benita, F.; López-Ramos, F.; Hernández-Luna, A. Bi-objective project portfolio selection in Lean Six Sigma. *Int. J. Prod. Econ.* **2017**, *186*, 81–88. [CrossRef]
25. Guo, Y.; Wang, L.; Li, S.; Chen, Z.; Cheng, Y. Balancing strategic contributions and financial returns: A project portfolio selection model under uncertainty. *Soft Comput.* **2018**, *22*, 5547–5559. [CrossRef]
26. Carazo, A.F.; Gómez, T.; Molina, J.; Hernández-Díaz, A.G.; Guerrero, F.M.; Caballero, R. Solving a comprehensive model for multiobjective project portfolio selection. *Comput. Oper. Res.* **2010**, *37*, 630–639. [CrossRef]
27. Kremmel, T.; Kubalík, J.; Biffl, S. Software project portfolio optimization with advanced multiobjective evolutionary algorithms. *Appl. Soft Comput.* **2011**, *11*, 1416–1426. [CrossRef]
28. Liesiö, J.; Punkka, A. Baseline value specification and sensitivity analysis in multiattribute project portfolio selection. *Eur. J. Oper. Res.* **2014**, *237*, 946–956. [CrossRef]
29. Hassanzadeh, F.; Nemati, H.; Sun, M. Robust optimization for interactive multiobjective programming with imprecise information applied to R&D project portfolio selection. *Eur. J. Oper. Res.* **2014**, *238*, 41–53.
30. Shou, Y.; Xiang, W.; Li, Y.; Yao, W. A multiagent evolutionary algorithm for the resource-constrained project portfolio selection and scheduling problem. *Math. Probl. Eng.* **2014**, *2014*, 302684. [CrossRef]
31. Dehouche, N. Non-profit project portfolio evaluation and selection: A multi-criteria approach. *Int. J. Appl. Manag. Sci.* **2015**, *7*, 338–363. [CrossRef]
32. Fliedner, T.; Liesiö, J. Adjustable robustness for multi-attribute project portfolio selection. *Eur. J. Oper. Res.* **2016**, *252*, 931–946. [CrossRef]
33. Perez, F.; Gomez, T. Multiobjective project portfolio selection with fuzzy constraints. *Ann. Oper. Res.* **2016**, *245*, 7–29. [CrossRef]
34. Danesh, D.; Ryan, M.J.; Abbasi, A. A systematic comparison of multi-criteria decision making methods for the improvement of project portfolio management in complex organisations. *Int. J. Manag. Decis. Mak.* **2017**, *16*, 280–320.

35. Ehsani, E.; Kazemi, N.; Olugu, E.U.; Grosse, E.H.; Schwindl, K. Applying fuzzy multi-objective linear programming to a project management decision with nonlinear fuzzy membership functions. *Neural Comput. Appl.* **2017**, *28*, 2193–2206. [CrossRef]
36. El-Kholany, M.M.; Abdelsalam, H.M. Multi-objective binary cuckoo search for constrained project portfolio selection under uncertainty. *Eur. J. Ind. Eng.* **2017**, *11*, 818–853. [CrossRef]
37. Liu, F.; Chen, Y.W.; Yang, J.B.; Xu, D.L.; Liu, W. Solving multiple-criteria R&D project selection problems with a data-driven evidential reasoning rule. *Int. J. Proj. Manag.* **2019**, *37*, 87–97.
38. Pérez, F.; Gómez, T.; Caballero, R.; Liern, V. Project portfolio selection and planning with fuzzy constraints. *Technol. Forecast. Soc. Chang.* **2018**, *131*, 117–129. [CrossRef]
39. Liu, Y.; Liu, Y.-K. Distributionally robust fuzzy project portfolio optimization problem with interactive returns. *Appl. Soft Comput.* **2017**, *56*, 655–668. [CrossRef]
40. Takami, M.A.; Sheikh, R.; Sana, S.S. A Hesitant Fuzzy Set Theory Based Approach for Project Portfolio Selection with Interactions under Uncertainty. *J. Inf. Sci. Eng.* **2018**, *34*, 65–79.
41. Naderi, B. The project portfolio selection and scheduling problem: Mathematical model and algorithms. *J. Optim. Ind. Eng.* **2013**, *6*, 65–72.
42. Wang, B.; Song, Y. Reinvestment Strategy-Based Project Portfolio Selection and Scheduling with Time-Dependent Budget Limit Considering Time Value of Capital. In Proceedings of the 2015 International Conference on Electrical and Information Technologies for Rail Transportation, Zhuzhou, China, 17–19 October 2015; Springer: Berlin/Heidelberg, Germany, 2016; pp. 373–381.
43. Martínez-Vega, D.A.; Cruz-Reyes, L.; Rangel-Valdez, N.; Santillán, C.G.; Sánchez-Solís, P.; Villafuerte, M.P. Project portfolio selection with scheduling: An evolutionary approach. *Int. J. Comb. Optim. Probl. Inform.* **2019**, *10*, 25–31.
44. Killen, C.P.; Jugdev, K.; Drouin, N.; Petit, Y. Advancing project and portfolio management research: Applying strategic management theories. *Int. J. Proj. Manag.* **2012**, *30*, 525–538. [CrossRef]
45. Jeng, D.J.-F.; Huang, K.-H. Strategic project portfolio selection for national research institutes. *J. Bus. Res.* **2015**, *68*, 2305–2311. [CrossRef]
46. Kopmann, J.; Kock, A.; Killen, C.P.; Gemünden, H.G. The role of project portfolio management in fostering both deliberate and emergent strategy. *Int. J. Proj. Manag.* **2017**, *35*, 557–570. [CrossRef]
47. Kaiser, M.G.; el Arbi, F.; Ahlemann, F. Successful project portfolio management beyond project selection techniques: Understanding the role of structural alignment. *Int. J. Proj. Manag.* **2015**, *33*, 126–139. [CrossRef]
48. Hoof, V.; Broeckx, F. Specification of Queuing Models: An Alternative Notation. *Int. J. Model. Simul.* **1982**, *2*, 67–70. [CrossRef]
49. Pajares, J.; López, A. New Methodological Approaches to Project Portfolio Management: The Role of Interactions within Projects and Portfolios. *Procedia—Soc. Behav. Sci.* **2014**, *119*, 645–652. [CrossRef]
50. Shariatmadari, M.; Nahavandi, N.; Zegordi, S.H.; Mohammad, H.S. Integrated resource management for simultaneous project selection and scheduling. *Comput. Ind. Eng.* **2017**, *109*, 39–47. [CrossRef]
51. Li, X.; Fang, S.-C.; Tian, Y.; Guo, X. Expanded model of the project portfolio selection problem with divisibility, time profile factors and cardinality constraints. *J. Oper. Res. Soc.* **2015**, *66*, 1132–1139. [CrossRef]
52. Rafiee, M.; Kianfar, F.; Farhadkhani, M. A multistage stochastic programming approach in project selection and scheduling. *Int. J. Adv. Manuf. Technol.* **2014**, *70*, 2125–2137. [CrossRef]
53. Sefair, J.A.; Méndez, C.Y.; Babat, O.; Medaglia, A.L.; Zuluaga, L.F. Linear solution schemes for Mean-SemiVariance Project portfolio selection problems: An application in the oil and gas industry. *Omega* **2018**, *68*, 39–48. [CrossRef]
54. Samphaiboon, N.; Yamada, T. Heuristic and exact algorithms for the precedence-constrained knapsack problem. *J. Optim. Theory Appl.* **2000**, *105*, 659–676. [CrossRef]
55. de Castro, L.; von Zuben, F.J. The Clonal Selection Algorithm with Engineering Applications. In Proceedings of the Workshop on Artificial Immune Systems and Their Applications, Las Vegas, NV, USA, 8–12 July 2000.
56. Swain, R.K.; Barisal, A.K.; Hota, P.K.; Chakrabarti, R. Short-term hydrothermal scheduling using clonal selection algorithm. *Int. J. Electr. Power Energy Syst.* **2011**, *33*, 647–656. [CrossRef]
57. Attwater, J.; Holliger, P. The cooperative gene. *Nature* **2012**, *491*, 48–49. [CrossRef] [PubMed]
58. Bohl, K.; Hummert, S.; Werner, S.; Basanta, D.; Deutsch, A.; Schuster, S.; Theißen, G.; Schroeter, A. Evolutionary game theory: Molecules as players. *Mol. BioSyst.* **2014**, *10*, 3066–3074. [CrossRef]
59. Furubayashi, T.; Ueda, K.; Bansho, Y.; Motooka, D.; Nakamura, S.; Mizuuchi, R.; Ichihashi, N. Emergence and diversification of a host-parasite RNA ecosystem through Darwinian evolution. *Elife* **2020**, *9*, e56038. [CrossRef]
60. Seifoddini, H.; Wolfe, P.M. Application of the similarity coefficient method in group technology. *IIE Trans.* **1986**, *18*, 271–277. [CrossRef]
61. Selim, H.M.; Askin, R.G.; Vakharia, A.J. Cell formation in group technology: Review, evaluation and directions for future research. *Comput. Ind. Eng.* **1998**, *34*, 3–20. [CrossRef]
62. Lu, P.; Zhang, Z. Critical nodes identification in complex networks via similarity coefficient. *Mod. Phys. Lett. B* **2022**, *36*, 2150620. [CrossRef]
63. Wu, L.; Shen, Y.; Niu, B.; Li, L.; Yang, C.; Feng, Y. Similarity coefficient-based cell formation method considering operation sequence with repeated operations. *Eng. Optim.* **2021**, *22*, 1–15. [CrossRef]

64. Yang, Z.; Qiao, T.; Ren, J.; Yuen, P.; Zhao, H.; Sun, G.; Marshall, S.; Benediktsson, J.A. Joint bilateral filtering and spectral similarity-based sparse representation: A generic framework for effective feature extraction and data classification in hyperspectral imaging. *Pattern Recognit.* **2018**, *77*, 316–328.
65. Erlich, Z.; Gelbard, R.; Spiegler, I. Data Mining by Means of Binary Representation: A Model for Similarity and Clustering. *Inf. Syst. Front.* **2002**, *4*, 187–197. [CrossRef]
66. Lozano, M.; García-Martínez, C. Hybrid metaheuristics with evolutionary algorithms specializing in intensification and diversification: Overview and progress report. *Comput. Oper. Res.* **2010**, *37*, 481–497. [CrossRef]
67. Scheibenpflug, A.; Wagner, S. An Analysis of the Intensification and Diversification Behavior of Different Operators for Genetic Algorithms. In Proceedings of the 14th International Conference on Computer Aided Systems Theory, Las Palmas, Spain, 10–15 February 2013.

mathematics

Article

A Statistical Comparison of Metaheuristics for Unrelated Parallel Machine Scheduling Problems with Setup Times

Ana Rita Antunes, Marina A. Matos, Ana Maria A. C. Rocha *, Lino A. Costa and Leonilde R. Varela

ALGORITMI Center, University of Minho, 4710-057 Braga, Portugal; b8769@algoritmi.uminho.pt (A.R.A.); mmatos@algoritmi.uminho.pt (M.A.M.); lac@dps.uminho.pt (L.A.C.); leonilde@dps.uminho.pt (L.R.V.)
* Correspondence: arocha@dps.uminho.pt

Abstract: Manufacturing scheduling aims to optimize one or more performance measures by allocating a set of resources to a set of jobs or tasks over a given period of time. It is an area that considers a very important decision-making process for manufacturing and production systems. In this paper, the unrelated parallel machine scheduling problem with machine-dependent and job-sequence-dependent setup times is addressed. This problem involves the scheduling of tasks on unrelated machines with setup times in order to minimize the makespan. The genetic algorithm is used to solve small and large instances of this problem when processing and setup times are balanced (Balanced problems), when processing times are dominant (Dominant P problems), and when setup times are dominant (Dominant S problems). For small instances, most of the values achieved the optimal makespan value, and, when compared to the metaheuristic ant colony optimization (ACOII) algorithm referred to in the literature, it was found that there were no significant differences between the two methods. However, in terms of large instances, there were significant differences between the optimal makespan obtained by the two methods, revealing overall better performance by the genetic algorithm for Dominant S and Dominant P problems.

Keywords: scheduling; unrelated parallel machines; sequence-dependent tasks; makespan; metaheuristics; genetic algorithm; statistical analysis

MSC: 90B36; 90C59; 90C27; 62P30; 68M20; 68Q25

Citation: Antunes, A.R.; Matos, M.A.; Rocha, A.M.A.C.; Costa, L.A.; Varela, L.R. A Statistical Comparison of Metaheuristics for Unrelated Parallel Machine Scheduling Problems with Setup Times. *Mathematics* **2022**, *10*, 2431. https://doi.org/10.3390/math10142431

Academic Editor: Ioannis G. Tsoulos

Received: 13 June 2022
Accepted: 7 July 2022
Published: 12 July 2022

1. Introduction

In the current 4th Industrial Revolution scenario, and with the underlying gradual transition of the use of exponential technologies and high-performance computing, companies, namely industrial ones, must be aware of the need to be able to progressively update their decision support tools, for instance, regarding manufacturing scheduling decision-making.

Thus, for an industrial company to remain successful and competitive, it is necessary to use effective and efficient methods and optimization algorithms, along with optimized production processes, in order to differentiate itself in the current global market. In this sense, the manufacturing scheduling decision support systems and underlying algorithms continue to play a crucial role by enabling manufacturing companies to obtain best-suited manufacturing schedules while avoiding unnecessary order cancellations or delays as well as the best use of production resources and costs; this has been a major concern of many researchers over the last decades.

Production management in an industrial environment faces several challenges due to dynamic changes in production and market volatility. Companies are increasingly looking to shorten delivery times as an economic measure due to increased competitiveness in emerging markets [1]. Industrial problems such as scheduling jobs or tasks are affected by customer requirements due to the variety of orders and speed of delivery [1,2].

Manufacturing scheduling is, therefore, a well-studied subject area and is considered a very important decision-making process for manufacturing and production systems. It aims to optimize one or more performance measures by allocating a set of resources to a set of jobs or tasks over a certain period of time [2].

Scheduling problems occur in different kinds of manufacturing environments, varying from a single stage of operation to multi-stage ones. In parallel machine scheduling problems, there is a need to assign tasks to machines coupled with sequencing problems. The scheduling problems in parallel machine environments have been the object of many studies in the last decades. However, cases on unrelated parallel machines are less investigated, especially when setup times are used [3–7].

Parallel machines are considered unrelated when the processing times of tasks depend on the machines to which they are assigned and when there is no relationship between the speeds of the machines [5]. The setup times are dependent since their values depend on both the work sequence and the configuration time matrix of each machine.

Many authors have considered the use of algorithms, heuristics, and metaheuristics to solve scheduling problems on unrelated parallel machines [3,6,8,9], although not all proposed approaches and underlying algorithms are equally effective and efficient in providing scheduling solutions. Yang et al. in 2022 [10] proposed an elite learning differential particle swarm optimization (DELPSO) in order to improve particle swarm optimization (PSO) for large problems. For this, two examples of guidance were applied to direct the update of each particle. The authors verified that the proposed model obtained better results when compared to the PSO. In another paper [11], a new approach was applied in order to improve the performance of multi-objective particle swarm optimizers (MOPSOs). The strategy used is called hybrid global leader selection (HGLSS), where Pareto dominance and density estimation are analyzed to verify the effectiveness of the proposed model. For this, the performance of the proposed approach was compared with nine multi-objective metaheuristics in solving several benchmark problems. The results showed an overcoming of the proposed algorithm in terms of the modified inverted generational distance indicator. The authors, Das and Suganthan [12], presented a literature review on the evolutionary algorithm (EA). In addition, an overview of works carried out on differential evolution (DE) was conducted, where subjects such as variants, multi-objective applications, constrained optimization problems, and theoretical studies of DE were addressed. Finally, engineering applications in which DE was applied were presented. In Pan et al. [13], a differential evolution (DE) algorithm was presented to solve a permutation flow shop scheduling problem in order to minimize the makespan. DE is a traditional continuous algorithm, and the lowest position value rule was presented in order to convert the continuous vector to a discrete work permutation.

This paper focuses on single-stage scheduling problems occurring in parallel machine environments. It is intended to evaluate the behavior of the genetic algorithm (GA) when applied to the scheduling problem of unrelated parallel machines in order to minimize the makespan of a set of tasks subject to varying setup times. A set of small and large instances of this problem will be used to assess the GA performance when compared to the solutions obtained by the metaheuristic ant colony optimization (ACOII) [8,9]. Then, statistical analysis of these two metaheuristics is performed.

In order to properly put forward the main contributions of this work, this paper is organized as follows. In the second section, a summarized introduction to manufacturing scheduling is provided. In Section 3, the unrelated parallel machine scheduling problem is briefly described. Section 4 presents the guidelines for the computational study to be carried out, the metaheuristic GA, and some implementation details. In Section 5, a comparative statistical analysis of the metaheuristics GA and ACOII is performed for small and large instances of scheduling problems, and the main conclusions and planned future work are summarized in Section 6.

2. Manufacturing Scheduling

2.1. General Overview

Scheduling plays a very important role in the decision-making process to be carried out in different decision-making environments, from manufacturing to information processing, transportation, and distribution, among other kinds of services [2]. In manufacturing, scheduling problems vary from single-stage or operations problems, occurring in single or parallel machine environments, up to multi-stage or operation ones, typically occurring in flow shop (FSP), job shop (JSP), or open shop scheduling problems (OSPs), to mention some of the most well-known [2]. The multi-stage scheduling problems tend to be more complex than the single-stage ones; for instance, FSPs have some complicated variants, namely, considering energy issues and material handling assumptions or stochastic data [14,15].

Manufacturing scheduling implies the allocation of jobs or tasks to resources or vice versa, its sequencing or order in time for being processed, along with the definition of the initial and final processing times for each job or task in a given resource, made available through a given manufacturing environment [16].

In [17], Brucker points out that the scheduling of a production system is dependent on several factors, such as manufacturing orders, available resources, available machines, and the operations to be performed by each one, and there may be processing times that are different from one machine to another one while satisfying and/or optimizing a single or complex performance measure or criteria.

The programming of parallel machines has been a growing research area since the first works carried out in the early 20th century [18]. This type of problem has, since then, received continuous interest from researchers due to its relevance to manufacturing environments [3,7–9].

Manufacturing scheduling to optimize configurations, either directly or indirectly, has been an important issue for different types of industries, including plastic, textile, and chemical industries, as well as for some service areas [3,7,9,19–25].

Allahverdi, in [26], presents a review of scheduling problems. In his work, the scheduling problems are further classified based on the underlying production environment as a single machine, parallel machines, flow shop, job shop, or open shop. It also classifies them according to the consideration and processing of information regarding their inclusion in family sets, besides the characterization of sequence-dependent jobs/tasks or machine configurations, which also affects production times and/or costs, among other characterization parameters that are also further explored in other publications, such as [3,21,22].

Thus, scheduling requires decisions about jobs/tasks and processing resources. The sequencing corresponds to a permutation of jobs/tasks or the order in which they might be processed on each resource or machine. On the other hand, the allocation of resources or machines refers to the choice of which one will process each job or task [27]. The scheduling problem aims to assign tasks to machines and define the periods that each task is processed on each machine in order to minimize and/or maximize an optimization criterion, usually expressed in the form of an objective function intended to be optimized [11].

The Brucker classification system, in [17], uses the nomenclature $\alpha \mid \beta \mid \gamma$, initially introduced by [28]. In this nomenclature, the α represents the scheduling classification factors related to the manufacturing environment, which usually include the type of manufacturing system and the number of machines. The manufacturing environment can range from a simple one, such as a single machine, up to more complex ones, occurring in flow shop, job shop, and open shop environments or flexible manufacturing environments, besides other kinds of manufacturing systems, for instance, taking place in different types of parallel machine environments. These manufacturing environments can have different complexity levels, not just according to the nature of the manufacturing systems' configuration and the underlying production flows themselves but further to other, additional characteristics, about a more or less widened set of conditions and constraints imposed in the scheduling problem, expressed by a corresponding β set of factors in the problem classification nomenclature. Moreover, a simple or combined set of performance measures can also be considered and expressed through the γ factor. Since its introduction, this notation has been used and reformulated by several authors, and many classifications have been added as other problems arise [22].

2.2. Scheduling Assumptions

The scheduling problem studied in this work occurs in a parallel-machines scheduling environment, where the processing times of the task are expressed through a task-machine tuple, which varies from one machine to another for processing a given task [29,30].

In scheduling problems arising in manufacturing environments, frequently, multiple machines are available for processing a set of tasks, and the processing times are often even more dependent on task sequences. Thus, there is a sequence-dependent setup time ($S_{i,j,k}$) whenever, after processing task j, preparation time $S_{i,j,k}$ is required before processing task k. Moreover, when these times are also dependent on the machines, index i is added [31–33].

The scheduling problem considers a set of tasks with setup times that are dependent on the machine used and sequence-dependent on the unrelated parallel machines set, with the goal of minimizing the maximum completion time or makespan. In scheduling theory, the makespan (C_{max}) is defined as the completion time of the final task of a job to be processed (when the job leaves the system). The scheduling problem is thus based on a set of N tasks that must be processed on a machine from a set of M unrelated parallel machines (R_M). The processing time of the tasks depends on the assigned machine, and there is no relationship between these machines as they are unrelated. The setup times are machine- and task-sequence-dependent ($S_{i,j,k}$). Each machine has its setup time matrix, and each matrix is different from the others for the remaining unrelated parallel machines.

The problem studied in this work is classified in the literature as $R_M \mid S_{i,j,k} \mid C_{max}$. Minimizing the makespan of a scheduling problem with identical parallel machines and sequence-dependent setup times is categorized as NP-hard. Therefore, the most complex problem of unrelated parallel machines is also considered NP-hard.

2.3. Review about Scheduling Sequence-Dependent Setups in Unrelated Parallel Machines

The scheduling problem occurring in unrelated parallel machines for processing sequence-dependent setup times is quite important as it can be found in several areas such as the electronics, steel, and textile industries [34–37].

Kim et al. [35] used the simulated annealing (SA) algorithm to solve a scheduling problem in the electronics industry. In their work, it was possible to conclude that the proposed SA method significantly outperformed a neighborhood search method regarding the total delay of jobs or tasks.

Tang and Wang [36] formulated a scheduling problem for the steel industry as a mixed nonlinear program and proposed the Tabu search (TS) heuristic to obtain satisfactory solutions. The results showed that their model and heuristic performed more efficiently and effectively than other manual planning approaches.

In the textile industry, Gendreau, Laporte, and Guimarães [34] applied a heuristic to the multiprocessor scheduling problem with sequence-dependent setup times; their results showed that their heuristic was faster than a TS-based one but, at the same time, provided solutions of almost similar quality.

Thus, the complexity of the scheduling problem of unrelated parallel machines has led to increased interest in heuristic procedures to find solutions in a reasonable time interval. Kim and Chen [38] proposed four research heuristics for the aforementioned problem. According to the authors, these heuristics can be easily applied to obtain practical production scheduling solutions. Ghirardi and Potts [39] also studied the problem of unrelated parallel machines for minimizing the makespan; the underlying heuristic used was an application of the recovering beam search technique. The computational results allowed them to generate approximate solutions for large instances of problems (up to 50 machines and 1000 jobs) in just a few minutes.

Another heuristic, called Meta-RaPS, was introduced by Rabadi, Moraga, and Al-Salem [7] to minimize the makespan in unrelated parallel machine problems with sequence-dependent setup times. The performance of the proposed heuristic was evaluated by comparing its solutions with those obtained by other existing heuristics for the same problem. The results showed that the Meta-RaPS found optimal solutions for small problems and performed better than other existing heuristics for larger problems.

For the same problem and makespan objective function, Arnaout et al. [8] introduced the ant colony optimization (ACO) approach. To evaluate the performance of the ACO, the authors compared their solutions with the ones obtained by other heuristics, for example, solutions based on Tabu search and a partitioning heuristic with those obtained by Meta-RaPS [7]. They concluded that the ACO outperformed the other algorithms.

Arnaout et al. [9] also proposed an improved ACO algorithm (ACOII) and mentioned achieving better performance than the previous version; further, the algorithm had the ability to solve more difficult combinatorial optimization problems by partitioning them into subproblems.

The minimization of the makespan is one of the most studied criterion in the production scheduling literature, whether in parallel or single machines. For example, Woo, Jung, and Kim, in [23], developed a mixed-integer linear programming (MILP) model to find the optimal solution to the scheduling problem in unrelated parallel machines with the aim of minimizing the makespan. They proposed a new rule based on a genetic algorithm with a chromosome that represents the sequence of assignment of jobs or tasks to a machine, with the scheduling of jobs/tasks for each machine being determined by a heuristic based on completion time during the chromosome decoding process.

Considering that the setup times are dependent on the work sequence, the authors of [6] presented a GA for minimizing the makespan when solving scheduling problems in unrelated parallel machines. Their GA algorithm was further compared with other algorithms found in the literature, and they concluded that their proposed GA outperformed existing ones.

More recently, the scheduling of unrelated parallel machines for green manufacturing purposes, with resource constraints, was proposed by Zheng and Wang [40]. This work aimed at minimizing the makespan and total carbon emissions; to solve the problem, a collaborative multi-objective fruit fly optimization algorithm (CMFOA) was proposed. The results showed that their multi-objective algorithm was able to obtain more and better non-dominated solutions than other existing algorithms in comparison.

Aydilek et al. [41] addressed a scheduling problem to minimize order delays, in which the setup times were independent of the processing times, through the application of algorithms of self-adaptive differential evolution and hybrid and simulated insertion algorithms. A scheduling problem with different approaches to setup times, aiming to minimize the makespan with the application of an enhanced version of the ACO algorithm, was studied in [24,25].

Abreu and Prata [42] presented a hybrid GA for solving the unrelated parallel machine scheduling problem with sequence-dependent setup times. A case study on the granite industry is presented, and the proposed approach outperformed three traditional dispatch rules presented in the current literature. Gedik et al. in [43] studied the non-preemptive unrelated parallel machine scheduling problem with job/task-sequence- and machine-dependent setup times in order to minimize the makespan. Their study provided a novel constraint programming (CP) model with two customized branching strategies that used CP's global constraints, interval decision variables, and domain filtering algorithms. According to the authors, in terms of average solution quality, the computational results indicated that their CP model slightly outperformed their analyzed contributions from state-of-art algorithms in solving small problem instances and was able to prove the optimality of 283 currently best-known solutions. It is also mentioned to be effective in finding good quality feasible solutions for larger problem instances. Fanjul-Peyro et al. in [44] studied the same problem with the same objective function, but they modeled the problem by means of two integer linear programming problems. One was based on a model previously proposed in the literature, and the other was based on packing problems. According to the authors, since their models were unable to solve medium-sized instances to optimality, they proposed three other metaheuristics for each of these two models. Their results showed that the proposed metaheuristics significantly outperformed the mathematical models. Recently, in [45], Arnaout addressed the unrelated parallel machine scheduling problem with setup times when minimizing the makespan through a worm optimization (WO) algorithm. The performance of the WO algorithm was evaluated by comparing its solutions to solutions of six other known metaheuristics.

Considering the previously presented literature review, most of the research addressed the scheduling problem with different objective functions and algorithm applications in different contexts and industrial environments.

The work underlying this paper was motivated by the work carried out by Arnaout et al. [9], where a comparison was made with the solutions obtained by the Meta-RaPS metaheuristic [46]. Based on data for small and large problem instances, this paper aims to propose a genetic algorithm for solving unrelated parallel machine scheduling problems with setup times for minimizing the makespan, C_{max}. Moreover, in this study, we intend to present an extended evaluation of the behavior of the GA when solving different types of case studies (small and large instances) of unrelated parallel machine scheduling problems with setup times.

3. The Scheduling Problem

This paper addresses the unrelated parallel machine scheduling problem considering the scheduling of N tasks that are available at time zero on M unrelated machines (R_M). The objective function is the makespan C_{max}, considering machine-dependent and sequence-dependent setup times $S_{i,j,k}$. This problem is classified in the literature as $R_M \mid S_{i,j,k} \mid C_{max}$, which is a generalization of the $P_M \mid \mid C_{max}$ problem of identical speeds for processing a set of tasks on the machines [2,47]. The unrelated parallel machine scheduling problem is known to be NP-hard [2] and can be formulated as a mixed-integer linear programming (MILP) model.

In the following, the problem assumptions are described:

- M is the number of parallel machines;
- N denotes the number of tasks to be scheduled;
- Each machine can only process one task at a time without preemption;
- For the initial time instant, which is at time zero, all tasks are available. No restrictions of precedence are imposed among tasks;
- For each machine i, each task j has a processing time, $p_{i,j}$;
- For each machine i, for processing tasks j just after tasks k, there is a setup time, $S_{i,j,k}$. The setup time is different for each machine;
- The objective is to minimize the makespan C_{max}. The term span is used to define the completion time of a machine, while the term makespan is used for the maximum span in the solution of the problem.

The mathematical programming model of the considered problem is presented below, which consists of finding an optimal solution to schedule a set of jobs or tasks in a set of unrelated parallel machines regarding the existence of sequence-dependent setup times, a similar model to the one used by Guinet in [8,48]. This MILP model includes binary variables ($x_{i,j,k} \in \{0,1\}$) indicating the assignment of tasks to machines and continuous variables denoting the completion times of tasks ($C_j \geq 0$ and $C_{max} \geq 0$).

$$Min\ C_{max} \tag{1}$$

subject to

$$\sum_{\substack{i=0 \\ i \neq j}}^{N} \sum_{k=1}^{M} x_{i,j,k} = 1, \qquad \forall j = 1, \ldots, N \tag{2}$$

$$\sum_{\substack{i=0 \\ i \neq j}}^{N} x_{i,h,k} - \sum_{\substack{j=0 \\ j \neq h}}^{N} x_{h,j,k} = 0, \quad \forall h = 1, \ldots, N, \; \forall k = 1, \ldots, M \tag{3}$$

$$C_j \geq C_i + \sum_{k=1}^{M} x_{i,j,k}\left(S_{i,j,k} + p_{j,k}\right) + HV\left(\sum_{k=1}^{M} x_{i,j,k} - 1\right), \quad \forall i = 0, \ldots, N, \; \forall j = 1, \ldots, N \tag{4}$$

$$\sum_{j=0}^{N} x_{0,j,k} = 1, \qquad \forall k = 1, \ldots, M \tag{5}$$

$$C_j \leq C_{max}, \qquad \forall j = 1, \ldots, N \tag{6}$$

$$x_{i,j,k} \in \{0,1\}, \; \forall i = 1, \ldots, N, \; \forall j = 1, \ldots, N, \; \forall k = 1, \ldots, M \tag{7}$$

$$C_0 = 0 \tag{8}$$

$$C_j \geq 0, \qquad \forall j = 1, \ldots, N \tag{9}$$

where

C_j: maximum completion time of task j;
$p_{j,i}$: processing time of task j in machine i;
$S_{i,j,k}$: setup time dependent on the processing sequence of task j after task k in machine i;
$S_{0,j,i}$: setup time for processing first task j on machine i;
$x_{k,j,i}$: 1 if task j is processed immediately after task k in machine i, and 0 otherwise;
$x_{0,j,i}$: 1 if task j is the first one to be processed in machine i, and 0 otherwise;
$x_{j,0,i}$: 1 if task j is the last task to be processed in machine i, and 0 otherwise;
HV: a large positive number (usually denoted by a capital M).

The objective function (1) intends to minimize the makespan, where C_{max} is the length of time that elapses from the start of jobs to the end of the last job. Constraints (2) ensure that each task is only scheduled once and processed by a single machine. Constraints (3) ensure that each task must not be preceded or succeeded by more than one task. Constraint (4) calculates the completion time and ensures that no task precedes or succeeds the same task. Constraint (5) ensures that only one task can be scheduled first on each machine. There is no need for additional constraints to ensure that only one task is scheduled last on each machine because Constraints (5) and (3) guarantee this. Constraints (6) express the makespan C_{max} as a variable that must be larger than any other job's completion time. Constraints (7) guarantee that the decision variable x is binary in all domains. Constraint (8) states that the completion time for dummy work is zero, and Constraint (9) ensures that the completion time is non-negative. Solving the scheduling problem described above enables optimal solutions to be obtained.

4. Computational Study

This section presents the way that this computational study will be conducted. Firstly, the scheduling data description will be presented. Then, the genetic algorithm will be briefly described, followed by the implementation details considered to achieve the goals of the research.

4.1. Scheduling Data Description

The data available on the Scheduling Research Virtual Center page (available online: https://sites.wp.odu.edu/schedulingresearch/wp-content/uploads/sites/99/2016/01/Rm-Cmax-ACO-Arnaout-2014.xlsx (accessed on 10 May 2021) [49] were used for comparison with the GA. This work is divided into two large groups: first, a comparative analysis of the solutions obtained by GA for small problems, and then, a statistical analysis of the solutions obtained with large problems.

The M and N values define the dimension of the problem in terms of the number of machines and tasks: small problems for some combinations of M in $\{2, 4, 6, 8\}$, and N in $\{6, 7, 8, 9, 10, 11\}$ were considered, and large problems were defined for some combinations of M in $\{2, 4, 6, 8, 10, 12\}$ and N in $\{20, 40, 60, 80, 100, 120\}$.

For each combination of the number of machines and tasks, there are 15 instances of problems. The study was carried out for three types of small and large problems: Balanced (when processing and setup times are balanced), Dominant P (when processing times are dominant), and Dominant S (when setup times are dominant).

4.2. Implementation Details

In this work, the implementation of the genetic algorithm [50–52] to solve the unrelated parallel machine scheduling problem with machine-dependent and job-sequence-dependent setup times is addressed. The *ga* function from MATLAB®, which implements the genetic algorithm, was used to solve this combinatorial problem. However, an implementation of a permutation to represent the solutions was previously defined in order to use the *ga* function. GA works with a population of chromosomes that represent the potential solution of the optimization problem and is adequate to tackle combinatorial problems since the constraints can be handled by permutation representations. For this scheduling problem, a solution is represented by a chromosome (a sequence of genes) that is a permutation of size $N + M - 1$. In this representation, a solution is a sequence of integer values that can occur only once. In a chromosome, the permutation values that are superior to N divide the chromosome into M subsequences that indicate the tasks and the corresponding order assigned to each machine. The genetic operators implemented to guarantee the feasibility of the solutions during the search were the order-based crossover and swap mutation. In the order-based crossover, the genetic material between two chromosomes is combined in order to generate offspring. Two random positions are selected, and the genes between them are swapped. Then, the remaining empty positions are filled with the other parent's genes while preventing repetitions. In the swap mutation, two positions are generated at random, and the genes in a chromosome are swapped. A stochastic uniform selection operator was used to select chromosomes for the application of genetic operators.

The population size and the maximum number of generations of GA were set to 50 (default value) and 15,000, respectively. The crossover fraction was 0.8, i.e., each generation of the order-based crossover was applied to 80% of the population. The swap mutation was applied to the same fraction of the population. GA parameter values were chosen, taking into account the ACOII parameter values used in [9] in order to guarantee a fair performance comparison between the algorithms. Due to the stochastic nature of the GA, 30 independent runs were performed; the value of MaxStallGenerations was 1000. The case studies were run in the MatLab® R2021a using a 11th Gen Intel(R) Core(TM) i7-1165G7 @ 2.80 GHz 2.80 GHz.

After obtaining the results using GA, the pandas and scipy.stats libraries [53,54] for Python version 3.8 were used to perform the statistical analysis.

Firstly, data were imported using the pandas' library by applying the *read_csv* function. Then, the function *groupby* was performed to identify the minimum value of the makespan, considering the same number of machines, tasks, and instances.

Thereafter, the library scipy.stats was used to implement the *shapiro*, *ttest_rel* and *wilcoxon* functions to evaluate if data followed a normal distribution and to apply a parametric t-test and non-parametric Wilcoxon test for the related samples, respectively.

In the case of data following a normal distribution, a t-test was considered to compare the metaheuristics GA and ACOII; otherwise, the Wilcoxon test was performed. After identifying the combinations of M and N values where there are differences between the two methods, the confidence interval for paired samples was computed, taking into consideration Equation (10) [55].

$$\overline{d} - t_{(n-1;\frac{\alpha}{2})} \times \frac{s_{\overline{d}}}{\sqrt{n}} < \mu_d < \overline{d} + t_{(n-1;\frac{\alpha}{2})} \times \frac{s_{\overline{d}}}{\sqrt{n}} \tag{10}$$

First, the differences between the makespan of ACOII and GA must be performed, defined as d. Moreover, $\overline{d}$ and $s_{\overline{d}}$ are the mean and standard deviation values of d, respectively. Furthermore, t is a quantile from the t-Student distribution, where n is the number of observations and α is the level of significance.

In order to visualize the obtained results, the seaborn library [56] was implemented to display the boxplot graphs using the *boxplot* function. Boxplots allow us to compare the distribution of the makespan values obtained by each algorithm for the different combinations of M and N.

5. Comparative Statistical Analysis

In this section, comparative statistical analysis is performed considering the makespan and the execution time obtained by the metaheuristics GA and ACOII when solving small and large instances arising in the scheduling problem on unrelated parallel machines with sequence-dependent setup times.

5.1. Small Problems

The comparison between both algorithms, GA and ACOII, was made considering the performance measures given by Equations (11) and (12) to identify which method achieved better results for small and large problems.

$$\gamma = \frac{C_{max}(Algorithm) - C_{max}(Optimal)}{C_{max}(Optimal)} \tag{11}$$

$$\delta = \frac{C_{max}(ACOII) - C_{max}(GA)}{C_{max}(GA)} \tag{12}$$

For the small instances, the optimal values are known when $M = 2$ and $N \in \{6, 7, 8, 9\}$, $M = 4$ and $N \in \{6, 7, 8\}$, and $M = 6$ and $N = 8$. Note that if the GA method, in the small instances, achieves the optimal value, then γ must be equal to zero. If γ is positive, it means that the proposed method achieved a worse value (greater value in terms of minimization) than the optimal value; otherwise, a better value was found (smaller). In terms of δ values, if it is positive, then GA achieved better results than ACOII; otherwise, ACOII achieved better results than GA.

Table 1 presents the average values of γ for the small instances where the optimal value is known. According to the results, in most of them, GA found the optimal value, except for the Dominant S ($M = 4$, $N = 7$) and Dominant P problems ($M = 4, N = 6$). For this Dominant S problem, the γ value is positive ($\gamma = 0.0005$), which means the result achieved is close to the optimal value. Furthermore, in the Dominant P problem, the γ value is negative ($\gamma = -0.00034$); thus, a better result than the optimal known value is found.

The δ values for the two algorithms are presented in Table 2 for Balanced, Dominant S, and Dominant P problems. In the Balanced instances, there are three cases where GA achieved better results ($\delta > 0$) and one case where ACOII had a better result ($\delta < 0$). In terms of Dominant S instances, there are three cases where GA had better results than ACOII and two cases where ACOII achieved better results than GA. Finally, in the Dominant P instances, there are also three cases where GA had the best performance and one case where ACOII performed better than GA. Overall, the GA method achieved better results when compared to ACOII, although these are very small differences.

Table 1. Average γ deviation from optimal solutions for small problems.

M	N	Balanced		Dominant S		Dominant P	
		ACOII	GA	ACOII	GA	ACOII	GA
2	6	0	0	0	0	0	0
	7	0	0	0	0	0	0
	8	0	0	0	0	0	0
	9	0	0	0	0	0	0
4	6	0	0	0	0	0	−0.00034
	7	0	0	0	0.0005	0	0
	8	0	0	0	0	0	0
6	8	0	0	NA	NA	NA	NA

Table 2. Average δ deviation from GA for small problems.

M	N	Balanced	Dominant S	Dominant P
2	10	0	−0.00026	−0.00026
	11	0.000469	0	0
4	6	0	0	0.000336
	7	0	−0.0005	0
	9	0	0.000117	0.00282
	11	−0.00055	0.000112	0
6	10	0.000271	0	0
	11	0.001328	0	0.000167
8	11	0	0.000351	0

Hypothesis testing has been used to test if there are statistically significant differences between the two methods (GA and ACOII). Parametric tests were first considered. Therefore, the underlying assumptions were tested, namely, the normality of data. If so, then the parametric *t*-test for comparing the means of two related samples was computed. Otherwise, a non-parametric Wilcoxon test was used. The hypotheses to be tested are:

H_0: There are no differences between the GA and ACOII results.

H_1: There are differences between the GA and ACOII results.

Considering the level of significance as 5%, if the *p*-value is greater than 0.05, it means that the null hypothesis (H_0) is not rejected. Otherwise, it is rejected.

For the small instances, the hypothesis test was performed for M and N values that presented some differences between the GA and ACOII methods (see Table 2). The *p*-values and the 95% confidence intervals (CI) are shown in Table 3. According to the results achieved, all the *p*-values were greater than 0.05, considering a significance level of 5%. Thus, the null hypothesis (H_0) is not rejected, and consequently, there are no statistically significant differences between GA and ACOII results for small instances. The same conclusion can be drawn using the confidence interval since the value zero belongs to all confidence intervals. In other words, the mean difference between the average δ deviations obtained for GA and ACOII can be equal to zero.

Table 4 shows the average execution times, in seconds, for each type of small instance, where the time increases as the number of concurrent machines and the number of tasks increase. It is possible to observe that for the Dominant P problems, a higher average execution time is required when the number of machines is 8. On the contrary, for the Dominant S problems, the lowest average execution time is observed for the number of machines equal to 2.

Table 3. Hypothesis test p-values and 95% confidence intervals for small problems.

M	N	Balanced		Dominant S		Dominant P	
		p-Value	CI	p-Value	CI	p-Value	CI
2	10	-	-	0.334	[−0.305;0.839]	0.334	[−0.305;0.839]
	11	0.334	[−1.048;0.382]	-	-	-	-
4	6	-	-	-	-	0.334	[−0.419;0.153]
	7	-	-	0.334	[−0.229;0.629]	-	-
	9	-	-	0.334	[0.210;0.076]	0.334	[−5.032;1.832]
	11	0.334	[−0.229;0.629]	0.334	[−0.210;0.076]	-	-
6	10	0.317	[−0.210;0.076]	-	-	-	-
	11	0.461	[−1.078;0.412]	-	-	0.751	[−0.509;0.376]
8	11	-	-	0.334	[0.419;0.153]	-	-

Table 4. Average GA execution time for Balanced, Dominant P, and Dominant S for small problems.

			Number of Tasks (N)					
			6	7	8	9	10	11
Number of Machines (M)	Balanced	2	11.44	11.76	12.03	12.34	12.71	12.79
		4	13.60	14.16	14.38	14.33	14.85	15.02
		6	-	-	15.16	15.77	16.13	15.85
		8	-	-	-	-	16.20	16.63
	Dominant P	2	10.97	11.11	11.39	12.24	12.54	12.66
		4	13.64	14.15	13.79	13.67	14.00	14.42
		6	-	-	14.77	15.88	16.19	16.32
		8	-	-	-	-	17.13	17.61
	Dominant S	2	11.34	11.81	12.01	11.35	11.69	12.33
		4	12.99	13.89	14.29	14.46	14.77	14.95
		6	-	-	14.54	15.37	15.01	15.13
		8	-	-	-	-	16.54	16.83

5.2. Large Problems

The same analysis was reproduced for large problems to compare GA and ACOII performances. The computed values for the average makespan deviation from GA (δ) are presented in Table 5. For the Balanced instances, negative values were obtained for all instances, which means that the ACOII method has better performance than GA. However, GA has better results in Dominant P and Dominant S instances, even though ACOII presents better results in Dominant P instances when the number of machines is equal to 12 and in Dominant S instances when $M = 12$ and $N = 40$.

Parametric and non-parametric tests were applied to statistically prove if there are significant differences between the average δ deviations achieved by the GA and ACOII algorithms for large instances. The hypotheses to be tested were the same as the ones previously defined for small instances. In terms of results, all p-values were less than 0.05; therefore, for the large instances, the null hypothesis is rejected, and it is possible to conclude that there are significant differences between GA and ACOII results considering the significance level of 5%. Furthermore, it is also possible to conclude that there are significant differences between the two algorithms for the significance level of 0.1% since all p-values are less than 0.001.

The next step was to compute the mean and median C_{max} values for each algorithm to identify which one achieved better results. For example, considering $M = 12$ and $N = 40$, in Dominant S instances, the mean values were 748.33 and 737.60, and the median values were 749.00 and 739.00 for GA and ACOII, respectively. Thereby, in this case, ACOII performed better than GA since it achieved lower mean and median C_{max} values. This analysis was conducted for all combinations of M and N. Overall, for Balanced instances, the ACOII results were significantly better than the GA results. On the other hand, for the Dominant P and Dominant S instances, the GA algorithm obtained better results than ACOII, except when $M = 12$ and $N = 40$ for Dominant S and $M = 12$ for Dominant P, where ACOII achieved better results.

Table 5. Average δ deviation from GA for large problems.

M	N	Balanced	Dominant S	Dominant P
2	20	−0.0052	0.6010	0.5968
	40	−0.0243	0.5839	0.5875
	60	−0.0312	0.5801	0.5823
	80	−0.0352	0.5771	0.5801
	100	−0.0374	0.5902	0.5929
	120	−0.0363	0.5883	0.5901
4	20	−0.0086	0.6038	0.5942
	40	−0.0281	0.6017	0.6017
	60	−0.0418	0.5909	0.5842
	80	−0.0456	0.5810	0.5783
	100	−0.0497	0.5803	0.5815
	120	−0.0544	0.5708	0.5755
6	20	−0.0064	0.6628	0.6659
	40	−0.0308	0.6185	0.6004
	60	−0.0499	0.5681	0.5698
	80	−0.0484	0.5864	0.5849
	100	−0.0663	0.5740	0.5666
	120	−0.0767	0.5463	0.5484
8	20	−0.0053	0.6594	0.6558
	40	−0.0491	0.5694	0.5703
	60	−0.0502	0.5904	0.5930
	80	−0.0764	0.5326	0.5282
	100	−0.0763	0.5577	0.5647
	120	−0.0978	0.5245	0.5113
10	20	−0.0192	0.6055	0.6072
	40	−0.0388	0.5704	0.5653
	60	−0.0723	0.5109	0.5082
	80	−0.0996	0.4983	0.4919
	100	−0.0996	0.5036	0.4976
	120	−0.1067	0.4945	0.5001
12	20	−0.0198	0.6008	−0.0187
	40	−0.0323	−0.0143	−0.0126
	60	−0.0917	0.4929	−0.0418
	80	−0.0924	0.5278	−0.0503
	100	−0.0905	0.5271	−0.0424
	120	−0.1261	0.4602	−0.0817

For a better understanding of the performance of the algorithms, boxplots showing the makespan (C_{max}) distribution in terms of N for different values of M are depicted in Figures 1–3 for Balanced, Dominant S, and Dominant P instances, respectively. Using this visualization, it is more perceptible that when the problem becomes more complex (that is, the number of tasks increases), the makespan value also increases. It can also be concluded that most of the GA results are better than the ACOII method.

In Table 6, the average execution time, in seconds, for the genetic algorithm when solving Balanced, Dominant P, and Dominant S large problems is presented. It can be seen that as the number of tasks increases for each number of concurrent machines, the execution time also increases. The lowest average execution time value is observed for problems with six machines. Conversely, the highest average execution time value is obtained for problems with four concurrent machines. It is also observed that the highest average execution time occurred for the Balanced problem with $M = 2$ and $N = 120$. For the Dominant P problem with $M = 2$ and $N = 20$, the shortest average execution time was obtained.

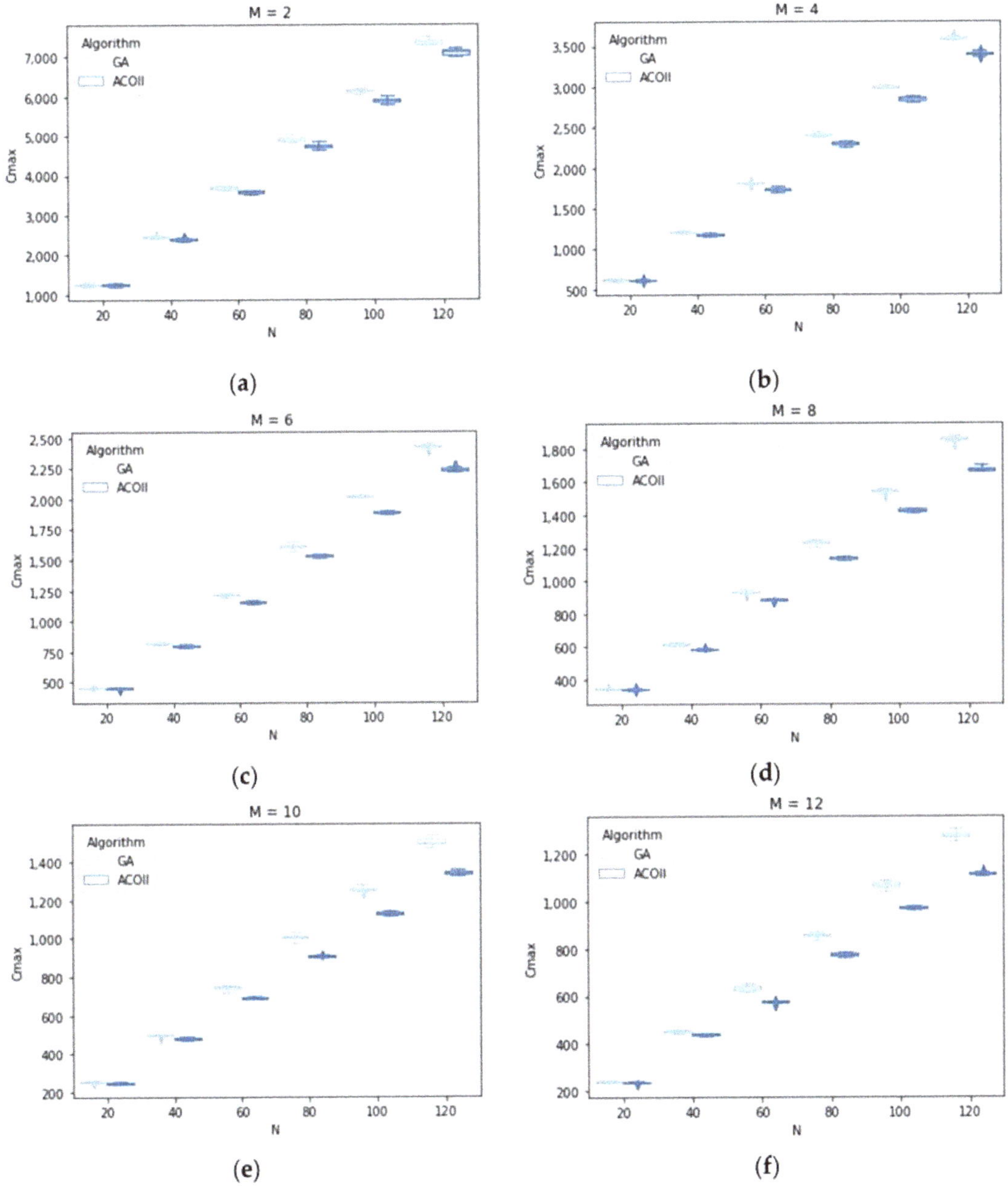

Figure 1. Makespan boxplot for Balanced large problems for (**a**) $M = 2$; (**b**) $M = 4$; (**c**) $M = 6$; (**d**) $M = 8$; (**e**) $M = 10$; (**f**) $M = 12$.

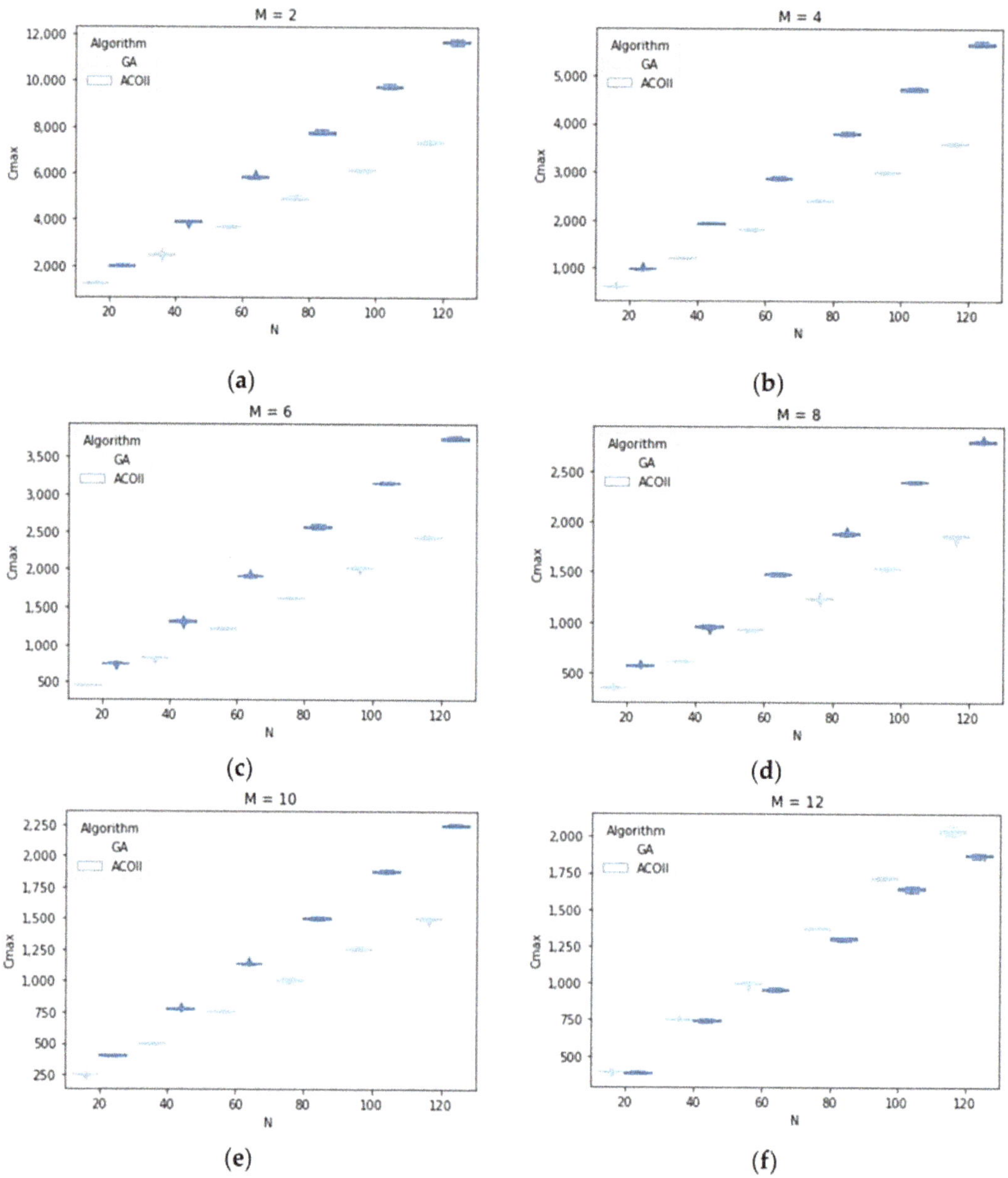

Figure 2. Makespan boxplot for Dominant P large problems for (**a**) $M = 2$; (**b**) $M = 4$; (**c**) $M = 6$; (**d**) $M = 8$; (**e**) $M = 10$; (**f**) $M = 12$.

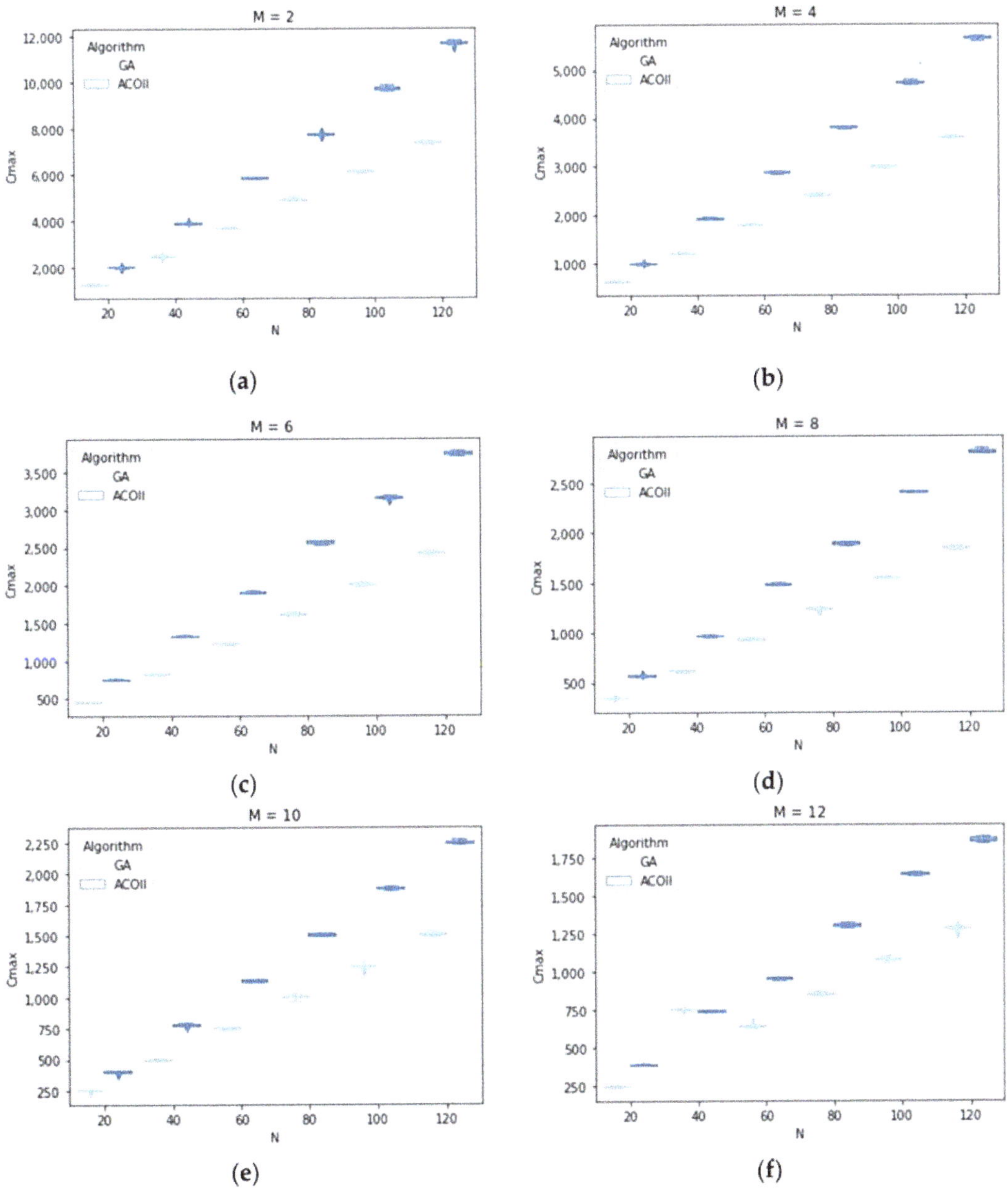

Figure 3. Makespan boxplot for Dominant S large problems for (**a**) $M = 2$; (**b**) $M = 4$; (**c**) $M = 6$; (**d**) $M = 8$; (**e**) $M = 10$; (**f**) $M = 12$.

Table 6. Average GA execution time for Balanced, Dominant P, and Dominant S for large problems.

			Number of Tasks (N)					
			20	**40**	**60**	**80**	**100**	**120**
Number of Machines (M)	Balanced	2	19.95	29.97	37.06	63.74	81.56	110.90
		4	22.24	35.65	50.56	60.74	85.44	94.31
		6	25.11	33.63	45.22	53.70	73.65	86.47
		8	26.74	35.38	41.47	50.82	60.51	71.67
		10	31.36	39.80	47.78	60.42	71.74	83.71
		12	31.20	40.78	50.03	53.02	61.63	69.19
	Dominant P	2	14.74	21.25	31.71	44.08	61.96	84.73
		4	16.93	23.22	30.66	41.56	51.82	65.73
		6	17.85	24.70	30.51	37.91	46.17	59.19
		8	25.12	33.38	42.34	50.66	59.03	70.64
		10	30.72	38.78	46.42	54.00	63.82	73.47
		12	32.47	40.64	49.22	57.69	65.36	73.28
	Dominant S	2	14.84	20.96	32.30	45.63	60.52	84.44
		4	17.17	23.37	32.06	42.02	49.55	64.90
		6	18.56	24.53	30.68	36.99	45.52	54.41
		8	24.99	33.62	41.27	49.82	57.57	67.14
		10	28.80	39.37	48.03	52.45	59.35	69.27
		12	31.11	36.81	44.94	51.85	59.18	64.69

6. Conclusions

Decision support systems are a much-needed factor for industries worldwide. For this, raw material production scheduling and mathematical algorithms are widely used methods to help companies make the best decisions. These decisions avoid unnecessary order cancellations or delays and help to better manage product production costs. In parallel machine scheduling problems, there is a need to assign tasks to machines, along with sequencing problems. In this article, a scheduling problem was developed in an environment of unrelated parallel machines. The machines are considered unrelated when the processing times of tasks depend on the machines to which they are assigned and when there is no relationship between the speeds of the machines. Two types of problems were applied, small and large problems (Balanced, Dominant S, and Dominant P), using the genetic algorithm to minimize the makespan.

In terms of small problems, in general, the proposed approach obtained good results in terms of the makespan, achieving the known optimal value. For Dominant S ($M = 4, N = 7$) and Dominant P ($M = 4, N = 6$) problems, the optimal solution was not found, despite the makespan of the solutions found being very close to the optimal value. Moreover, a new comparison was made with the ACOII method, and there are nine values where the genetic algorithm achieved better results and only four values where ACOII performed better. However, after applying hypothesis testing, there were no significant differences between GA and ACOII results, considering a level of confidence of 95%. The average execution time was the longest for the Dominant S problems when $M = 8$ and $N = 11$. However, the shortest average execution time was obtained for Dominant P problems for $M = 2$ and $N = 6$.

For large problems, the optimal solution is unknown, and thus, a comparison was made between the solutions obtained by the genetic algorithm and the ACOII method. For the Balanced problems, the genetic algorithm exhibited the worst performance. On the contrary, for both Dominant S and Dominant P problems, the genetic algorithm performed better, except when $M = 12$ and $N = 40$ and for $M = 12$, respectively. Furthermore, it was proved that there are significant differences between the average makespan values of the genetic algorithm and the ACOII methods. Moreover, it was also possible to observe that the highest average execution time occurred for the Balanced problem with $M = 2$

and $N = 120$. For the Dominant P problem with $M = 2$ and $N = 20$, the shortest average execution time was obtained.

In conclusion, GA has the ability to effectively solve small and large scheduling problems when minimizing the makespan of unrelated parallel machines with sequence-dependent setup times; better performance for the GA was observed in the comparative statistical analysis between the two metaheuristics, especially for the large problems.

In the future, we intend to study the effect of using different chromosome representations and genetic operators in GA performance and to consider other objectives such as completion time and tardiness.

Author Contributions: Conceptualization, L.R.V., A.M.A.C.R. and L.A.C.; methodology, A.M.A.C.R., L.R.V., L.A.C., A.R.A. and M.A.M.; software, A.M.A.C.R., L.A.C., A.R.A. and M.A.M.; validation, A.R.A., A.M.A.C.R., L.A.C., L.R.V. and M.A.M.; formal analysis, A.R.A., A.M.A.C.R., M.A.M. and L.A.C.; investigation, L.R.V., A.R.A., A.M.A.C.R., L.A.C. and M.A.M.; resources, A.M.A.C.R., L.A.C. and L.R.V.; data curation, A.R.A., A.M.A.C.R., L.A.C., L.R.V. and M.A.M.; writing—original draft preparation, A.R.A., L.R.V. and M.A.M.; writing—review and editing, A.R.A., A.M.A.C.R., L.A.C., L.R.V. and M.A.M.; visualization, A.R.A., A.M.A.C.R., L.A.C., L.R.V. and M.A.M.; supervision, A.M.A.C.R., L.A.C. and L.R.V.; project administration, A.M.A.C.R., L.A.C. and L.R.V.; funding acquisition, A.M.A.C.R., L.A.C. and L.R.V. All authors have read and agreed to the published version of the manuscript.

Funding: The project is funded by the FCT—Fundação para a Ciência e Tecnologia through the R&D Units Project Scope UIDB/00319/2020 and EXPL/EME-SIS/1224/2021 and PhD grant UI/BD/150936/2021.

Institutional Review Board Statement: Not applicable.

Informed Consent Statement: Not applicable.

Data Availability Statement: Not applicable.

Conflicts of Interest: The authors declare no conflict of interest.

References

1. Schuh, G.; Reuter, C.; Prote, J.-P.; Brambring, F.; Ays, J. Increasing data integrity for improving decision making in production planning and control. *CIRP Ann.* **2017**, *66*, 425–428. [CrossRef]
2. Pinedo, M.L. *Scheduling Theory, Algorithms, and Systems*, 5th ed.; Springer: New York, NY, USA, 2016.
3. Santos, A.S.; Madureira, A.M.; Varela, M.L.R. An ordered heuristic for the allocation of resources in unrelated paral-lel-machines. *Int. J. Ind. Eng. Comput.* **2015**, *6*, 145–156.
4. Su, L.-H.; Cheng, T.; Chou, F.-D. A minimum-cost network flow approach to preemptive parallel-machine scheduling. *Comput. Ind. Eng.* **2013**, *64*, 453–458. [CrossRef]
5. Tan, Z.; Chen, Y.; Zhang, A. Parallel machines scheduling with machine maintenance for minsum criteria. *Eur. J. Oper. Res.* **2011**, *212*, 287–292. [CrossRef]
6. Vallada, E.; Ruiz, R. A genetic algorithm for the unrelated parallel machine scheduling problem with sequence dependent setup times. *Eur. J. Oper. Res.* **2011**, *211*, 612–622. [CrossRef]
7. Rabadi, G.; Moraga, R.J.; Al-Salem, A. Heuristics for the Unrelated Parallel Machine Scheduling Problem with Setup Times. *J. Intell. Manuf.* **2006**, *17*, 85–97. [CrossRef]
8. Arnaout, J.-P.; Rabadi, G.; Musa, R. A two-stage Ant Colony Optimization algorithm to minimize the makespan on unrelated parallel machines with sequence-dependent setup times. *J. Intell. Manuf.* **2009**, *21*, 693–701. [CrossRef]
9. Arnaout, J.-P.; Musa, R.; Rabadi, G. A two-stage Ant Colony optimization algorithm to minimize the makespan on unrelated parallel machines—part II: Enhancements and experimentations. *J. Intell. Manuf.* **2012**, *25*, 43–53. [CrossRef]
10. Yang, Q.; Guo, X.; Gao, X.-D.; Xu, D.-D.; Lu, Z.-Y. Differential Elite Learning Particle Swarm Optimization for Global Numerical Optimization. *Mathematics* **2022**, *10*, 1261. [CrossRef]
11. Leung, M.-F.; Coello, C.A.C.; Cheung, C.-C.; Ng, S.-C.; Lui, A.K.-F. A Hybrid Leader Selection Strategy for Many-Objective Particle Swarm Optimization. *IEEE Access* **2020**, *8*, 189527–189545. [CrossRef]
12. Das, S.; Suganthan, P.N. Differential Evolution: A Survey of the State-of-the-Art. *IEEE Trans. Evol. Comput.* **2011**, *15*, 4–31. [CrossRef]
13. Pan, Q.-K.; Tasgetiren, M.F.; Liang, Y.-C. A discrete differential evolution algorithm for the permutation flowshop scheduling problem. *Comput. Ind. Eng.* **2008**, *55*, 795–816. [CrossRef]

14. Ho, M.H.; Hnaien, F.; Dugardinr, F. Exact method to optimize the total electricity cost in two-machine permutation flow shop scheduling problem under Time-of-use tariff. *Comput. Oper. Res.* **2022**, *144*, 105788. [CrossRef]
15. Foumani, M.; Razeghi, A.; Smith-Miles, K. Stochastic optimization of two-machine flow shop robotic cells with con-trollable inspection times: From theory toward practice. *Robot. Comput.-Integr. Manuf.* **2020**, *61*, 101822. [CrossRef]
16. Artiba, A.; Elmaghraby, S.E. *The Planning and Scheduling of Production Systems*; Springer Science & Business Media: Berlin, Germany, 1996. [CrossRef]
17. Brucker, P. Due-date scheduling. In *Scheduling Algorithms*; Springer: Berlin/Heidelberg, Germany, 2001. [CrossRef]
18. McNaughton, R. Scheduling with Deadlines and Loss Functions. *Manag. Sci.* **1959**, *6*, 1–12. [CrossRef]
19. Bülbül, K.; Şen, H. An exact extended formulation for the unrelated parallel machine total weighted completion time problem. *J. Sched.* **2016**, *20*, 373–389. [CrossRef]
20. Nikabadi, M.S.; Naderi, R. A hybrid algorithm for unrelated parallel machines scheduling. *Int. J. Ind. Eng. Comput.* **2016**, *7*, 681–702. [CrossRef]
21. Reddy, M.S.; Ratnam, C.; Agrawal, R.; Varela, M.L.; Sharma, I.; Manupati, V. Investigation of reconfiguration effect on makespan with social network method for flexible job shop scheduling problem. *Comput. Ind. Eng.* **2017**, *110*, 231–241. [CrossRef]
22. Varela, M.L.R.; Silva, S.D.C. An ontology for a model of manufacturing scheduling problems to be solved on the web. In Proceedings of the International Conference on Information Technology for Balanced Automation Systems, Porto, Portugal, 23–25 June 2008; Springer: Boston, MA, USA, 2008; pp. 197–204.
23. Woo, Y.-B.; Jung, S.; Kim, B.S. A rule-based genetic algorithm with an improvement heuristic for unrelated parallel machine scheduling problem with time-dependent deterioration and multiple rate-modifying activities. *Comput. Ind. Eng.* **2017**, *109*, 179–190. [CrossRef]
24. Xu, L.; Wang, Q.; Huang, S. Dynamic order acceptance and scheduling problem with sequence-dependent setup time. *Int. J. Prod. Res.* **2015**, *53*, 1–12. [CrossRef]
25. Zhang, S.; Wong, T. Studying the impact of sequence-dependent set-up times in integrated process planning and scheduling with E-ACO heuristic. *Int. J. Prod. Res.* **2015**, *54*, 4815–4838. [CrossRef]
26. Allahverdi, A. The third comprehensive survey on scheduling problems with setup times/costs. *Eur. J. Oper. Res.* **2015**, *246*, 345–378. [CrossRef]
27. Baker, K.R.; Trietsch, D. *Principles of Sequencing and Scheduling*; John Wiley & Sons, Inc.: Hoboken, NJ, USA, 2009. [CrossRef]
28. Graham, R.L.; Lawler, E.L.; Lenstra, J.K.; Kan, A.H.G.R. *Optimization and Approximation in Deterministic Se-Quencing and Scheduling*; Elsevier: Amsterdam, The Netherlands, 1979; Volume 5, pp. 287–326.
29. Blazewicz, J.; Domschke, W.; Pesch, E. The job shop scheduling problem: Conventional and new solution techniques. *Eur. J. Oper. Res.* **1996**, *93*, 1–33. [CrossRef]
30. Błażewicz, J.; Machowiak, M.; Węglarz, J.; Kovalyov, M.Y.; Trystram, D. Scheduling Malleable Tasks on Parallel Processors to Minimize the Makespan. *Ann. Oper. Res.* **2004**, *129*, 65–80. [CrossRef]
31. Lin, S.-W.; Lu, C.-C.; Ying, K.-C. Minimization of total tardiness on unrelated parallel machines with sequence- and machine-dependent setup times under due date constraints. *Int. J. Adv. Manuf. Technol.* **2010**, *53*, 353–361. [CrossRef]
32. Zeidi, J.R.; MohammadHosseini, S. Scheduling unrelated parallel machines with sequence-dependent setup times. *Int. J. Adv. Manuf. Technol.* **2015**, *81*, 1487–1496. [CrossRef]
33. Jungwattanakit, J.; Reodecha, M.; Chaovalitwongse, P.; Werner, F. A comparison of scheduling algorithms for flexible flow shop problems with unrelated parallel machines, setup times, and dual criteria. *Comput. Oper. Res.* **2009**, *36*, 358–378. [CrossRef]
34. Gendreau, M.; Laporte, G.; Guimarães, E.M. A divide and merge heuristic for the multiprocessor scheduling problem with sequence dependent setup times. *Eur. J. Oper. Res.* **2001**, *133*, 183–189. [CrossRef]
35. Kim, D.-W.; Kim, K.-H.; Jang, W.; Chen, F.F. Unrelated parallel machine scheduling with setup times using simulated annealing. *Robot. Comput. Manuf.* **2002**, *18*, 223–231. [CrossRef]
36. Tang, L.; Wang, X. Simultaneously scheduling multiple turns for steel color-coating production. *Eur. J. Oper. Res.* **2009**, *198*, 715–725. [CrossRef]
37. Pfund, M.; Fowler, J.W.; Gupta, J.N.D. A Survey Of Algorithms For Single And Multi-Objective Unrelated Parallel-Machine Deterministic Scheduling Problems. *J. Chin. Inst. Ind. Eng.* **2004**, *21*, 230–241. [CrossRef]
38. Kim, D.-W.; Na, D.-G.; Chen, F.F. Unrelated parallel machine scheduling with setup times and a total weighted tardiness objective. *Robot. Comput. Manuf.* **2003**, *19*, 173–181. [CrossRef]
39. Ghirardi, M.; Potts, C. Makespan minimization for scheduling unrelated parallel machines: A recovering beam search approach. *Eur. J. Oper. Res.* **2005**, *165*, 457–467. [CrossRef]
40. Zheng, X.-L.; Wang, L. A Collaborative Multiobjective Fruit Fly Optimization Algorithm for the Resource Constrained Unrelated Parallel Machine Green Scheduling Problem. *IEEE Trans. Syst. Man, Cybern. Syst.* **2016**, *48*, 790–800. [CrossRef]
41. Aydilek, A.; Aydilek, H.; Allahverdi, A. Minimising maximum tardiness in assembly flowshops with setup times. *Int. J. Prod. Res.* **2017**, *55*, 7541–7565. [CrossRef]
42. Abreu, L.; Prata, B. A Hybrid Genetic Algorithm for Solving the Unrelated Parallel Machine Scheduling Problem with Sequence Dependent Setup Times. *IEEE Lat. Am. Trans.* **2018**, *16*, 1715–1722. [CrossRef]
43. Gedik, R.; Kalathia, D.; Egilmez, G.; Kirac, E. A constraint programming approach for solving unrelated parallel machine scheduling problem. *Comput. Ind. Eng.* **2018**, *121*, 139–149. [CrossRef]

44. Fanjul-Peyro, L.; Perea, F.; Ruiz, R. Models and matheuristics for the unrelated parallel machine scheduling problem with additional resources. *Eur. J. Oper. Res.* **2017**, *260*, 482–493. [CrossRef]
45. Arnaout, J.-P. A worm optimization algorithm to minimize the makespan on unrelated parallel machines with sequence-dependent setup times. *Ann. Oper. Res.* **2019**, *285*, 273–293. [CrossRef]
46. Amaral, G.; Costa, L.; Rocha, A.M.A.C.; Varela, L.; Madureira, A. Application of the Simulated Annealing Algorithm to Minimize the makespan on the Unrelated Parallel Machine Scheduling Problem with Setup Times. In *International Conference on Hybrid Intelligent Systems*; Springer: Cham, Switzerland, 2019; pp. 398–407. [CrossRef]
47. Rabadi, G. (Ed.) *Heuristics, Metaheuristics and Approximate Methods in Planning and Scheduling*; Springer: Berlin, Germany, 2016; Volume 236.
48. Guinet, A. Textile production systems: A succession of non-identical parallel processor shops. *J. Oper. Res. Soc.* **1991**, *42*, 655–671. [CrossRef]
49. Scheduling Research Virtual Center Homepage. Available online: www.SchedulingResearch.com (accessed on 17 January 2022).
50. Holland, J.H. *Adaptation in Natural and Artificial Systems*, 1st ed.; University of Michigan Press: Ann Arbor, MI, USA, 1975.
51. Katoch, S.; Chauhan, S.S.; Kumar, V. A review on genetic algorithm: Past, present, and future. *Multimed. Tools Appl.* **2021**, *80*, 8091–8126. [CrossRef]
52. Lambora, A.; Gupta, K.; Chopra, K. Genetic Algorithm—A Literature Review. In Proceedings of the 2019 International Conference on Machine Learning, Big Data, Cloud and Parallel Computing (COMITCon), Faridabad, India, 14–16 February 2019; pp. 380–384.
53. Teoh, T.T.; Rong, Z. Python for Data Analysis. In *Artificial Intelligence with Python*; Springer: Singapore, 2022; pp. 107–122. [CrossRef]
54. Bonald, T.; de Lara, N.; Lutz, Q.; Charpentier, B. Scikit-network: Graph analysis in python. *J. Mach. Learn Res.* **2020**, *21*, 1–6.
55. Montgomery, D.C.; Runger, G.C. *Applied Statistics and Probability for Engineers*; Wiley: Hoboken, NJ, USA, 2018; p. 710.
56. Waskom, M.L. Seaborn: Statistical data visualization. *J. Open Source Softw.* **2021**, *6*, 3021. [CrossRef]

mathematics

Article

A Family of Hybrid Stochastic Conjugate Gradient Algorithms for Local and Global Minimization Problems

Khalid Abdulaziz Alnowibet [1], Salem Mahdi [2], Ahmad M. Alshamrani [1], Karam M. Sallam [3] and Ali Wagdy Mohamed [4,*]

1 Statistics and Operations Research Department, College of Science, King Saud University, P.O. Box 2455, Riyadh 11451, Saudi Arabia
2 Department of Mathematics & Computer Science, Faculty of Science, Alexandria University, Alexandria 21544, Egypt
3 School of IT and Systems, University of Canberra, Canberra, ACT 2601, Australia
4 Operations Research Department, Faculty of Graduate Studies for Statistical Research, Cairo University, Giza 12613, Egypt
* Correspondence: aliwagdy@staff.cu.edu.eg

Abstract: This paper contains two main parts, Part I and Part II, which discuss the local and global minimization problems, respectively. In Part I, a fresh conjugate gradient (CG) technique is suggested and then combined with a line-search technique to obtain a globally convergent algorithm. The finite difference approximations approach is used to compute the approximate values of the first derivative of the function f. The convergence analysis of the suggested method is established. The comparisons between the performance of the new CG method and the performance of four other CG methods demonstrate that the proposed CG method is promising and competitive for finding a local optimum point. In Part II, three formulas are designed by which a group of solutions are generated. This set of random formulas is hybridized with the globally convergent CG algorithm to obtain a hybrid stochastic conjugate gradient algorithm denoted by HSSZH. The HSSZH algorithm finds the approximate value of the global solution of a global optimization problem. Five combined stochastic conjugate gradient algorithms are constructed. The performance profiles are used to assess and compare the rendition of the family of hybrid stochastic conjugate gradient algorithms. The comparison results between our proposed HSSZH algorithm and four other hybrid stochastic conjugate gradient techniques demonstrate that the suggested HSSZH method is competitive with, and in all cases superior to, the four algorithms in terms of the efficiency, reliability and effectiveness to find the approximate solution of the global optimization problem that contains a non-convex function.

Keywords: global optimization; unconstrained minimization; numerical approximations of gradients; meta-heuristics; stochastic parameters; conjugate gradient methods; efficient algorithm; performance profiles; comparisons; testing

MSC: 90C26

Citation: Alnowibet, K.A.; Mahdi, S.; Alshamrani, A.M.; Sallam, K.M.; Mohamed, A.W. A Family of Hybrid Stochastic Conjugate Gradient Algorithms for Local and Global Minimization Problems. *Mathematics* **2022**, *10*, 3595. https://doi.org/10.3390/math10193595

Academic Editors: Humberto Rocha and Ana Maria Rocha

Received: 23 August 2022
Accepted: 24 September 2022
Published: 1 October 2022

1. Introduction

The major goal of this paper is to find the local and global minima of a convex and non-convex function. The local and global minimization problems are defined as follows.

Definition 1. *A local minimum $x_{lo} \in \mathbb{S}$ of the function f, $f : \mathbb{S} \to \mathbb{R}$ is an input element with $f(x_{lo}) \le f(x)$ for all x neighboring x_{lo}. If $\mathbb{S} \subseteq \mathbb{R}^n$, it is formulated by*

$$\forall x_{lo}\ \exists \varepsilon > 0 : f(x_{lo}) \le f(x)\ \forall x \in \mathbb{S}, \|x - x_{lo}\| \le \varepsilon. \tag{1}$$

Definition 2. *The point $x_{gl} \in \mathbb{S}$ is called the global minimizer of the function f; $f : \mathbb{S} \rightarrow \mathbb{R}$ such that $f(x_{gl}) \leq f(x)$ $\forall x \in \mathbb{S}$. When $S \subseteq \mathbb{R}^n$, then the problem can be formulated by*

$$\min_{x \in \mathbb{S}} f(x) : \mathbb{S} \rightarrow \mathbb{R}, \tag{2}$$

In both problems (formulae) $\mathbb{S} \subseteq \mathbb{R}^n$ is the range in which we find the global minimizer of $f(x)$. $f(x)$ is continuously differentiable.

Global optimization (GO) attempts to find the approximate solution of the objective function are shown in Problem (2).

However, this task can be difficult since the knowledge about f is usually only local. On the other hand, the fastest algorithms (LO) prefer to find a local point since these algorithms are not capable of finding the global solution at each run.

The bottom line is that the core difference between the GO methods and the LO algorithms is as follows: the GO methods focus on solving Problem (2) over the given set, while the task of the LO methods is to solve (1). Consequently, solving Problem (1) is relatively simple by using deterministic (classical) local optimization methods. On the contrary, finding the global optimum of Problem (2) is an NP-hard problem.

Challenging problems arise in different application fields, for example, technical sciences, industrial engineering, economics, networks, chemical engineering, etc. See [1–11].

Recently, many optimization algorithms have been proposed to deal with these problems. The thoughts of those suggested methods rely on the standard of meta-heuristic strategies (random search).

There are different classifications for meta-heuristic methods [12].

Mohamed et al. [7] presented a brief description of these classifications.

In random algorithms, the minimization technique relies partly on probability.

In contrast, in the deterministic algorithms, a guessing scale is not utilized. Hence, deterministic techniques need an exhaustive examination over the research domain of function f to find the approximate solution to Problem (2) at each run. Otherwise, they fail in this task.

Therefore, finding the approximate solution to Problem (2) by using random techniques can be proved by the asymptotic convergence probability. See [13–15].

There are many deterministic methods that have been proposed for dealing with the local optimization problems. See, for example, Refs. [16–20].

The most popular deterministic method is the CG method [18]. CG methods are exceedingly utilized to find the local minimizer of Problem (1) [21].

However, the CG algorithms have a numerical weakness, so their subsequent actions might be low if a little step is created away from the local point. Hence, for solving this issue, a line-search technique is combined with the CG technique to create a globally convergent algorithm [22,23].

Therefore, many conjugant gradient line-search methods are suggested; see, for example, refs. [18,24–28].

The CG method is an efficient and inexpensive technique to deal with Problem (1).

The CG method is an iterative algorithm. Therefore, the candidate solutions are generated by the following recursive formula.

$$x_{k+1} = x_k + \alpha_k d_k, \tag{3}$$

where the step size $\alpha_k > 0$, and the directions d_k are created by the following formula:

$$d_{k+1} = -g_{k+1} + \beta_k d_k, d_0 = -g_0. \tag{4}$$

where g_k denotes the gradient vector of the function f at the point x_k.

Several versions of the CG methods are suggested. The core difference between those CG algorithms relies on choosing the parameter β_k [18,27–29]. The main features of the

CG method are as follows: it has low memory requirements, it is strongly local, and it has global convergence properties [30].

Many authors presented several studies to analyze the CG method; see, for example, Refs. [31,32].

In 1964, the authors of [33] applied the CG methods to nonlinear problems, and they proposed the following parameter.

$$\beta_k^{FR} = \frac{||g_{k+1}||^2}{||g_k||^2}. \tag{5}$$

The authors of [34,35] established the global convergence of the scheme defined in (5); they used an exact line search and an inexact line search respectively.

However, the author of [36] showed that there are some cases that have some strays; these jamming occurrences happen when the search directions d_k are almost orthogonal to the gradient vector g_k [18].

The authors of [37,38] presented a modification of the parameter β_k^{FR} for treating the noise event denoted in [36]. Hence, they proposed the following parameter.

$$\beta_k^{PRP} = \frac{y_k^T g_{k+1}}{||g_k||^2}, \tag{6}$$

where $y_k = g_{k+1} - g_k$. When a noise occurs $g_{k+1} \approx g_k$, $\beta_k^{PRP} \approx 0$, and $d_{k+1} \approx -g_{k+1}$, i.e., when jamming happens, the search direction d_k is no longer perpendicular to the gradient vector g_k, but it is aligned with the vector $-g_k$. This built-in restart advantage of the β_k^{PRP} parameter usually has better quick convergence when compared to the parameter β_k^{FR} [18].

The authors of [39] proposed an approach closely related to β_k^{PRP}, and it is defined as follows.

$$\beta_k^{HS} = \frac{y_k^T g_{k+1}}{d_k^T y_k}. \tag{7}$$

in the case that step-size α_k is found by an exact line search algorithm. Hence, by (4) and the orthogonality situation $g_{k+1}^T y_k = 0$, the following can be obtained:

$$d_k^T y_k = (g_{k+1} - g_k)^T d_k = -d_k^T g_k = ||g_k||^2. \tag{8}$$

Therefore, $\beta_k^{HS} = \beta_k^{PRP}$ when the step size α_k is calculated by an exact line search method. Other fundamentals formulas of the parameter β_k which contain one term are listed as follows.

$$\beta_k^{LS} = \frac{g_{k+1}^T y_k}{-d_k^T g_k}. \tag{9}$$

Formula (9) was proposed by [40].

$$\beta_k^{DY} = \frac{||g_{k+1}||^2}{y_k^T d_k}. \tag{10}$$

Formula (10) was proposed by Dai and Yuan [41]. It is noteworthy that when the f is quadratic and step size α_k is selected to reduce f along d_k, the options of the parameter β_k mentioned above are alike for the generic nonlinear function.

Different alternatives have fully different convergence possessions [18].

Many version of the parameter β_k have been proposed in two- and three terms; see, for example, Refs. [32,42–50].

For example, in the following two approaches, we present some modifications to obtain a new CG method. See Section 2.

$$\beta_k^{HZ} = \frac{(y_k^T g_k)(d_{k-1}^T y_k) - 2||y_k||^2 (d_{k-1}^T g_k)}{(d_{k-1}^T y_k)^2}. \tag{11}$$

Formula (11) was proposed by [30].

$$\beta_k^{MHZ} = \frac{(y_k^T g_k)(d_{k-1}^T y_k) - 2||y_k||^2 (d_{k-1}^T g_k)}{\max\{\sigma ||y_k||^2 ||d_k||^2, (d_{k-1}^T y_k)^2\}}, \tag{12}$$

where $\sigma > 0.5$ is a constant. Formula (12) was proposed by [49]. The denominator $(d_{k-1}^T y_k)^2$ in the β_k^{HZ} is modified to $\max\{\sigma \|y_k\|^2 \|d_k\|^2, (d_{k-1}^T y_k)^2\}$ in the β_k^{MHZ}. This procedure may help the d_k stay in a trusted area automatically beneath each iteration [49]. Furthermore, in a situation $\sigma ||y_k||^2 ||d_k||^2 < (d_{k-1}^T y_k)^2$, β_k^{MHZ} decreases to β_k^{HZ} with α_k calculated to satisfy the inexact line search. Moreover, β_k^{HZ} decreases to β_k^{HS} under the exact line search.

Consequently, by using a line search method, the CG method can satisfy the following descent condition:

$$g_k^T d_k \leq -C \|g_k\|^2, \tag{13}$$

where $C > 0$ is a constant.

The sufficient descent condition (13) has a core task in the convergence analysis of the algorithms. See [17,30–32,35,41,49,51,52].

However, the CG method has a numerical obstacle; its sub-sequential phases might be low if a little step is created away from the intended point [49].

Recently, the authors of [48,49] proved that the CG algorithm includes powerful convergence features if it satisfies the trust-region feature that is determined by

$$||d_k|| < C_v ||g_k||, \tag{14}$$

where $C_v > 0$ is a constant. It is shown, therefore, that the trust-region property can enable the search direction d_k to be bounded in the trust radius [49]. Numerous researchers proposed many CG algorithms that give perfect results and powerful convergence properties. See [30,48,49,51].

The selection of the right step size α_k can help the CG algorithms to achieve global convergence.

The exact line search is defined as follows:

$$f(x_k + \alpha_k d_k) = \min_{\alpha \geq 0} \theta(\alpha) = f(x_k + \alpha d_k). \tag{15}$$

It is clear that in big-scale problems, the exact line search cannot be used.

Therefore, there are many techniques to achieve this task. Formula (15), for example, the weak Wolfe–Powell algorithm (WWP), is a popular technique, and it is exceedingly utilized. The WWP technique is designed to find the step size α_k to satisfy the following inequalities:

$$f(x_k + \alpha_k d_k) \leq f(x_k) + \delta \alpha_k g_k^T d_k, \tag{16}$$

and

$$g(x_k + \alpha_k d_k)^T d_k \geq \sigma g_k^T d_k, \tag{17}$$

where $\delta \in (0, 0.5)$ and $\sigma \in (\delta, 1)$ are constants.

Inequality (16) is named the Armijo condition, and the WWP line search decreases to strong Wolfe–Powell (SWP) by substituting Inequality (17) with the following inequality:

$$|g(x_k + \alpha_k d_k)^T d_k| \leq -\sigma g_k^T d_k, \tag{18}$$

Generally, under the WWP line search, it is assumed that the gradient $g(x)$ is Lipschitz continuous in the convergence analysis. Therefore, the following inequality is satisfied:

$$||g(x) - g(y)|| \leq L||x - y||, \tag{19}$$

with L is a constant $\forall\, x, y \in \mathbb{R}^n$.

In fact, the CG technique with the line search methods has proven notability in solving the local optimization problem [18,27,28]. However, in trying to solve Problem (2), the CG method fails to achieve this task per run because it is trapped to a local point. To prevent sticking in a local point, random parameters are used [53].

We can summarize the essence of the above discussions as follows.

Recently, there have been many and many proposed approaches presented to improve the performance of deterministic methods, such as CG methods, gradient descent methods, Newton methods, etc. Those new approaches are designed to deal with the local optimization problems. See, for example, Refs. [16–20].

On the other hand, a plentiful number of stochastic approaches are suggested to deal with the global optimization problems. See, for example, Refs. [1,2,4,5,7,54].

Therefore, to gain the features of both deterministic and stochastic methods, many studies presented several ideas and suggestions to combine deterministic and stochastic techniques to obtain a new technique that is efficient and effective in solving Problem (2). Numerical outcomes demonstrated that the interbreed between classical and stochastic techniques has been hugely successful. See [55–59].

This work focuses on solving the local and global minimization problems. So, the first part of this study trades with Problem (1) by suggesting a new modified CG method, while the second part of this paper presents a new random approach that includes three formulae by which the candidate solutions are generated randomly.

Therefore, the new proposed stochastic approach is combined with the new modified CG method that is proposed in the first part of this paper to obtain a new hybrid stochastic conjugate gradient algorithm that solves Problem (2). The new hybrid stochastic conjugate gradient algorithm has four formulae by which the candidate solutions are created. One of the four formulae is a purely deterministic formula, the second one is a mixture of deterministic and stochastic parameters, and the other two formulas contain parameters generated randomly. The bottom line is that we can claim that the main merit that makes the new hybrid algorithm capable of finding the approximate solution to the global minimum of a non-convex function comes from the hybridization of random and non-random parameters.

Consequently, the contribution of this paper is divided into two parts.

Part I presents the following contributions.

- A new modified CG technique is proposed and added with a line search for obtaining a globally convergent algorithm that solves Problem (1). It is abbreviated by SHZ.
- The convergence analysis of the SHZ algorithm is designed.
- The gradient vector is estimated by using a numerical approximation approach (DFF); step-size h (interval) is randomly.
- The convergence analysis of the DFF method is designed.
- The four FR, SH, HZ and MZH methods are designed like the SHZ algorithm to solve Problem (1).
- Numerical experiments of the five SHZ FR, SH, HZ and MZH algorithms are analyzed by using the performance profiles.

Part II presents the following contributions.

◇ Stochastic parameters are designed (SP).
◇ The five SHZ, FR, SH, HZ and MZH algorithms are hybridized with the SP technique to obtain five hybrid algorithms; HSSHZ, HSFR, HSSH, HSHZ and HSMZH. These five algorithms solve Problem (2).
◇ Numerical experiments of the five HSSHZ, HSFR, HSSH, HSHZ and HSMZH algorithms are analyzed by using the performance profiles.

Consequently, the remainder of the study is arranged as follows.

Part I contains the following sections: Section 2 presents a new modified CG- SHZ technique with its convergence analysis.

In Section 3, the approximate value of the gradient vector is calculated by using the numerical differentiation. Section 4 presents the numerical investigations of the local minimization problem. Part II contains the following sections: Section 5 presents a random approach for unconstrained global optimization. Section 6 presents the hybridization of the conjugate gradient method with stochastic parameters. The numerical experiments of Problem (2) are presented in Section 7. Some concluding remarks are given in Section 8.

Part I: Local Minimization Problem

In this part, a new modified CG technique is presented, the convergence analysis of this technique is designed, the numerical differentiation approach is utilized to calculate the approximate values of the first derivative, the five algorithms are designed to solve Problem (1), and their numerical experiments are analyzed by using the performance profiles.

2. Suggested CG Method

Recently, the authors of [49] suggested a new MHZ-CG method, relying on the study which was proposed by the authors of [30]. The MHZ method contains the sufficient descent and the trust-region features independent of a line search technique. The parameter of the MHZ is defined by (12).

Therefore, the story in this section begins with the authors of [30] who proposed a new CG-HZ method, where the parameter of the HZ method is defined by (11). The parameter β_k^{HZ} can ensure that $\boldsymbol{d}_k$ satisfies the following inequality:

$$\boldsymbol{d}_k^T g_k \leq -\frac{7}{8}||g_k||^2, \tag{20}$$

where (20) is proved by [30]. If the step size α_k is calculated by the true line search, then β_k^{HZ} decreases to the β_k^{HS} that was proposed by [39] because $d_k^T g_k = 0$ is true [49].

Hence, for obtaining the global convergence for a general function, Hager and Zhang [30] dynamically adjusted the down limitation of β_k^{HZ} by

$$\boldsymbol{d}_k = -g_k + \beta_k^{HZ^+}\boldsymbol{d}_{k-1}, \boldsymbol{d}_0 = -g_0, \tag{21}$$

$\beta_k^{HZ^+} = \max\{\beta^{HZ}, r_k\}, r_k = \frac{-1}{||d_{k-1}|| \min\{r, ||g_{k-1}||\}}$, where $r > 0$ is a constant.

Many researchers have suggested several modifications and refinements to improve the performance of the CG-HZ algorithm. The latest version of the CG-HZ method was offered by [49]. Yuan et al. [49] presented some modifications to the HZ-CG method, and the result was obtaining the new CG-MHZ algorithm.

The CG-MHZ algorithm contains a sufficient condition and the trust-region feature. The research direction of the MHZ-CG technique is designed as follows:

$$\boldsymbol{d}_k = -g_k + \beta_k^{MHZ}\boldsymbol{d}_{k-1}, \boldsymbol{d}_0 = -g_0, \tag{22}$$

where the β^{MHZ} is defined by (12).

In this paper, the MHZ method is extended and modified to obtain a new proposed method called the SHZ method such that the SHZ method has a sufficient condition and the trust-region feature. This method is defined as follows:

$$\boldsymbol{d}_k = -g_k + \beta_k^{SHZ}\boldsymbol{d}_{k-1}, \boldsymbol{d}_0 = -g_0, \tag{23}$$

$$\beta_k^{SHZ} = \frac{(\boldsymbol{y}_k^T g_k)(\boldsymbol{d}_{k-1}^T \boldsymbol{y}_k) - 2||\boldsymbol{y}_k||^2(\boldsymbol{d}_{k-1}^T g_k)}{\max\{\vartheta||\boldsymbol{y}_k||^2||\boldsymbol{d}_k||^2, (\boldsymbol{d}_{k-1}^T \boldsymbol{y}_k)^2\}}, \tag{24}$$

where the $\vartheta = \max\{\rho, R_k\}$, the ρ and R_k are defined as follows. The parameter ρ is changed randomly at each iteration and its values are taken from the range $[0.8, 2)$ and $R_k = \triangle f \triangle x$. The values of $\triangle f$ and $\triangle x$ are calculated by

$$\triangle f = |f_0 - f_{Itr}|, \tag{25}$$

where Itr is the number of iterations, and after the Itr number of iterations, f_{Itr} and $\triangle f$ are computed. Then, we set $f_0 = f_{Itr}$, while $\triangle x$ is defined by

$$\triangle x = \|x_{k+1} - x_k\|, \text{for } k = 0, 1, \ldots, Itr. \tag{26}$$

Hence, when $\vartheta = \sigma$, β_k^{SHZ} inevitably reduces to one of the following methods $\{\beta_k^{MHZ}, \beta_k^{HZ}, \beta_k^{HS}\}$ as follows.

If $\vartheta = \sigma$ and $\delta\|y_k\|^2\|d_k\|^2 > (d_{k-1}^T y_k)^2$, the β_k^{SHZ} reduces to the β_k^{MHZ}. Otherwise, β_k^{SHZ} reduces to β_k^{HZ} or to β_k^{HS} under the exact line search [49]. This procedure gives the advantages of the MHZ, HZ and HS methods to the proposed SHZ method. In other words, the SHZ algorithm gains the characteristics of the three MHZ, HZ and HS algorithms. This is why the SHZ algorithm is superior to the four other MHZ, HZ, HS and FR methods.

Note: The authors of [49] imposed that the $\sigma > 0.5$ is a constant, while the parameter ϑ is modified dynamically at each iteration.

Convergence Analysis of Algorithm 1

In this section, we present the features of Algorithm 1. We also present the convergence analysis of this algorithm, and we show that the search direction d_k that is defined by Formula (23) satisfies the sufficient descent condition and the trust-region merit, which are defined by Formulae (13) and (14), respectively.

Algorithm 1 A conjugate gradient method (CG-SHZ).

Input: $f : \mathbb{R}^n \to \mathbb{R}, f \in C^1, \gamma \in (0,1)$,$k = 0$, a starting point $x_k \in \mathbb{R}^n$ and $\varepsilon > 0$.
Output: $x^* = x_{loc}$ the local minimizer of f, $f(x^*)$, the value of f at x^*

```
Set d_0 = -g_0 and k := 0.
while ||g_k|| > ε. do
    compute α_k to satisfy (16) and (17).
    Calculate a new point x_{k+1} = x_k + α_k d_k.
    compute f_k = f(x_{k+1}), g_k = g(x_{k+1})
    Set k = k + 1.
    calculate the search direction d_k by (23).
end while
return x_ac the local minimizer and its function value f_ac
```

Two sensible hypotheses are assumed as follows.

Hypothesis 1. *We suppose that Problems (1) and (2) contain an objective function $f(x)$ with the following characteristics: continuity and differentiability properties.*

Hypothesis 2. *In some neighborhood $\aleph$ of the level set*

$$\ell = \{x \in \mathbb{R}^n : f(x) \le f(x_0)\},$$

the gradient vector $g(x)$ is Lipschitz continuous. This means that there is a fixed real number $L < \infty$ such that

$$\|g(x) - g(y)\| \le L\|x - y\|,$$

for all $x, y \in \aleph$.

Lemma 1. *Suppose that the sequence $\{x_k\}$ is obtained by Algorithm 1. If $d_k^T y_k \neq 0$, then*

$$g_k^T d_k \leq -c\|g_k\|^2, \tag{27}$$

and

$$||d_k|| \leq r_v \|g_k\|, \tag{28}$$

where $c = 1 - \frac{7}{9\vartheta} > 0$, $\vartheta = \max\{\rho, R_k\}$, ρ is taken randomly from $\in [\frac{8}{10}, 2)$ at each iteration of Algorithm 1, $0 \leq R_k < \infty$, and $r_v = (1 + \frac{3}{\vartheta})$ is the trust-region radius.

Proof. If $k = 0$, $d_0 = -g_0$, then $g_0^T d_0 = -||g_0||^2$ and $||d_0|| = \|g_0\|$, which indicates (27) and (28) by picking $c \in (0, 1]$ and $r_v \in [1, \infty)$.

Merging (23) with (24), the result is obtaining the following:

$$g_k^T d_k = \frac{(y_k^T g_k)(d_{k-1}^T y_k)(g_k{}^T d_{k-1}) - 2||y_k||^2 (g_k{}^T d_{k-1})^2}{\max\{\vartheta\|y_k\|^2\|d_{k-1}\|^2, (d_{k-1}^T y_k)^2\}} - \|g_k\|^2. \tag{29}$$

The following inequality $u^T v \leq \frac{1}{2}(||u||^2 + \|v\|^2)$ is applied to the first term of the numerator of Inequality (29), where $u = d_{k-1} g_k^T y_k$, $v = y_k g_k{}^T d_{k-1}$, and it is clear that $u^T v \leq \frac{7}{9}(||u||^2 + \|v\|^2)$ is right.

Therefore, the following inequality obtains

$$g_k^T d_k = \frac{(y_k^T g_k)(d_{k-1}^T y_k)(g_k{}^T d_{k-1}) - 2||y_k||^2 (g_k{}^T d_{k-1})^2}{\max\{\vartheta\|y_k\|^2\|d_{k-1}\|^2, (d_{k-1}^T y_k)^2\}} - \|g_k\|^2 \leq$$

$$-\|g_k\|^2 + \frac{\frac{7}{9}||y_k||^2\|g_k\|^2||d_{k-1}||^2 + \frac{7}{9}||y_k||^2 (g_k{}^T d_{k-1})^2 - 2||y_k||^2 (g_k{}^T d_{k-1})^2}{\max\{\vartheta\|y_k\|^2\|d_{k-1}\|^2, (d_{k-1}^T y_k)^2\}} =$$

$$-\|g_k\|^2 + \frac{\frac{7}{9}||y_k||^2\|g_k\|^2||d_{k-1}||^2 - \frac{11}{9}||y_k||^2 (g_k{}^T d_{k-1})^2}{\max\{\vartheta\|y_k\|^2\|d_{k-1}\|^2, (d_{k-1}^T y_k)^2\}} \leq$$

$$-\|g_k\|^2 + \frac{\frac{7}{9}||y_k||^2\|g_k\|^2||d_{k-1}||^2}{\max\{\vartheta\|y_k\|^2\|d_{k-1}\|^2, (d_{k-1}^T y_k)^2\}} \leq (\frac{7}{9\vartheta} - 1)\|g_k\|^2,$$

such that

$$\max\{\vartheta\|y_k\|^2\|d_{k-1}\|^2, (d_{k-1}^T y_k)^2\} \geq \vartheta\|y_k\|^2\|d_{k-1}\|^2, \tag{30}$$

where $\vartheta = \max\{\rho, R_k\}$. Since $\vartheta \geq \frac{8}{10}$ and $c = 1 - \frac{7}{9\vartheta} > 0$, (27) is true.

By using (30), it is obvious that

$$\|d_k\| = \left\| - g_k + \frac{(y_k^T g_k)(d_{k-1}^T y_k) - 2||y_k||^2 (d_{k-1}^T g_k)}{\max\{\vartheta\|y_k\|^2\|d_k\|^2, (d_{k-1}^T y_k)^2\}} d_{k-1} \right\| \leq$$

$$\| - g_k\| + \frac{||y_k||^2\|g_k\|\|d_{k-1}\|^2 + 2\|y_k\|^2\|g_k\|\|d_{k-1}\|^2}{\vartheta\|y_k\|^2\|d_{k-1}\|^2} = (1 + \frac{3}{\vartheta})\|g_k\|$$

Consequently, (28) is met, where $r_v \in [1 + \frac{3}{\vartheta}, \infty)$. The proof is complete. □

Corollary 1. *According to Formula (28) of Lemma 1, the following formula is met.*

$$\sum_{k=0}^{\infty} \frac{\|g_k\|^4}{\|d_k\|^2} = \infty. \tag{31}$$

Proof. Since $\|d_k\| \leq r_v \|g_k\|^2$, where $1 < r_v < \infty$, then $\|d_k\|^2 \leq r_v^2 \|g_k\|^4$, therefore, $\frac{\|d_k\|^2}{\|g_k\|^4} \leq r_v^2$, hence $\frac{\|g_k\|^4}{\|d_k\|^2} \geq \frac{1}{r_v^2}$. Now, the final expression is summed as $k \to \infty$. The result is obtaining the following inequality: $\sum_{k=0}^{\infty} \frac{\|g_k\|^4}{\|d_k\|^2} \geq \sum_{k=0}^{\infty} \frac{1}{r_v^2} = \frac{1}{r_v^2} \sum_{k=0}^{\infty} 1 = \infty$. Therefore, (31) is met. □

Under the assumptions, we give a helpful lemma that was basically proved by Zoutendijk [60] and Wolfe [61,62].

Lemma 2. *Assume that the x_0 is the initial point by which Assumption 1 is satisfied. Regarding any algorithm of Formula (23), d_k is a descent direction, and α_k satisfies the standard Wolfe conditions (16) and (17). Hence, the following inequality is met:*

$$\sum_{k=0}^{\infty} \frac{(g_k{}^T d_k)^2}{\|d_k\|^2} < \infty \tag{32}$$

Proof. It tracks Formula (17), such that

$$d_k^T y_k = d_k^T (g_{k+1} - g_k) \geq (\sigma - 1) g_k^T d_k. \tag{33}$$

On the other hand, the Lipschitz condition (19) implies

$$(g_{k+1} - g_k)^T d_k \leq \alpha_k L \|d_k\|^2. \tag{34}$$

The above two inequalities give

$$\alpha_k \geq \frac{\sigma - 1}{L} . \frac{g_k{}^T d_k}{\|d_k\|^2}, \tag{35}$$

which with (16) implies that

$$f_k - f_{k+1} \geq c \frac{(g_k{}^T d_k)^2}{\|d_k\|^2}, \tag{36}$$

where $c = \frac{\delta(1-\sigma)}{L}$. By summing (36) and with the observation that f is limited below, we see that (32) holds, which concludes the proof. □

Theorem 1. *Suppose that Hypotheses 1 and 2 hold, and by utilizing the outcome of Corollary 1, the sequence $\{g_k\}$ that is generated by Algorithm 1 satisfies the following:*

$$\liminf_{k \to \infty} \|g_k\| = 0, \tag{37}$$

Proof. By contradiction, suppose that (37) is not true; then, for some $\epsilon > 0$, the following inequality is true:

$$\|g_k\| \geq \epsilon. \tag{38}$$

Hence, with inequality (38) and (27), we obtain

$$g_k{}^T d_k \leq -c \|g_k\|^2 \leq -\epsilon^2. \tag{39}$$

Then, we have

$$\frac{g_k{}^T d_k}{\|d_k\|} \leq \frac{-\epsilon^2}{\|d_k\|};$$

$$\frac{g_k{}^T d_k}{\|d_k\|} \geq \frac{\epsilon^4}{\|d_k\|^2},$$

and by summing the final expression, we obtain

$$\sum_{k=0}^{\infty} \frac{(g_k{}^T d_k)^2}{||d_k||^2} \geq \sum_{k=0}^{\infty} \frac{\epsilon^4}{||d_k||^2} = \infty. \tag{40}$$

Therefore, the above leads to a contradiction with (32). So, (37) is met. □

Note 1: The search direction d_k that is defined by Formula (23) satisfies the sufficient descent condition which is defined by Formula (13).

Note 2: Lemma 1 guarantees that Algorithm 1 has a sufficient descent property and the trust-region feature automatically.

Note 3: Theorem 1 confirms that the series $\{g_k\}$ that is obtained by Algorithm 1 approaches to 0 as long as $k \to \infty$.

In the next section, the numerical differentiation approach is discussed by which the first derivative is estimated and the step size α_k is computed.

3. Numerical Differentiation

We now turn our attention to the numerical approximation to compute the approximate value of the gradient vector. In precept, it can be possible to find an analytic form for the first derivative for any continuous and differentiable function. However, in some cases, the analytic form is very complicated. The numerical approximation of the derivative may be sufficient for some purposes.

In this paper, the values of the α_k, g_k and the direction d_k are computed by using the numerical differentiation method. Moreover, we have another step size and research directions that are generated randomly.

Several suggested methods have given fair outcomes for computing the gradient vector values numerically. See [63–67].

The common approaches by which the first derivative is computed are the finite difference approximation methods. Therefore, the first derivative $f'(x)$ can be estimated by the following numerical differentiation formula:

$$D_f f(x_i) = \frac{f(x_{i+1}) - f(x_i)}{x_{i+1} - x_i} = \frac{f(x_i + h) - f(x_i)}{h}, \tag{41}$$

where h is limited and little, but it is not necessarily infinitesimally small.

Reasonably, if the value of the h is small, the approximated value of the first derivative may improve. The forward difference and the central difference are the familiar and common methods used in many studies; see for example, [68–72].

The Taylor series can be used to derive these formulas. Thus, 3, 4 and 5 points can be utilized to derive these formulas, but it will be more costly than utilizing 2 points. The central difference method is known to include aspects of both accuracy and precision [73] but it needs $2n$ function evaluations against the forward-difference approximation approach, which needs n function evaluations for each iteration. So, in this study, the forward-difference approximation approach is used, because it is a cheap method and it has sensible precision [66,68].

The advantage of the finite difference approximation approaches relies on choosing the fit values of the h.

Error approximation of the first derivative is discussed in the next section.

Therefore, the discussion of the error analysis guides us to define an appropriate finite-difference interval for the forward-difference approximation that balances the truncation error that grows from the error in the Taylor formula, and the magnitude error that is obtained from noise during computing the function values [66].

3.1. Error Analysis

Formula (41) contains the forward-difference approximation form that is used to estimate the first derivative of the function f. Its errors are proportional to some power of the values of h. Therefore, it appears that the errors go on to reduce if h is reduced. However, it is a part of the problem since it is assumed only the truncation error yielded by truncating the high-order terms in the Taylor series expansion and does not take into account the round-off error induced by quantization. The round-off error is beside the truncation error; all of them are discussed in this section as follows.

Regarding this goal, suppose that the function values $f(x), f(x+h)$, are quantized to $\theta_1 = f(x+h) + \epsilon_1, \theta_0 = f(x) + \epsilon_0$, with the sizes of the round-off errors ϵ_1 and ϵ_0 all being smaller than some positive number ε, that is $|\epsilon_j| \le \varepsilon$; with $j = 0, 1$.

Hence, the total error of the forward difference approximation defined by (41) is derived by

$$D_f f(x) = \frac{\theta_1 - \theta_0}{h} = \frac{f(x+h) + \epsilon_1 - f(x) - \epsilon_0}{h} = f'(x) + \frac{\epsilon_1 - \epsilon_0}{h} + \frac{T_f}{2}h. \tag{42}$$

Hence,

$$\left|D_f f(x) - f'(x)\right| \le \left|\frac{\epsilon_1 - \epsilon_0}{h}\right| + \left|\frac{T_f}{2}\right|h \le \frac{2\varepsilon}{h} + \frac{|T_f|}{2}h, \tag{43}$$

with $T_f = f''(x)$. Therefore, the upper bound of the error is illustrated by the right-hand side of Formula (43). The maximum limited of error contains two expressions; the first comes from the rounding error and in inverse proportion to step-size h, whilst the second comes from the truncation error and in direct proportion to h. These two parts can be formulated as a function $\phi(h)$ with respect to h as follows $\phi(h) = \frac{2\varepsilon}{h} + \frac{|T_f|}{2}h$. Now, if we find the minimizer h^* of the function $\phi(h)$, then the value $\phi(h^*)$ is the upper bound of the total error. Hence $\frac{d\phi(h)}{dh} = \frac{-2\varepsilon}{h^2} + \frac{|T_f|}{2} = 0$, then

$$h^* = 2\sqrt{\frac{\varepsilon}{|T_f|}} = 2\sqrt{\frac{\varepsilon}{|f''(x)|}}. \tag{44}$$

Therefore, it can be concluded that as we create small values of h, the round-off error might grow, whilst the truncation error reduces. It is called the "step-size dilemma".

Consequently, there have to be some optimal values of the h^* for the forward difference approximation formula, as derived analytically in (44). However, Formula (44) is only of theoretical value and cannot be used practically to determine h^* because we do not have any information about the second derivative and, therefore, we cannot estimate the values of T_f.

Therefore, there are many approaches which have been presented to deal with the step-size dilemma.

Recently, Shi et al. [66] proposed a bisection search for finding a finite-difference interval for a finite-difference method. Their approach was presented to balance the truncation error that grows from the error in the Taylor formula and the measurement error obtained from noise in the function evaluation. According to their numerical experience, the finite-difference interval h^* are bounded between the following ranges $[2 \times 10^{-4}, 6.32 \times 10^{-1}]$, $[2.72 \times 10^{-4}, 8.26 \times 10^{0}]$ and $[8.44 \times 10^{-3}, 3.94 \times 10^{0}]$ by using the forward and central differences to estimate the values of the first derivative of the f.

Additionally, the authors of [68] gave a study of the theoretical and practical comparison of the approximate values of the gradient vector in derivative-free optimization. These authors analyzed some approaches for approximating gradients of noisy functions utilizing only function values; those techniques include a finite difference.

The values of the finite difference interval are as follows $10^{-8} \le h^* \le 1$.

According to the earlier investigations, the core of the difference between all approaches is to determine the step size h. Hence, the value of the step size is ranged between this range $h^* \in [1, 12 \times 10^{-10}]$.

In this paper, the h is designed in a way that makes its values generated randomly. Additionally, the values of the h are connected to the function values per iteration to cover this domain, thus the feature here is that the value of h is modified per iteration randomly.

Therefore, a fresh approach to define the h^* is presented in the following section.

3.2. Selecting a Step-Size h

The forward difference approach is a cheap method compared to the different techniques.

The forward difference approach has shown promising results for minimizing noisy black-box functions [66].

Depending on the hypotheses which are listed in Section 2, let x_0 be any starting point, thus function f satisfies the following $f_0 \geq f_1 \geq \dots \geq f_k$, for $k = 0, 1, 2, \dots$. The numerical outcomes that are given in the past papers denote that the values of step-size h belong to the following range $[10^{-10}, \leq 1]$.

Therefore, the next Algorithm 2 is created to generate the values of the h^* randomly from the intervals $[0.1,\ 10^{-8}]$.

Algorithm 2 Algorithm for calculating the values of h^*.

Step 1: At each iteration k, we generate a set random values between 10^{-2}, and 10^{-7}, and this set of random values is denoted by $L_\epsilon = \{l_{\epsilon_1}, l_{\epsilon_2}, \dots, l_{\epsilon_{10}}\}$.
Step 2: The minimum and maximum of the set L_ϵ are extracted, respectively, as follows $M_\epsilon = \min\{l_{\epsilon_i}{:}i=1,2,\dots,10\}$, $N_\epsilon = \max\{l_{\epsilon_i}{:}i=1,2,\dots,10\}$ and set $M_f = M_\epsilon^{-1}$.
Step 3: The function value f is calculated at each k; $f_k = f(x_k)$.

Now we determine two cases according to the function values of the $|f_k|$ as follows.

Case 1: If $|f_k| \in [10^{-1}, \infty)$, the value of the h is determined by

$$h_k = \begin{cases} \sqrt{\frac{N_\epsilon}{M_f}} & \text{if } |f_k| > M_f, \\ \sqrt{\frac{M_\epsilon}{|f_f|}} & \text{otherwise.} \end{cases} \tag{45}$$

Case 2: If $|f_k| \in [0, 10^{-1})$, the value of the h is determined by a random way from the range $[10^{-4}, 10^{-8}]$.

Example: In this example, we show how the above algorithm is run.

Let us suppose that the point x_0 has four different values as starting points with four different values of f, for example, $f_0 = f(x_0) = \{10^{10}, 10^6, 10^3, 10^{-1}\}$ and suppose we generate the set L_ϵ as random values between 10^{-1}, and 10^{-7} such that $L_\epsilon = \{1.50 \times 10^{-4},\ 5.10 \times 10^{-6},\ 1.01 \times 10^{-6},\ 1.40 \times 10^{-2},\ 1.78 \times 10^{-7},\ 1.92 \times 10^{-5},\ 1.09 \times 10^{-3},\ 2.77 \times 10^{-4},\ 2.99 \times 10^{-04},\ 5.15 \times 10^{-4}\}$, $M_\epsilon = 1.78 \times 10^{-7}$; hence, $M_f = 5.618 \times 10^6$, since $f_0 = 10^{10} > M_f = 5.618 \times 10^6$, then we set $F_0 = M_f = 5.618 \times 10^6$ and $h_1 = 2\sqrt{\frac{M_\epsilon}{M_f}} = 2\sqrt{\frac{1.78 \times 10^{-7}}{5.618 \times 10^6}} = 3.56 \times 10^{-7}$. If $f_0 = 10^6$, $f_0 = 10^6 < M_f = 5.618 \times 10^6$, and then $h_1 = 2\sqrt{\frac{M_\epsilon}{F_0}} = 2\sqrt{\frac{5.618 \times 10^6}{10^6}} = 8.438 \times 10^{-7}$, and $f_0 = \{10^3 < M_f 5.618 \times 10^6$, we set $F_0 = 10^3$, then $h_1 = 2\sqrt{\frac{M_\epsilon}{F_0}} = 2\sqrt{\frac{5.618 \times 10^6}{10^3}} = 2.6683 \times 10^{-5}$.

Finally, if $f_0 = 10^{-1}$, then $h_1 = 2\sqrt{\frac{5.618 \times 10^6}{10^{-1}}} = 2.67 \times 10^{-3}$.

The above example shows how Case 1 is implemented by using Formula (45).

Regarding Case 2 when $0 \leq |f_k| < 0.1$, the value of the h_k is taken randomly from the range $[10^{-4}, 10^{-8}]$.

3.3. Estimating Gradient Vector

The forward finite difference (DFF) is utilized to compute the approximate value of the gradient vector of function f at $\boldsymbol{x} \in \mathbb{R}^n$ by

$$[\boldsymbol{DFF}]_i = \frac{f(x + \boldsymbol{h}e_i) - f(x)}{h}, \text{for } i = 1, 2, \ldots, n. \tag{46}$$

where $h > 0$ is the finite difference interval defined in Section 3.2, and $e_i \in \mathbb{R}^n$ is the ith column of the identity matrix.

Therefore, $\boldsymbol{g}(\boldsymbol{x}) \approx \boldsymbol{DFF}(\boldsymbol{x})$, is the approximate value of the gradient vector of function f at point $\boldsymbol{x}$.

Therefore, the step size φ_k is defined in the following.

The function $f(x)$ is estimated by utilizing Taylor's expansion up to the linear term around the point $\boldsymbol{x}_k$, for each iteration k. Then we have

$$f(\boldsymbol{x}_k + \boldsymbol{p}) \approx f(\boldsymbol{x}_k) + \boldsymbol{g}(\boldsymbol{x}_k)^T \boldsymbol{p}.$$

We define the quadratic model of $f(x)$ at $\boldsymbol{x}_k$ as

$$m_k(p) = \frac{1}{2}\left(f(\boldsymbol{x}_k) + \boldsymbol{g}(\boldsymbol{x}_k)^T \boldsymbol{p}\right)^2 = \frac{1}{2} f(\boldsymbol{x}_k)^2 + f(\boldsymbol{x}_k)\boldsymbol{g}(\boldsymbol{x}_k)^T \boldsymbol{p} + \frac{1}{2}\boldsymbol{p}^T \boldsymbol{g}(\boldsymbol{x}_k)\boldsymbol{g}(\boldsymbol{x}_k)^T \boldsymbol{p}.$$

Set $\boldsymbol{p} = -\varphi \boldsymbol{g}(\boldsymbol{x}_k)$ where φ is the step size along the $-\boldsymbol{g}(\boldsymbol{x}_k)$. The optimal value of the φ is picked by solving the following subproblem: $\min_{\varphi \in \mathbb{R}} m_k(\varphi) = \frac{1}{2} f(\boldsymbol{x}_k)^2 - \varphi f(\boldsymbol{x}_k)\boldsymbol{g}(\boldsymbol{x}_i)^T \boldsymbol{g}(\boldsymbol{x}_k) + \frac{1}{2}\varphi^2 (\boldsymbol{g}(\boldsymbol{x}_k)^T \boldsymbol{g}(\boldsymbol{x}_k))^2$. This gives

$$\varphi_k = \frac{f(\boldsymbol{x}_k)}{\|\boldsymbol{g}(\boldsymbol{x}_k)\|^2}. \tag{47}$$

Therefore,

$$\|\boldsymbol{g}(\boldsymbol{x}_k)\|^2 = \frac{f(\boldsymbol{x}_k)}{\varphi_k}, \varphi_k \neq 0, \tag{48}$$

where $\boldsymbol{g}(\boldsymbol{x}_k) \approx \boldsymbol{DFF}(\boldsymbol{x}_k)$.

3.4. Convergence Analysis of DFF

The condition which is usually utilized in the convergence analysis of first-order methods with inexact gradient (DFF) vectors is defined by

$$||\boldsymbol{DFF}(\boldsymbol{x}) - \boldsymbol{g}(\boldsymbol{x})|| \leq C||\boldsymbol{g}(\boldsymbol{x})||, \tag{49}$$

for some $0 \leq C < 1$. This condition is introduced by [74,75] and it is called a norm condition. This condition denotes that the $\boldsymbol{g}(\boldsymbol{x}) \approx DFF(x)$ is a descent direction for the function f [68].

However, condition (49) cannot be applied, unless we know $\|\boldsymbol{g}(\boldsymbol{x})\|$; therefore, this condition might be hard or impossible to verify.

There are many authors who have attempted to deal with this issue; see, for example, Refs. [68,76–79]. Byrd et al. [76] suggested a practical approach to estimate $\|g(x_k)\|$, and they utilized it to guarantee some approximation of (49). Cartis and Scheinberg [77] and Paquette and Scheinberg [79] replaced condition (49) by

$$\|DFF(x) - \boldsymbol{g}(\boldsymbol{x})\| \leq k\alpha_k ||\boldsymbol{g}(\boldsymbol{x})||, \tag{50}$$

where $k > 0$, and convergence rate analysis were derived for a line search method that has access to deterministic function values in [77] and stochastic function values (with additional assumptions) in [79]. Berahas et al. [68] established conditions under which (49) holds. For the forward finite differences method (DFF), they set $h^* = 2\sqrt{\frac{M_\epsilon}{L}}$.

Therefore, we present the following

Theorem 2. *Under Assumptions 1 and 2 of Section 2, let* $\boldsymbol{DFF}(\boldsymbol{x})$ *denote the forward finite difference approximation to the gradient* $\boldsymbol{g}(\boldsymbol{x})$. *Then, for all* $\boldsymbol{x} \in \mathbb{R}^n$, *the following inequality is true:*

$$\Big| \|DFF(\boldsymbol{x}_k)\|_\infty - \|g(\boldsymbol{x}_k)\|_\infty \Big| \leq |f(\boldsymbol{x}_k)_{h_i} - f(\boldsymbol{x}_k)| + \frac{f(\boldsymbol{x}_k)}{\varphi_k}, \varphi_k \neq 0, \tag{51}$$

where the value of the φ_k *is estimated by (47). We know that* $\|\boldsymbol{X}\|_\infty$ *and* $\|\boldsymbol{X}\|$ *are the norm infinity and the 2-norm, respectively, and they are defined by*

$$\|\boldsymbol{X}\|_\infty = \max_{1 \leq i \leq n} |x_i|, \tag{52}$$

$$\|\boldsymbol{X}\| = \sqrt{\sum_i x_i^{\,2}}, \tag{53}$$

and then

$$\|\boldsymbol{X}\|_\infty = \max_{1 \leq i \leq n} |x_i| \leq \sqrt{\sum_i x_i^{\,2}}. \tag{54}$$

According to (46) which defines the gradient approximation by forward differences, the vector of $[\boldsymbol{DFF}(\boldsymbol{x}_k)]_i$ is described by $[\boldsymbol{DFF}(\boldsymbol{x}_k)]_i = \frac{1}{h}[f(\boldsymbol{x}_k + e_i h) - f(\boldsymbol{x}_k)]_i$, wherer $i = 1, 2, \ldots, n$, then

$$\|\boldsymbol{DFF}(\boldsymbol{x}_k)\|_\infty = \max_{1 \leq i \leq n} \left| \left[\frac{f(\boldsymbol{x}_k + e_i h) - f(\boldsymbol{x}_k)}{h} \right]_i \right| = \frac{1}{h} \max_{1 \leq i \leq n} |[f(\boldsymbol{x}_k + e_i h) - f(\boldsymbol{x}_k)]_i|,$$

and therefore, the next inequality is true

$$\|DFF(\boldsymbol{x}_k)\|_\infty = \frac{1}{h} \max_{1 \leq i \leq n} |[f(\boldsymbol{x}_k + e_i h) - f(\boldsymbol{x}_k)]_i| \leq |f(\boldsymbol{x}_k)_{h_i} - f(\boldsymbol{x}_k)|. \tag{55}$$

By using (48), (51), (54) and (55), we obtain $\Big| \|\boldsymbol{DFF}(\boldsymbol{x}_k)\|_\infty - \|g(\boldsymbol{x}_k)\|_\infty \Big| \leq \|\boldsymbol{DFF}(\boldsymbol{x}_k)\|_\infty + \|g(\boldsymbol{x}_k)\|_\infty \leq |f(\boldsymbol{x}_k)_{h_i} - f(\boldsymbol{x}_k)| + \|g(\boldsymbol{x}_k)\|^2 = |f(\boldsymbol{x}_k)_{h_i} - f(\boldsymbol{x}_k)| + \frac{f(\boldsymbol{x}_k)}{\varphi_k}, \varphi_k \neq 0.$

Therefore, the theorem holds.

4. Numerical Experiments of Part I

All experiments were run on a PC with Intel(R) Core(TM) i5-3230M CPU@2.60GHz 2.60 GHz with RAM 4.00 GB of memory on a Windows 10 operating system. The five methods were coded by utilizing MATLAB version 8.5.0.197613 (R2015a) and the machine epsilon was about 10^{-16}.

The model optimization test problems are categorized into two types. The first type is the test problems that contain a convex function, while the second type include a non-convex function. Both kinds of test problems are listed in Tables 1–8 such that the second type of the test problem is referred to by $*$. Columns 1–4 of Table 1 give the data of the test problems as follows: the abbreviation of the function f is given on Column 1, the number of variables n is listed on Column 2, the exact function value $f(x^*)$ at the global point x^* is presented on Column 3, and the exact value of the norm of the gradient $\|g(x^*)\|$ vector is given by Column 4, where the mark "$-$" denotes that the value of the norm of the gradient $\|g(x^*)\|$ for the convex function satisfies the stopping criterion $\|g(x^*)\| < 10^{-6}$. Columns 5–8 are as Columns 1–4.

The data in Table 1 are taken from [56].

The numerical results for the local minimizers of all test problems are listed in Tables 2–8. Columns 1–2 and 8–9 contain the abbreviation of the function f and the number of the variables n, respectively. Columns 3–7 contain the abbreviation of each algorithm of the five algorithm SHZ, MHZ, HZ, HS and FR, which present the number of worst iterations, number of worst function evaluations, number of best iterations, number of best function evaluations, average of time (CPU), average of the number of iterations and average of the number of function evaluations, respectively. Columns 10–14 are similar to Columns 3–7.

Note 1: It is worth noting that the full name for each test function is mentioned in Appendix A according to the reference in which the test problem is.

Note 2: F denotes that the algorithm has failed to find the local minimizer of the function f according to the stopping criteria of Algorithm 1 which are listed in Section 4.1 below.

Table 1. List of both kinds of test problems.

f	n	$f(x^*)$	$\|g(x^*)\|$	f	n	$f(x^*)$	$\|g(x^*)\|$
Rn	10, 30, 50, 80, 100	0	-	Zn	10, 30, 50, 80, 100	0	-
PW	8, 32, 84, 120	0	-	SP	10, 30, 80, 100	0	-
Tr	10, 30, 60, 80	$\frac{-n(n+4)(n-1)}{6}$	-	Su	10, 30, 50, 80, 100	0	-
CV	4	0	-	BR	2	0.397887	-
DJ	3	0	-	BO	2	0	-
Ma	2	0	-	$S5^*$	4	−10.1532	3.2×10^{-5}
$S7^*$	4	−10.4029	-	$S10^*$	4	−10.5364	3×10^{-5}
GP^*	2	3	2×10^{-6}	Ras^*	2	−2	2.5×10^{-6}
$Bh1^*$	2	0	2.4×10^{-5}	SH^*	2	−186.7309	2×10^{-6}
$P8^*$	3	0	-	$P16^*$	5	0	1.2×10^{-6}
CB^*	2	−1.0316285	2×10^{-5}	$H3^*$	3	−3.86278	2×10^{-5}
$H6^*$	6	−3.32237	6×10^{-5}	HM^*	2	0	1.1×10^{-8}
Le^*	10	0	2.1×10^{-6}				

Table 2. The number of worst iterations.

f	n	SHZ	MHZ	HZ	HS	FR	f	n	SHZ	MHZ	HZ	HS	FR
Rn	10	2915	3740	5705	5080	5185	Rn	30	2270	3555	5170	5140	5050
Rn	50	2605	3805	5705	5290	5145	Rn	80	2750	4010	5795	5150	5890
Rn	100	2820	2950	5050	5930	5840	Zn	10	145	170	225	210	195
Zn	30	1075	995	1825	1575	1425	Zn	50	2295	2600	4180	3645	3515
Zn	80	5335	4900	9255	8610	7345	Zn	100	9095	7490	9905	9905	9905
PW	8	1470	2230	7120	3980	970	PW	32	2135	4515	9700	9700	2075
PW	84	3345	6575	9885	9885	2145	PW	120	4385	7750	9920	9920	4495
SP	10	15	25	25	30	25	SP	30	15	25	30	30	30
SP	80	15	30	25	35	25	SP	100	15	30	30	35	30
Tr	10	575	160	135	355	155	Tr	30	2830	1765	2055	9680	2280
Tr	60	9840	9840	9840	9840	9840	Tr	100	9880	9905	9905	9905	9905
Su	100	155	155	190	200	185	Su	80	140	135	175	190	185
Su	50	115	95	130	130	130	Su	30	75	80	90	95	80
Su	10	45	40	45	40	40	BR	2	75	75	70	65	200
CV	4	2070	1745	1760	2455	5705	DJ	3	15	15	35	40	30
BO	2	35	35	40	40	35	Ma	2	80	105	65	F	140
$S5^*$	4	115	445	150	750	155	$S7^*$	4	200	275	220	1500	215
$S10^*$	4	100	250	205	620	120	GP^*	2	6670	6670	6670	6670	6670
Ras^*	2	30	175	1665	280	220	$Bh1^*$	2	35	50	400	70	75
SH^*	2	6670	6670	6670	6670	6670	$P8^*$	4	20	8000	8000	1880	4730
$P16^*$	5	20	8000	8000	1880	4730	CB^*	2	25	25	115	25	150
$H3^*$	3	415	655	1300	365	7500	$H6^*$	6	445	1425	2190	8575	565
HM^*	2	25	30	25	25	25	Le^*	10	1105	1575	1815	1025	1200

Table 3. The number of worst function evaluations.

f	*n*	SHZ	MHZ	HZ	HS	FR	*f*	*n*	SHZ	MHZ	HZ	HS	FR
Rn	10	32,065	41,140	290,955	55,880	57,035	*Rn*	30	70,370	110,205	160,270	159,340	156,550
Rn	50	132,855	194,055	290,955	269,790	262,395	*Rn*	80	222,750	324,810	469,395	417,150	477,090
Rn	100	284,820	297,950	510,050	598,930	589,840	*Zn*	10	1595	1870	2475	2310	2145
Zn	30	33,325	30,845	56,575	48,825	44,175	*Zn*	50	117,045	132,600	213,180	185,895	179,265
Zn	80	432,135	396,900	749,655	697,410	594,945	*Zn*	100	918,595	756,490	1,000,405	1,000,405	1,000,405
PW	8	13,230	20,070	64,080	35,820	8730	*PW*	32	70,455	148,995	320,100	320,100	68,475
PW	84	284,325	558,875	840,225	840,225	182,325	*PW*	120	530,585	937,750	1,200,320	1,200,320	543,895
SP	10	165	275	275	330	275	*SP*	30	465	775	930	930	930
SP	80	1215	2430	2025	2835	2025	*SP*	100	1515	3030	3030	3535	3030
Tr	10	6325	1760	1485	3905	1705	*Tr*	30	87,730	54,715	63,705	300,080	70,680
Tr	60	600,240	600,240	600,240	600,240	600,240	*Tr*	100	800,280	1,000,405	1,000,405	1,000,405	1,000,405
Su	100	15,655	15,655	19,190	20,200	18,685	*Su*	80	11,340	10,935	14,175	15,390	14,985
Su	50	5865	4845	6630	6630	6630	*Su*	30	2325	2480	2790	2945	2480
Su	10	495	440	495	440	440	*BR*	2	225	225	210	195	600
CV	4	10,350	8725	8800	12,275	28,525	*DJ*	3	60	60	140	160	120
BO	2	105	105	160	120	105	*Ma*	2	240	315	195	F	420
*S5**	4	575	2225	750	3750	775	*S7**	4	1000	1375	1100	7500	1075
*S10**	4	500	1250	1025	3100	600	*GP**	2	20,010	20,010	20,010	20,010	20,010
*Ras**	2	90	525	4995	840	660	*Bh1**	2	105	150	1200	210	225
*SH**	2	20,010	20,010	20,010	20,010	20,010	*P8**	4	100	40,000	40,000	9400	23,650
*P16**	5	100	40,000	40,000	9400	23,650	*CB**	2	75	75	345	75	450
*H3**	3	1660	2620	5200	1460	30,000	*H6**	6	3115	9975	15,330	60,025	3955
*HM**	2	75	90	75	75	75	*Le**	10	12,155	17,325	19,965	11,275	13,200

Table 4. The number of best iterations.

f	*n*	SHZ	MHZ	HZ	HS	FR	*f*	*n*	SHZ	MHZ	HZ	HS	FR
Rn	10	360	460	490	520	510	*Rn*	30	230	485	190	590	420
Rn	50	705	375	490	650	440	*Rn*	80	230	400	920	125	460
Rn	100	240	275	885	670	705	*Zn*	10	60	60	115	80	75
Zn	30	245	330	875	875	810	*Zn*	50	570	905	2235	1765	1885
Zn	80	935	1565	4080	4365	3495	*Zn*	100	1670	2545	6345	6045	5095
PW	8	180	175	2080	375	225	*PW*	32	280	2610	1250	390	280
PW	84	510	3525	4115	520	410	*PW*	120	535	2745	2765	395	435
SP	10	5	10	10	20	10	*SP*	30	10	10	10	20	10
SP	80	10	10	10	20	15	*SP*	100	10	10	10	20	15
Tr	10	85	85	65	80	55	*Tr*	30	735	1370	230	350	220
Tr	60	9840	9840	9840	9840	430	*Tr*	100	9880	9905	9905	9905	9905
Su	100	70	80	95	95	75	*Su*	80	65	55	75	100	70
Su	50	50	55	60	70	50	*Su*	30	40	40	40	45	40
Su	10	20	25	20	25	20	*BR*	2	15	15	15	15	10
CV	4	275	275	690	370	600	*DJ*	3	10	10	10	20	10
BO	2	15	15	20	20	20	*Ma*	2	30	20	20	F	15
*S5**	4	15	20	20	25	125	*S7**	4	15	15	15	30	100
*S10**	4	15	15	20	15	100	*GP**	2	25	180	170	60	165
*Ras**	2	10	20	95	15	45	*Bh1**	2	20	20	30	25	75
*SH**	2	390	255	840	155	20010	*P8**	4	10	15	5	15	125
*P16**	5	10	15	5	15	125	*CB**	2	15	15	20	15	45
*H3**	3	5	5	5	5	15	*H6**	6	50	50	50	50	175
*HM**	2	10	15	15	15	30	*Le**	10	65	40	105	70	550

Table 5. The number of best function evaluations.

f	*n*	SHZ	MHZ	HZ	HS	FR	*f*	*n*	SHZ	MHZ	HZ	HS	FR
Rn	10	3960	5060	24,990	5720	5610	*Rn*	30	7130	15,035	5890	18,290	13,020
Rn	50	35,955	19,125	24,990	33,150	22,440	*Rn*	80	18,630	32,400	74,520	10,125	37,260
Rn	100	24,240	27,775	89,385	67,670	71,205	*Zn*	10	660	660	1265	880	825
Zn	30	7595	10,230	27,125	27,125	25,110	*Zn*	50	29,070	46,155	113,985	90,015	96,135
Zn	80	75,735	126,765	330,480	353,565	283,095	*Zn*	100	168,670	257,045	640,845	610,545	514,595
PW	8	1620	1575	18,720	3375	2025	*PW*	32	9240	86,130	41,250	12,870	9240
PW	84	43,350	299,625	349,775	44,200	34,850	*PW*	120	64,735	332,145	334,565	47,795	52,635
SP	10	55	110	110	220	110	*SP*	30	310	310	310	620	310
SP	80	810	810	810	1620	1215	*SP*	100	1010	1010	1010	2020	1515
Tr	10	935	935	715	880	605	*Tr*	30	22,785	42,470	7130	10,850	6820
Tr	60	600,240	600,240	600,240	600,240	26,230	*Tr*	100	800,280	1,000,405	1,000,405	1,000,405	1,000,405
Su	100	7070	8080	9595	9595	7575	*Su*	80	5265	4455	6075	8100	5670
Su	50	2550	2805	3060	3570	2550	*Su*	30	1240	1240	1240	1395	1240
Su	10	220	275	220	275	220	*BR*	2	45	45	45	45	30
CV	4	1375	1375	3450	1850	3000	*DJ*	3	40	40	40	80	40
BO	2	45	45	80	60	60	*Ma*	2	90	60	60	F	45
*S5**	4	75	100	100	125	125	*S7**	4	75	75	75	150	100
*S10**	4	75	75	100	75	100	*GP**	2	75	540	510	180	165
*Ras**	2	30	60	285	45	45	*Bh1**	2	60	60	90	75	75
*SH**	2	1170	765	2520	465	20,010	*P8**	4	50	75	25	75	125
*P16**	5	50	75	25	75	125	*CB**	2	45	45	60	45	45
*H3**	3	15	15	15	15	15	*H6**	6	300	300	300	300	175
*HM**	2	30	45	45	45	30	*Le**	10	715	440	1155	770	550

Table 6. The average of time.

f	*n*	SHZ	MHZ	HZ	HS	FR	*f*	*n*	SHZ	MHZ	HZ	HS	FR
Rn	10	1.463	1.441	3.316	2.436	3.215	*Rn*	30	2.771	3.702	6.831	6.934	6.816
Rn	50	5.571	6.123	13.288	12.286	13.258	*Rn*	80	10.149	11.273	19.529	21.934	22.283
Rn	100	14.761	15.973	29.596	29.495	32.934	*Zn*	10	0.083	0.091	0.139	0.137	0.115
Zn	30	1.336	1.365	2.477	3.220	2.290	*Zn*	50	5.445	5.297	11.689	11.293	10.938
Zn	80	24.818	27.532	58.547	55.403	50.910	*Zn*	100	53.552	51.210	107.859	104.313	109.531
PW	8	0.493	1.231	4.760	0.985	0.309	*PW*	32	1.783	6.812	21.873	6.496	1.085
PW	84	8.779	38.644	76.499	16.061	4.562	*PW*	120	16.955	72.473	113.171	29.623	7.631
SP	10	0.011	0.016	0.020	0.026	0.017	*SP*	30	0.021	0.032	0.033	0.042	0.036
SP	80	0.060	0.099	0.096	0.165	0.096	*SP*	100	0.075	0.137	0.125	0.191	0.148
Tr	10	0.183	0.084	0.068	0.144	0.069	*Tr*	30	2.948	2.891	1.982	24.737	1.256
Tr	60	63.990	73.812	80.235	58.588	70.106	*Tr*	100	90.259	130.122	134.463	145.078	135.992
Su	100	4.542	4.706	4.736	5.194	5.127	*Su*	80	2.288	2.753	2.839	2.948	2.716
Su	50	0.780	0.799	0.842	0.921	0.889	*Su*	30	0.294	0.265	0.298	0.296	0.247
Su	10	0.051	0.043	0.038	0.041	0.036	*BR*	2	0.022	0.024	0.022	0.019	0.045
CV	4	0.568	0.505	0.762	0.774	6.317	*DJ*	3	0.008	0.009	0.013	0.020	0.013
BO	2	0.014	0.014	0.016	0.016	0.016	*Ma*	2	0.026	0.026	0.017	F	0.019
*S5**	4	0.107	0.321	0.166	0.297	0.162	*S7**	4	0.231	0.204	0.180	0.554	0.273
*S10**	4	0.124	0.180	0.208	0.432	0.194	*GP**	2	6.068	7.927	5.410	11.203	3.164
*Ras**	2	0.021	0.091	1.355	0.109	0.120	*Bh1**	2	0.019	0.030	0.184	0.039	0.043
*SH**	2	13.341	11.597	13.226	12.487	17.294	*P8**	4	0.011	0.307	0.234	0.247	0.155
*P16**	5	0.128	0.276	3.452	0.129	3.886	*CB**	2	0.014	0.015	0.060	0.015	0.058
*H3**	3	0.103	0.411	0.203	0.114	0.400	*H6**	6	0.224	0.902	0.205	1.064	0.164
*HM**	2	0.016	0.021	0.015	0.015	0.016	*Le**	10	0.501	0.513	0.836	0.553	0.612

Table 7. The average of number of iterations.

f	n	SHZ	MHZ	HZ	HS	FR	f	n	SHZ	MHZ	HZ	HS	FR
Rn	10	1469.3	1479.3	3114.8	2517.2	2952.5	*Rn*	30	1273.4	1523.8	2877.9	2867.5	2721.5
Rn	50	1375	1530.6	3114.8	2746.4	2851.1	*Rn*	80	1379.2	1535.2	2524.7	2885.5	2593.2
Rn	100	1403.9	1421.2	2821	2839	2809.7	*Zn*	10	104.61	108.92	168.04	155.2	142.16
Zn	30	654.02	674.22	1229.3	1187.3	1078.8	*Zn*	50	1491.3	1479.2	2947.1	2914.6	2817
Zn	80	3378.6	3298.6	6529	6519.2	6185.1	*Zn*	100	5125.6	4818.4	9066.1	8703.7	9048.1
PW	8	697.16	1731.6	5774.2	1339.6	441.47	*PW*	32	1042.5	3675	8852.9	2993.3	660.1
PW	84	1665.2	5767.7	9547.3	2632.5	817.55	*PW*	120	1774.8	6851.4	9424	2964.5	897.55
SP	10	10.294	16.765	16.471	23.824	18.529	*SP*	30	10.784	17.941	18.627	25.294	21.176
SP	80	11.176	19.118	19.216	26.471	19.412	*SP*	100	10.588	19.314	18.725	26.765	20.98
Tr	10	283.24	125.88	100.59	182.06	96.961	*Tr*	30	1713.5	1610.5	1053	7268.8	649.41
Tr	60	9840	9840	9840	9840	9117.5	*Tr*	100	9880	9905	9905	9905	9905
Su	100	116.08	112.35	152.06	147.94	141.86	*Su*	80	96.373	99.02	134.9	137.25	117.35
Su	50	76.667	72.353	95.98	95.686	89.51	*Su*	30	58.725	54.314	70.392	69.608	57.255
Su	10	32.451	29.706	31.961	33.627	29.706	*BR*	2	36.961	32.255	35.686	31.667	49.902
CV	4	704.41	634.9	1114.1	1015.2	3365.8	*DJ*	3	11.078	11.373	21.176	31.275	18.824
BO	2	24.608	23.824	29.02	28.235	27.451	*Ma*	2	58.824	60.882	39.902	F	40.882
*S5**	4	52	106.5	76.75	255.25	66.75	*S7**	4	73.25	73	61.25	387.75	76.5
*S10**	4	40.25	49.75	57.75	172.75	51.75	*GP**	2	2053.8	3273.3	2340.3	4125	1381
*Ras**	2	21.5	68.25	802.25	86.25	88.25	*Bh1**	2	28.5	33.25	119.5	38	42.5
*SH**	2	5927.5	5855	6184.5	6344.3	6670	*P8**	4	13.25	771.5	575.25	603.75	367.5
*P16**	5	13.25	771.5	575.25	603.75	367.5	*CB**	2	19.75	19	50.75	19.75	47.5
*H3**	3	92.25	201.25	225.5	104.75	450	*H6**	6	108.25	250	130.25	580.5	103.75
*HM**	2	19.75	20.25	19	19.25	19	*Le**	10	303.5	323.25	550.75	375	379

Table 8. The average of number of function evaluations.

f	n	SHZ	MHZ	HZ	HS	FR	f	n	SHZ	MHZ	HZ	HS	FR
Rn	10	16,163	16,273	158,855	27,689	32,477	*Rn*	30	39,476	47,239	89,216	88,894	84,366
Rn	50	70,125	78,060	158,855	140,065	145,405	*Rn*	80	111,717	124,351	204,501	233,725	210,052
Rn	100	141,796	143,539	284,919	286,741	283,780	*Zn*	10	1151	1198	1848	1707	1564
Zn	30	20,275	20,901	38,109	36,805	33,444	*Zn*	50	76,055	75,440	150,300	148,645	143,665
Zn	80	273,669	267,189	528,851	528,057	500,993	*Zn*	100	517,684	486,662	915,674	879,076	913,862
PW	8	6274	15,584	51,968	12,057	3973	*PW*	32	34,404	121,275	292,147	98,780	21,783
PW	84	141,542	490,258	811,517	223,758	69,492	*PW*	120	214,751	829,016	1,140,306	358,706	108,603
SP	10	113	184	181	262	204	*SP*	30	334	556	578	784	657
SP	80	905	1549	1557	2144	1572	*SP*	100	1069	1951	1891	2703	2119
Tr	10	3116	1385	1107	2003	1067	*Tr*	30	53,119	49,925	32,644	225,333	20,132
Tr	60	600,240	600,240	600,240	600,240	556,165	*Tr*	100	800,280	1,000,405	1,000,405	1,000,405	1,000,405
Su	100	11,724	11,348	15,358	14,942	14,328	*Su*	80	7806	8021	10,927	11,118	9506
Su	50	3910	3690	4895	4880	4565	*Su*	30	1821	1684	2182	2158	1775
Su	10	357	327	352	370	327	*BR*	2	111	97	107	95	150
CV	4	3522	3175	5571	5076	16,829	*DJ*	3	44	46	85	125	75
BO	2	74	72	116	85	82	*Ma*	2	177	183	120	F	123
*S5**	4	260	533	384	1276	334	*S7**	4	366	365	306	1939	383
*S10**	4	201	249	289	864	259	*GP**	2	6161	9820	7021	12,375	4143
*Ras**	2	65	205	2407	259	265	*Bh1**	2	86	100	359	114	128
*SH**	2	17,783	17565	18,554	19,033	20,010	*P8**	4	66	3858	2876	3019	1838
*P16**	5	66	3858	2876	3019	1838	*CB**	2	59	57	152	59	143
*H3**	3	369	805	902	419	1800	*H6**	6	758	1750	912	4064	671
*HM**	2	59	61	57	58	57	*Le**	10	3339	3556	6058	4125	4169

The stopping criteria of Algorithm 1 are as follows.

4.1. Stopping Criteria of Algorithm 1

Since this section focuses in finding a local minimizer of all test problems, the stopping criteria of Algorithm 1 can be defined as follows.

According to the discussions of the convergence analysis which are mentioned in the previous sections, the stopping criterion of Algorithm 1 is, if $\|g(x_k)\| \leq \varepsilon_1$ is satisfied, Algorithm 1 stops, where $\varepsilon_1 \in [10^{-6}, 10^{-8}]$. However, the exact value of the gradient vector is unknown since the value of the gradient vector is estimated by Formula (46); therefore, this condition is replaced by $\|\boldsymbol{DFF}_k\| \leq \varepsilon_2$ or FEs $= n10^4$, i.e., if one of them is met, Algorithm 1 stops, where $\varepsilon_2 \in [10^{-7}, 10^{-9}]$, FEs denotes the maximum function evaluations and n is the number variables of the f.

In the following section, the performance profile is presented as an easy tool to compare the performance of our proposed method versus other methods in finding local minimizers of convex or non-convex functions regarding the worst and best numbers of iterations and function evaluations, the average of CPU time and the average of iterations and function evaluations, respectively.

4.2. Performance Profiles

The performance profile is the best tool for testing the performance of the proposed algorithms [80–84].

In this paper, the five algorithms' performance evaluation standards are as follows: the worst and best numbers of iterations and function emulations, and the average of the CPU time, iterations and function emulations. They are abbreviated as itr.w, itr.be, FEs.w, FEs.be, time.a, itr.a and EFs.a, respectively. In the remainder of the paper, the set Fit will be used to denote the seven criteria; Fit $= \{\text{itr.w}, \text{itr.be}, \text{FEs.w}, \text{FEs.be}, \text{time.a}, \text{itr.a}, \text{EFs.a}\}$.

Therefore, the numerical outcomes are presented in the form of performance profiles, as depicted in [82]. The most important characteristic of the performance profiles is that they can be shown in one figure by plotting for the different solvers a cumulative distribution function $\rho_s(\tau)$.

The performance ratio is defined by first setting $r_{p,s} = \frac{t_{p,s}}{\min\{t_{p,s}: s \in S\}}$, where $p \in P$, P is a set of test problems, S is the set of solvers, and $t_{p,s}$ is the value obtained by solver s on test problem p.

Then, define $\rho_s(\tau) = \frac{1}{|P|}\text{size}\{p \in P : r_{p,s} \leq \tau\}$, where $|P|$ is the number of test problems.

The value of $\rho_s(1)$ is the probability that the solver will win over the remaining ones, i.e., it will yield a value lower than the values of the remaining ones.

In the following, the performance profiles are utilized to evaluate the performance of the five methods: SHZ, MHZ, HZ, SH and FR.

Therefore, in this paper, the term $t_{p,s}$ indicates one element of the set Fit, $|P| = 46$ is the number of test problems. We have 46 unconstrained test problems, 14 of which include non-convex functions. The group of solvers $S = \{SHZ, MHZ, HZ, SH, FR\}$ finds the local minimizers of the 46 test problems; therefore, the values of the Fit are taken from the results of the 46 test problems as follows.

Each solver s of the set S is run 51 times for each of the 46 problems; at each run, every element of the set Fit has owned its value. So, they are analyzed in the following.

$$r_{p,s} = \begin{cases} \frac{\text{fit}_{p,s}}{\min\{\text{fit}_{p,s}: s \in S\}} & \text{if the s pass to solve the p,} \\ \infty & \text{otherwise,} \end{cases} \tag{56}$$

where $\text{fit}_{p,s}$ is an element of the Fit for the test problem p by using the solver s.

Note: Formula (56) means that if the final result, obtained by a solver $s \in S$, satisfies Inequality (57), then the first branch of (56) is computed. Otherwise, we set $r_{p,s} = \infty$.

$$\|DFF_k\| \leq \varepsilon_2, \tag{57}$$

where $\varepsilon_2 \in [10^{-5}, 10^{-9}]$.

Therefore, the performance profile of solver s is defined as follows:

$$\delta(r_{p,s}, \tau) = \begin{cases} 1 & \text{if } r_{p,s} \leq \tau, \\ 0 & \text{otherwise,} \end{cases} \tag{58}$$

Therefore, the performance profile for solver s is then given by the following function:

$$\rho_s(\tau) = \frac{1}{|P|}\left\{ \sum_{p \in P} \delta(r_{p,s}, \tau) \right\}, \ \tau \geq 1. \tag{59}$$

As we mentioned above, $|P| = 46$ and $\tau \in [1, 60]$.

By definition of $\text{Fit}_{p,s}$, $\rho_s(1)$ denotes the fraction of test problems for which solver s performs the best. In general, $\rho_s(\tau)$ can be explained as the probability for solver $s \in S$ that the performance ratio $r_{p,s}$ is within a factor τ of the best possible ratio. Additionally, the essential characteristic of performance profiles is that they present data on the proportional performance of numerous solvers [82,83].

The numerical outcomes of the five methods are analyzed by using the performance profiles as follows. Figures 1–4 show the performance profiles of the set solvers S, for each element of the set Fit, respectively.

The performance profile depicted on the left of Figure 1 (in the term itr.w) compares the five techniques for a set of the 46 test problems.

The SHZ method has the best performance for the 46 test problems; this means that our suggested approach is capable of finding a local minimizer to the 46 test problems as fast as, or faster than, the other four approaches.

For instance, if $\tau = 1$, the SHZ technique is capable of finding the local minimizer for 65% of problems versus the 33%, 20%, 20% and 13% of a set of test problems solved by the MHZ, HS, FR and HZ methods, respectively.

In general, the term itr.w, $\tau = 60$ displays that all test problems are solved by SHZ against 96% of test problems solved by the MHZ, HZ and FR methods respectively, while 93% of test problems are solved by the HS method. At $\tau \geq 400$, all test problems are solved by the MHZ, HZ and FR methods respectively, while 98% of test problems are solved by the HS.

The right graph of Figure 1 shows that the method SHZ is capable of finding the local minimum of all test problems regarding term FEs.w.

The rest of Figures 2–4 show that the SHZ algorithm is superior to the four algorithms regarding the rest of the terms of the set Fit.

Therefore, the SHZ technique includes the characteristics of efficiency, reliability and effectiveness in solving Problem (1) compared to the other four methods.

Note: The power of the SHZ technique comes from the fact that the SHZ method gains the features of the four methods MHZ, HZ and HS, as we mentioned in Section 2.

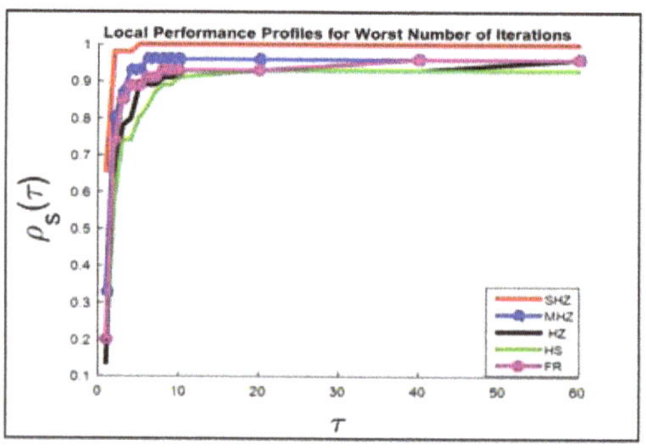

Figure 1. *Cont.*

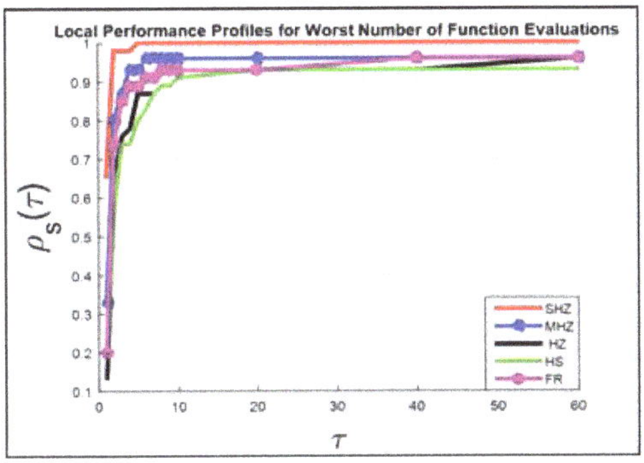

Figure 1. Plotting the results of the terms itr.w and FEs.w for 5 algorithms.

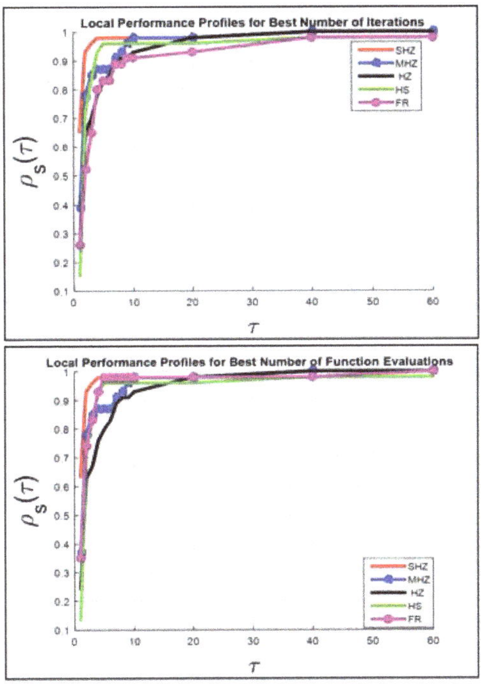

Figure 2. Plotting the results of the terms itr.be and FEs.be for 5 algorithms.

Part II: Global Minimization Problem

It is worth mentioning that the final results of Part I for the second set of test problems contain some global minimizers at some runs for some non-convex functions. This means that the pure CG technique could not find the global minimizer of the second type of test problems for each run because it is a local method.

Therefore, to make this method capable of solving Problem (2) per run, the random technique is proposed and it is added to the CG approach to gain a new PS-CG hybrid technique that solves Problem (2). In many studies, the numerical outcomes indicated that the interbreed between a classical method and a random technique is very successful in overcoming the weakness of these methods. See [55–59].

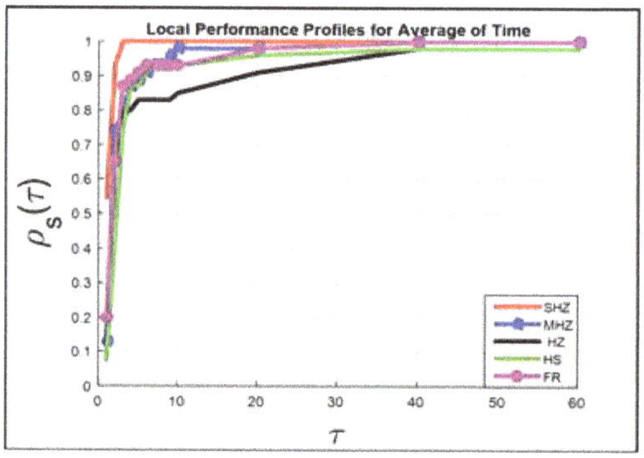

Figure 3. Plotting the results of the term time.a "CPU" for 5 algorithms.

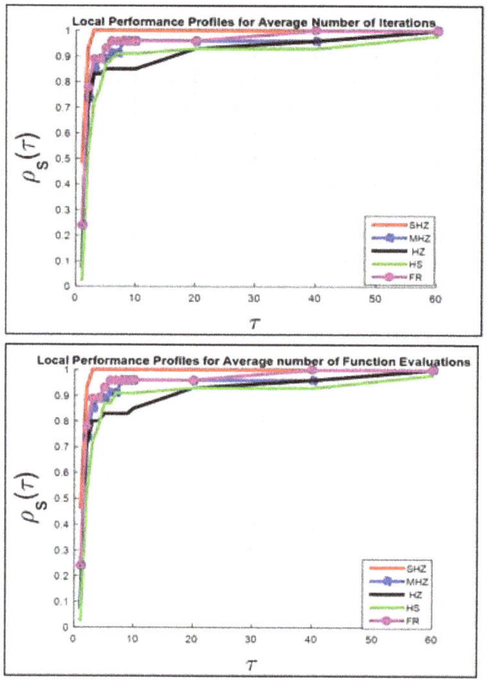

Figure 4. Plotting the results of the terms itr.a and FEs.a for 5 algorithms.

Consequently, this part of the paper seeks to solve Problem (2).

Therefore, each method of the five CG methods mentioned in Part I is hybridized with the stochastic technique to obtain five algorithms to try to solve Problem (2).

In the next section, a stochastic technique is presented.

5. Random Technique

In this section, a new random parameter "SP" is presented. This stochastic technique contains three different formulas by which three different points are generated. This set of formulas is combined with the CG method to obtain a new algorithm that solves Problem (2).

Random Parameters (SP Technique)

Step 1: The first point is computed as follows, generate $V_k \sim [-1,1]^n$ is as a random vector, set $\gamma_k = 10^{\psi_k}$, $\psi_k \in [0.01,1)$, where the interval $[0.01,1)$ is divided into Itr of fractions and at every iteration k, the parameter ψ_k takes one value of the Itr and then computes $\lambda_k = \frac{(1+\gamma_k)^{|V_i|}}{\gamma_k} SV_i$ as a research direction with the step lengths, where $i = 1,2,\ldots,n$, n is the of number variables, Itr is the number of iterations, and SV_i denotes the signs of the $\boldsymbol{V}$ and is defined by

$$SV_i = \begin{cases} -1 & \text{if } V_i < 0, \\ 1 & \text{otherwise.} \end{cases} \tag{60}$$

Thus, a point is calculated as follows:

$$\boldsymbol{x}_1 = \boldsymbol{x}_{ac} + \boldsymbol{\lambda}_k, \tag{61}$$

where $\boldsymbol{x}_{ac}$ is the best point obtained yet, and then we compute $f_1 = f(x_1)$.

Step 2: The second point is defined by

$$\boldsymbol{x}_2 = \boldsymbol{x}_{ac} + \eta_k \boldsymbol{B}_k, \tag{62}$$

where $B_k = \varphi_k \boldsymbol{d}_k$, φ_k is defined by (47), $\eta_k \in (0,2)$ is a random number, and the $\boldsymbol{d}_k$ is defined by (23). Then, we compute $f_2 = f(\boldsymbol{x}_2)$.

Step 3: This point is defined by

$$\boldsymbol{x}_3 = \boldsymbol{X}_w + \frac{1}{2}\boldsymbol{Dx}, \tag{63}$$

where $\boldsymbol{Dx} = \frac{(1+\mu_k)^{|V_i|-1}}{\mu_k+0.1} SV_i$, $\mu_k = |f_{ac}|^2$, f_{ac} is the function value at the point x_{ac} that has been accepted, and $\boldsymbol{X}_w$ is a stochastic variable picked from the feasible range of the objective function. This means that for $\boldsymbol{X}_w \sim [a,b]^n$, a and b are the lower and upper bounds of the feasible range, respectively, and the random vector $\boldsymbol{V}$ with its signs SV_i is defined by the first step.

Therefore, we calculate $f_3 = f(\boldsymbol{x}_3)$.

For finding the global minimizer of a non-convex function, the above stochastic technique is used since Algorithm 1 is not capable of finding the global solution at each run. In other words, in some runs, Algorithm 1 fails to find the global solution to this function due to it sticking to a local point.

In the following example, we show how the SP algorithm is run.

Example: This example shows how the three steps of the SP algorithm are implemented. We use the first test problem of the list of the test problems that are listed in Appendix A. $R_2(x) = 100(x_1^2 - x_2)^2 + (x_1 - 1)^2$, to facilitate an explanation of the mechanism of using the Sp algorithm (Formulas (61)–(63)), we use the following easy information about the function $R_2(x)$, $n = 2$ is the number of the variables, $x_{ac} = [2;-1]$, or $x_{ac} = [2;1]$, where x_{ac} represents the best solution has been accepted so far or the starting point; hence, the function values at the two points are $R_2(x_{ac}) = 100(2^2+1)^2 + (2-1)^2 = 2500 + 1 = 2501$ and $R_2(x_{ac}) = 100(2^2-1)^2 + (2-1)^2 = 900 + 1 = 901$.

Supposing $Itr = 5$ is the number of iterations, the interval $[0.01;1)$ is divided into five fractions with step size $\frac{1-0.01}{5} = 0.198$, and thus the set of this fractions is $A = \{0.01,\ 0.208,\ 0.406,\ 0.604,\ 0.802\}$, let k be 3 which means the algorithm is at the third iteration. Then, $\psi_3 = 0.406$, $\gamma_3 = 10^{\psi_3} = 10^{0.406} = 2.5468$. Let V_3 be $[-0.5;1]$, then $\lambda_3 = \left[\frac{(1+2.5468)^{|-0.5|}}{2.5468} \times -1;\ \frac{(1+2.5468)^{|1|}}{2.5468} \times 1\right] = \left[-\frac{1.8833}{2.5468}; \frac{3.5468}{2.5468}\right] = \left[-0.73948; 1.3926\right]$.

Therefore, the new solution is computed by Formula (61) as follows.

$x_1 = x_{ac} + \lambda_3 = [2;-1] + [-0.73948;\ 1.3926] = [1.2605;\ 0.3926]$ or $x_1 = x_{ac} + \lambda_3 = [2;1] + [-0.73948; 1.3926] = [1.2605;\ 2.3926]$.

The function values at both points are as follows.

$R_2(x_1) = 100(1.2605^2 - 0.3926)^2 + (1.2605 - 1)^2 = 143.1 + 0.06786 = 143.17$ or $R_2(x_1) = 100(1.2605^2 - 2.3926)^2 + (1.2605 - 1)^2 = 64.6 + 0.06786 = 64.668$.

Therefore, $R_2(x_1) < R_2(x_{ac})$; this means the solution that is generated by Formula (61) reduces the function value.

In the following, we explain how the candidate solution is generated by Formula (62).

Let $M_\epsilon = 1.2 \times 10^{-6}$. By using Formula (45), we obtain $h_3 = 4.381 \times 10^{-5}$ as the step size h (a random interval) to the difference approximations method, and then we have $x_{h_1} = [x_{ac}(1) + h_3;\ x_{ac}(2)] = [2 + 4.381 \times 10^{-5}; -1]$, $x_{h_2} = [x_{ac}(1);\ x_{ac}(2) + h_3] = [2;\ -1 + 4.381 \times 10^{-5}]$.

Therefore, the values of the function at the three points x_{ac}, x_{h_1} and x_{h_2} are listed in the following.

$R_2(x_{ac}) = 2501$, $R_2(x_{h_1}) = 2501.175$ and $R_2(x_{h_2}) = 2500.956$.

We compute the approximate value of the gradient vector by Formula (46) as follows:

$$DFF(x_{ac}) = \left[\frac{2501.175 - 2501}{4.381 \times 10^{-5}}; \frac{2500.956 - 2501}{4.381 \times 10^{-5}}\right] = \left[3994.522;\ -1004.34\right],$$

$\varphi_3 = \frac{2501}{\|DFF\|^2} = 0.0002$, where φ_3 is defined by (47).

We consider $d_3 = -g(x_{ac}) \approx \left[-3994.522;\ 1004.34\right]$ because we do not have information about the value of the d_2 in this illustration example.

Now, we apply Formula (62), as follows $B_3 = \varphi_3 d_2 = [-0.799;\ 0.201]$, we take $\eta_3 = 0.971$ as a random number from the range $(0,\ 2)$, then $x_2 = [2; -1] + 0.971 \times [-0.799;\ 0.201] = [1.2242;\ -0.80483]$, the function value at the point x_2 is $R_2(x_2) = 530.66$.

We note that the $R_2(x_2) = 530.66 < R_2(x_{ac}) = 2501$, i.e., the function value is reduced by the point x_2.

In the following, we explain how the candidate solution is generated by Formula (63).

$\mu_3 = |f_{ac}|^2 = 2501^2 = 6{,}255{,}001$, $\boldsymbol{Dx} = \left[\frac{(1+6{,}255{,}001)^{|-0.5|} - 1}{6{,}255{,}001 + 0.1} \times -1; \frac{(1+6{,}255{,}001)^{|1|} - 1}{6{,}255{,}001 + 0.1} \times 1\right] = \left[-\frac{2501-1}{6{,}255{,}001.1}; \frac{6{,}255{,}002-1}{6{,}255{,}001.1}\right] = \left[-0.0004;\ 0.999\right]$. $\boldsymbol{X}_w = [-3.095;\ 8.701]$ is as a random vector picked from the range $[-5, 10]^2$, and then $x_3 = [-3.095;\ 8.701] + \frac{1}{2}[-0.0004;\ 0.999] = [-3.095;\ 8.701] + [-0.0002;\ 0.4995] = [-3.0952;\ 9.2005]$.

We compute the function value at the point x_3; $R_2(x_3) = 100((-3.0952)^2 - 9.2005)^2 + (-3.0952 - 1)^2 = 14.422 + 16.771 = 31.193$.

We note that the $R_2(x_3) = 31.193 < R_2(x_{ac}) = 2501$. Therefore, the point x_3 minimizes the function value.

According to the above example that illustrates the mechanism of Formulas (61)–(63), we deduce the following results.

Remark 1. *Formulas (3), (61) and (62) are the main formulas which are used in the new hybrid proposed algorithm that is described in Section 6. However, Formula (63) is used when $\Delta f = 0$ that is defined by Formula (25); in this case, Algorithm 3 reaches a critical point, thus if this point is the approximate value of the global minimizer point of the f, then Algorithm 3 stops according to the condition in Line 4 or Line 1 of Algorithm 3. Otherwise, the candidate solution is generated by Formula (63); see Section 6. Consequently, in this example, at iteration $k = 3$, the result which is obtained by Formula (63) cannot be taken into account due to the $\Delta f \neq 0$.*

Remark 2. *All Formulas (61)–(63) minimize the function value from any starting point.*

6. Hybridization of the CG Method with Stochastic Parameters

When a stochastic method as a global optimization algorithm is combined with a globally convergent method (deterministic method), the result is a global optimization algorithm [55,56].

Therefore, the SP technique is hybridized with each of the five conjugate gradient methods SHZ, MHZ, HZ, HS and FR to obtain five techniques.

Our proposed algorithm is called a hybrid stochastic CG method abbreviated by HSSHZ that solves Problem (2). However, Algorithm 3 represents five alternative algorithms when the SHZ method is hybridized with the PS technique, then we obtain a new algorithm abbreviated by HSSHZ. When we combine any method of MHZ, HZ, HS or FR, we obtain four other abbreviations of algorithms as follows: HSMHZ, HSHZ, HSHS and HSFR, respectively.

In general, the outputs of this paper are five algorithms that solve Problem (2), where the best one is the HSSHZ algorithm as illustrated by the numerical experiments section of Part II.

In the following, Algorithm 1 is combined with SP technique to obtain Algorithm 3.

The SP method permits conducting an exhaustive wipe of the search range to guarantee that the global minimizer point is visited at least once per run.

Algorithm 3 Hybrid stochastic CG method.

Input: $f : \mathbb{R}^n \to \mathbb{R}, f \in C^1, f_{ac} = f_{cg}$ gained by Algorithm 1 and $\varepsilon > 0$.
Output: $x_{gl} = x_{ac}$ the global minimizer of f, $f(x_{gl})$, the value of f at x_{gl}.

1: **while** $|f_{ac} - f^*| > \varepsilon$ or FEs$< n10^4$ **do**
2:     f_{cg} is a function value f gained by Algorithm 1.
3:     $f_{ac} = \min\{f_{cg}, f_1, f_2\}$ and x_{ac} the best point gives the f_{ac}.
4:     **if** $|f_{ac} - f^*| \leq \varepsilon$ **then**
5:         Stop.
6:     **end if**
7:     **if** $\triangle f == 0$ **then**
8:         calculate the x_3 and the $f_3 = f(x_3)$ by Formula (63).
9:         **if** $f_3 < f_{ac}$ **then**
10:             the x_3 is accepted, compute the $x_{ac} \to x_3$, $f_{ac} \to f_3$, and go to Line 1.
11:         **else**
12:             generate another point x_3 by Formula (63).
13:         **end if**
14:     **else**
15:         go to Line 1.
16:     **end if**
17: **end while**
18: **return** x_{ac} the best point and its function value f_{ac}

A Mechanism Running Algorithm 3

As we mentioned above, Algorithm 3 is a combination of two methods; the first is a CG method of the five techniques CG $= \{SHZ, MHZ, HZ, SH, FR\}$ that are discussed in Part I, and the second is a random method is depicted by Section 5. The point x_{cg} is obtained by Algorithm 1 and it will be an input to Algorithm 3.

Algorithm 3 begins with Line 1 that is the stopping standard of the algorithm. Therefore, Algorithm 3 ends if one of the following standards is satisfied: The first standard is $|f_{ac} - f^*| \leq \varepsilon$, and the second standard is FEs$\geq n10^4$, where f_{ac} the best value of the function f is gained, the f^* is the true solution, $\varepsilon = 10^{-6}$, FEs is the number of function evaluations, and FEs $= n10^4$ is a stopping standard indicated by [85,86].

In Line 3, the best value of f is selected from the three values of the function f_{cg}, f_1 and f_2, and indicated by f_{ac}, the three values of the function f are calculated by Algorithms (1), (61) and (62), respectively, and x_{ac} indicates this.

In Line 4, if $|f_{ac} - f^*| \leq \varepsilon$ is fulfilled, the algorithm ends. The standard that is listed in Line 7 gives the algorithm an opportunity to flee from the local points. Consequently, if $\triangle f = 0$, then the algorithm has reached a crucial point. Therefore, if the norm of the gradient vector is 0 or $\approx$0, this point is either a local point or the global point. According to the above actions, the hybrid algorithm has been granted sequential opportunities to escape out of a snare (a local point). Thus, the procedures in Lines 8–12 are eligible for helping the algorithm to flee this snare, especially since the second stopping standard guarantees that most of the research domain is scanned.

The numerical outcomes of the five methods are given in the next section.

7. Numerical Experiments of Part II

The numerical results for the second test problems (non-convex functions) are presented, and these results are obtained by Algorithm 3.

The performance profiles tool that is described in Part I is used here for assessing the achievement of Algorithm 3 that contains five alternatives of algorithms as we mentioned above in Section 6.

The numerical results of the second type of the test problems are listed in Tables 9–15. Columns 1–2 and 8–9 contain the abbreviation of the function f and the number of the unknowns n, respectively. Columns 3–7 contain the abbreviation of each algorithm of the five algorithm HSSHZ, HSMHZ, HSHZ, HSHS and HSFR, which present the number of worst iterations, number of worst function evaluations, number of best iterations, number of best function evaluations, average of time (CPU), average of number of iterations and average of number of function evaluations, respectively. Columns 10–14 are similar to Columns 3–7.

Note: F denotes that the algorithm has failed to find the local minimizer of the function f according to the stopping criteria of Algorithm 3 which are listed in Section 6.

Table 9. The number of worst iterations.

f	n	HSSHZ	HSMHZ	HSHZ	HSHS	HSFR	f	n	HSSHZ	HSMHZ	HSHZ	HSHS	HSFR
$S5^*$	4	3150	55	85	F	F	$S7^*$	4	10,000	F	10,000	F	F
$S10^*$	4	710	F	3020	F	F	HM^*	2	40	100	95	75	180
H^*	3	300	590	1155	465	1270	H^*	6	50	500	300	9550	F
CB^*	2	55	145	15	200	90	$P8^*$	4	20	15	15	550	10
$P16^*$	5	755	835	3280	F	7300	SH^*	2	100	115	200	250	190
$Bh1^*$	2	205	F	F	F	F	Ras^*	2	1310	F	F	F	F
GP^*	2	20	F	F	300	F	Le^*	10	2470	1430	F	F	3100

Table 10. The number of worst function evaluations.

f	n	HSSHZ	HSMHZ	HSHZ	HSHS	HSFR	f	n	HSSHZ	HSMHZ	HSHZ	HSHS	HSFR
$S5^*$	4	12,600	220	340	F	F	$S7^*$	4	40,000	F	40,000	F	F
$S10^*$	4	2840	F	12,080	F	F	HM^*	2	120	200	190	150	360
H^*	3	900	1770	3465	1395	3810	H^*	6	300	3000	1800	57,300	F
CB^*	2	110	290	30	400	180	$P8^*$	4	80	60	60	2200	40
$P16^*$	5	3775	4175	16,400	F	36,500	SH^*	2	200	230	400	500	380
$Bh1^*$	2	410	F	F	F	F	Ras^*	2	2620	F	F	F	F
GP^*	2	40	F	F	600	F	Le^*	10	24,700	14,300	F	F	31,000

Table 11. The number of best iterations.

f	n	HSSHZ	HSMHZ	HSHZ	HSHS	HSFR	f	n	HSSHZ	HSMHZ	HSHZ	HSHS	HSFR
$S5^*$	4	50	35	55	F	F	$S7^*$	4	750	F	520	F	F
$S10^*$	4	20	F	70	F	F	HM^*	2	15	10	10	10	5
H^*	3	50	60	85	20	130	H^*	6	50	100	100	50	F
CB^*	2	15	10	10	50	10	$P8^*$	4	5	5	5	50	5
$P16^*$	5	150	35	80	F	40	SH^*	2	10	10	10	50	10
$Bh1^*$	2	20	F	F	F	F	Ras^*	2	10	F	F	F	F
GP^*	2	15	F	F	50	F	Le^*	10	400	120	F	F	395

Table 12. The number of best function evaluations.

f	n	HSSHZ	HSMHZ	HSHZ	HSHS	HSFR	f	n	HSSHZ	HSMHZ	HSHZ	HSHS	HSFR
$S5^*$	4	200	140	220	F	F	$S7^*$	4	3000	F	2080	F	F
$S10^*$	4	80	F	280	F	F	HM^*	2	45	20	20	20	10
H^*	3	150	180	255	60	390	H^*	6	300	600	600	300	F
CB^*	2	30	20	20	100	20	$P8^*$	4	20	20	20	200	20
$P16^*$	5	725	175	400	F	200	SH^*	2	20	20	20	100	20
$Bh1^*$	2	40	F	F	F	F	Ras^*	2	20	F	F	F	F
GP^*	2	30	F	F	100	F	Le^*	10	4000	1200	F	F	3950

Table 13. The average of time.

f	n	HSSHZ	HSMHZ	HSHZ	HSHS	HSFR	f	n	HSSHZ	HSMHZ	HSHZ	HSHS	HSFR
$S5^*$	4	0.720	0.050	0.046	F	F	$S7^*$	4	7.368	F	13.249	F	F
$S10^*$	4	0.151	F	0.885	F	F	HM^*	2	0.017	0.031	0.028	0.021	0.053
H^*	3	0.186	0.353	0.409	0.271	0.361	H^*	6	0.057	0.194	0.143	4.712	F
CB^*	2	0.018	0.025	0.010	0.049	0.030	$P8^*$	4	0.014	0.015	0.009	0.116	0.007
$P16^*$	5	0.319	0.135	0.683	F	1.606	SH^*	2	0.028	0.037	0.050	0.084	0.039
$Bh1^*$	2	0.039	F	F	F	F	Ras^*	2	0.261	F	F	F	F
GP^*	2	0.015	F	F	0.078	F	Le^*	10	1.221	0.627	F	F	2.087

Table 14. The average of number of iterations.

f	n	HSSHZ	HSMHZ	HSHZ	HSHS	HSFR	f	n	HSSHZ	HSMHZ	HSHZ	HSHS	HSFR
$S5^*$	4	416.7	47	67	F	F	$S7^*$	4	5928.7	F	8648	F	F
$S10^*$	4	131.3	F	589	F	F	HM^*	2	24.3	29	34.8	25.8	50.8
H^*	3	213.3	268.8	373.3	247	333.8	H^*	6	50	205	177.5	2382.5	F
CB^*	2	26	23.3	12.8	57.5	36.8	$P8^*$	4	12	11	10.5	267.5	6.3
$P16^*$	5	376	171.25	878.8	F	2208.5	SH^*	2	30.3	39	48.5	65	41.5
$Bh1^*$	2	74.3	F	F	F	F	Ras^*	2	346.7	F	F	F	F
GP^*	2	18	F	F	62.5	F	Le^*	10	1012.7	506	F	F	1380.3

Table 15. The average of number of function evaluations.

f	n	HSSHZ	HSMHZ	HSHZ	HSHS	HSFR	f	n	HSSHZ	HSMHZ	HSHZ	HSHS	HSFR
$S5^*$	4	1666.7	188	268	F	F	$S7^*$	4	23,714.7	F	34,592	F	F
$S10^*$	4	525.3	F	2356	F	F	HM^*	2	73	58	69.5	51.5	101.5
H^*	3	640	806.3	1119.8	741	1001.3	H^*	6	300	1230	1065	14,295	F
CB^*	2	52	46.5	25.5	115	73.5	$P8^*$	4	48	44	42	1070	25
$P16^*$	5	1880	856.3	4393.8	F	11,042.5	SH^*	2	60.7	78	97	130	83
$Bh1^*$	2	148.7	F	F	F	F	Ras^*	2	693.3	F	F	F	F
GP^*	2	36	F	F	125	F	Le^*	10	10,126.7	5060	F	F	13,802.5

The performance profiles for the five algorithms are analyzed as follows.

Figures 5–8 show the performance profiles of the five set solvers S regarding the set standard Fit that is mentioned in Section 4.2.

The performance profiles which are drawn on the left of Figure 5 (in the term itr.w) compares 5 methods for the 14 test problems.

The HSSHZ technique has a good achievement (for the term itr.w) for all test problems, which indicates that the HSSHZ technique is capable of solving Problem (2) as fast as or faster than the four techniques.

For instance, if $\tau = 1$, the HSSHZ algorithm solves 71% of the 14 test problems against 14%, 14%, 7% and 0%, of the 14 test problems solved by the HSMHZ, HSHZ, HSFR and HSHS algorithms, respectively.

In general, for the term itr.w, $\tau \geq 60$ exhibits that the second type of the test problems are solved by HSSHZ, while 64%, 71%, 43% and 50% of test problems are solved by the HSMHZ, HSHZ, HSHS and HSFR algorithms respectively.

Figures 5–8 demonstrate that the performance of the HSSHZ technique is better than the performance of the four techniques regarding the seven standards listed in the set Fit, respectively.

Therefore, the HSSHZ technique includes the characteristics of efficiency, reliability and effectiveness in finding the global minimizer of the non-convex function f compared to the other four methods.

It is worth observing that the power of the HSSHZ algorithm comes from the fact that the SHZ method gains the features of the four methods, MHZ, HZ, HS and FR, as mentioned in Section 2.

Note 1: In Algorithm 3, a run is considered successful if Inequality (64) is met.

$$|f_{ac} - f^*| \leq 10^{-5}, \tag{64}$$

where f^* is the exact global solution that is listed in Columns 3 and 7 of Table 1, respectively, and the f_{ac} is the final result obtained by Algorithm 3.

Note 2: Formula (56) means if the final result f_{ac}, obtained by Algorithm 3 satisfies Inequality (64), then the first branch of (56) is computed; otherwise, we set $r_{p,s} = \infty$.

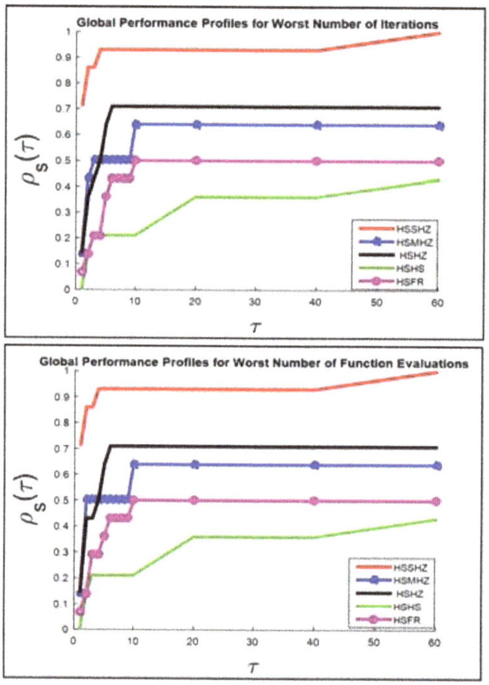

Figure 5. Plotting the results of the terms itr.w and FEs.w for 5 algorithms.

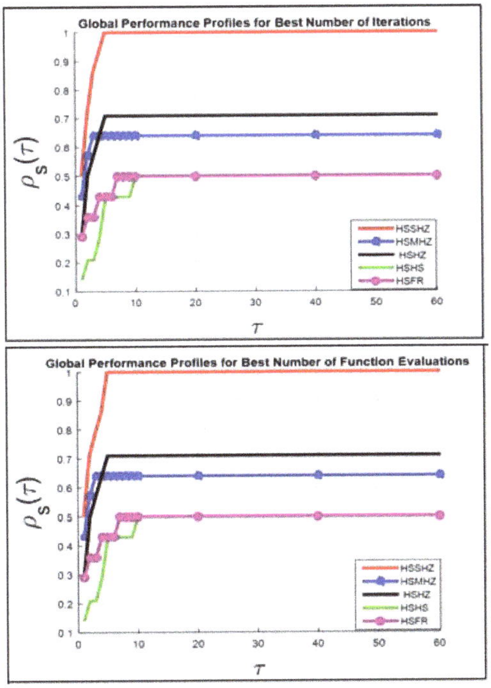

Figure 6. Plotting the results of the terms itr.be and FEs.be for 5 algorithms.

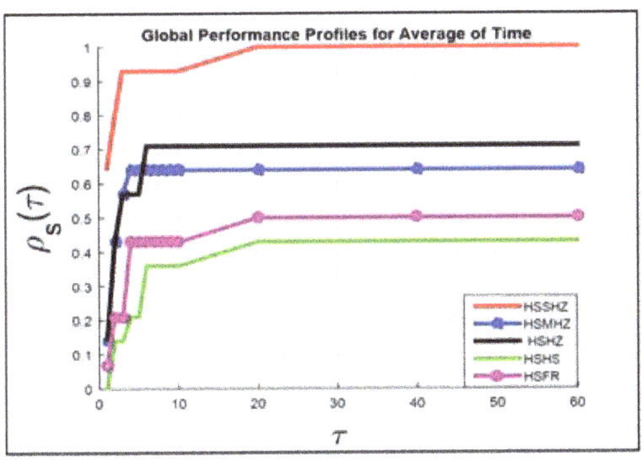

Figure 7. Plotting the results of the term time.a "CPU" for 5 algorithms.

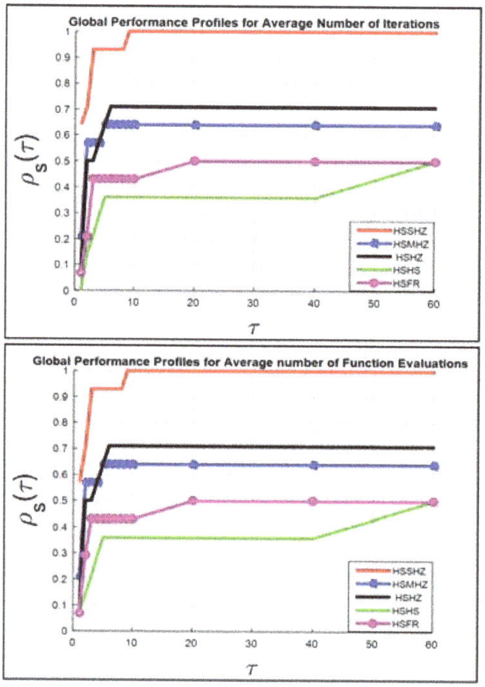

Figure 8. Plotting the results of the terms itr.a and FEs.a for 5 algorithms.

8. Conclusions and Future Work

A new modified CG algorithm is suggested, named SHZ. The SHZ finds the local minimizers of unconstrained optimization problems. The modernized formulae of the SHZ algorithm are more complicated than previous approaches; nevertheless, the numerical experiments of the SHZ are very strong. The convergence analysis of the SHZ algorithm is designed. We also analyzed the gradient approximation $g(x) \approx$ DFF constructed by finite differences (the forward differences method). This method includes a new approach for selecting the fit value of the h according to the value of the objective function and it is updated dynamically at each iteration. The numerical results demonstrate that the performance of the SHZ method is positively competitive with the other four conjugate gradient methods based on performance profiles.

Comparing the final results of the gradient vector that were obtained by the method DFF to the exact values of the gradient vector demonstrates that the fresh technique succeeded in picking the right value of h. The proposed random approach recreates a critical role to make the SHZ method capable of finding the global minimizers of unconstrained optimization test problems, especially when the objective function is non-convex.

It can be worth observing that the power of the HSSHZ algorithm comes from the fact that the SHZ method gains the characteristics of the four methods, MHZ, HZ, HS and FR.

The suggested approach can be improved and modified to deal with constrained, multi-objective optimization problems, and it will be used for image restorations.

Author Contributions: Conceptualization, K.A.A.; Data curation, A.M.A.; Formal analysis, A.M.A. and K.M.S.; Funding acquisition, K.A.A.; Investigation, A.W.M.; Methodology, S.M.; Project administration, K.A.A. and K.M.S.; Software, S.M.; Supervision, A.W.M.; Validation, A.M.A.; Writing—original draft, S.M. All authors have read and agreed to the published version of the manuscript.

Funding: The research is funded by Researchers Supporting Program at King Saud University (Project# RSP-2021/305).

Data Availability Statement: Not applicable.

Acknowledgments: The authors present their appreciation to King Saud University for funding the publication of this research through Researchers Supporting Program (RSP-2021/305), King Saud University, Riyadh, Saudi Arabia.

Conflicts of Interest: The authors declare no conflict of interest.

Appendix A. List of Test Problems

1 R_n: Rosenbrock functions [57,87,88]

$$\min_x \left\{ \sum_{i=1}^{n-1} \left[100(x_i^2 - x_{i+1})^2 + (x_i - 1)^2\right] \right\}.$$

Range of starting points $-5 < x_i < 10, i = 1, 2, \ldots, n$.
Global minima: $f(x^*) = 0$ at $x^* = (1, 1, \ldots, 1)$.

2 Z_n: Zakharov functions [57,80,87,88]

$$\min_x \left\{ \sum_{i=1}^{n} x_i^2 + \left(\sum_{i=1}^{n} 0.5 i x_i \right)^2 + \left(\sum 0.5 i x_i \right)^4 \right\}.$$

Range of starting points $-5 < x_i < 10, i = 1, 2, \ldots, n$.
Global minima: $f(x^*) = 0$ at $x^* = (0, 0, \ldots, 0)$.

3 PW: Powell function [80]

$$\min_x \left\{ \sum_{i=1}^{\frac{n}{4}} \left[(x_{4i-3} + 10x_{4i-2})^2 + 5(x_{4i-1} - x_{4i})^2 + (x_{4i-2} - 2x_{4i-1})^4 + 10(x_{4i-3} - x_{4i})^4 \right] \right\}.$$

Range of starting points $-600 < x_i < 600, i = 1, 2, \ldots, n$.
Global minima: $f(x^*) = 0$ at $x^* = (0, 0, \ldots, 0)$.

4 SP: Sphere function [89]

$$\min_x \left\{ \sum_{i=1}^{n} x_i^2 \right\}.$$

Range of starting points $-10 \le x_i \le 10, i = 1, 2, \ldots, n$.
Global minima: $f(x^*) = 0$ at $x^* = (0, 0, \ldots, 0)$.

5 Tr: Trid function [80]

$$\min \sum_{i=1}^{n} (x_i - 1)^2 - \sum_{i=2}^{n} x_i x_{i-1}.$$

Range of starting points $-n^2 < x_i < n^2, i = 1, 2, \ldots, n$.
Global minima : $f(x^*) = \frac{n(n+4)(n-1)}{6}$. at $x^* = i(n + 1 - i)$

6: Sum Squares function [90]

$$\min_x \left\{ \sum_{i=1}^{n} i x_i^2 \right\}.$$

Range of starting points $-100 < x_i < 100, i = 1, 2, \ldots, n$.
Global minima: $f(x^*) = 0$ at $x^* = (0, 0, \ldots, 0)$.

7 CV: Colville function [57,80,91]

$$\min_x \left\{ 100\left(x_1^2 - x_2\right)^2 + \left(x_1 - 1\right)^2 + \left(x_3 - 1\right)^2 + 90\left(x_3^2 - x_4\right)^2 + 10.1\left(\left(x_2 - 1\right)^2 + \left(x_4 - 1\right)^2\right) + 19.8\left(x_2 - 1\right)\left(x_4 - 1\right)^2 \right\}.$$

Range of starting points $-10 < x_i < 10, i = 1, 2, \ldots, n$.
Global minima: $f(x^*) = 0$ at $x^* = (1, 1, 1, 1)$.

8 BR: Branin function [57,92,93]

$$\min_x \left\{ (x_2 - \frac{5.1}{4\pi^2}x_1^2 + \frac{5}{\pi}x_1 - 6)^2 + 10(1 - \frac{1}{8\pi}cos(x_1)) + 10 \right\}.$$

Range of starting points $-5 < x_i < 15, i = 1, 2$.
Only one global minima: $f(x^*) = 0.397887$. at $x^* = \{(-\pi, 12.275), (9.42478, 2.475), (\pi, 2.275)\}$.

9 DJ: De Joung function [57,87,88]

$$\min_x \left\{ x_1^2 + x_2^2 + x_3^2 \right\}.$$

Range of starting points $-5 < x_i < 15, i = 1, 2, 3$.
Number of local minima: no local minima.
Global minima: $f(x^*) = 0$ at $x^* = (0, 0, 0)$.

10 BO: Booth function [89]

$$\min_x \left\{ (x_1 + 2x_2 - 7)^2 + (2x_1 + x_2 - 5)^2 \right\}.$$

Range of starting points $-10 < x_i < 10, i = 1, 2, \ldots, n$.
Global minima: $f(x^*) = 0$ at $x^* = (1, 3)$.

11 Ma: Matyas function [90]

$$\min_x \left\{ 0.26\left(x_1^2 + x_2^2\right) - 0.48x_1x_2 \right\}.$$

Range of starting points $-10 < x_i < 10, i = 1, 2, \ldots, n$.
Global minima: $f(x^*) = 0$ at $x^* = (0, 0)$.

12 Sm*: Shekel functions [57,80,87,88,92–94]

$$\min_x \left\{ -\sum_{j=1}^{m} \left(\sum_{i=1}^{4} (x_i - A_{ij})^2 + c_j \right)^{-1} \right\}.$$

where $c = 0.1[1, 2, 2, 4, 4, 6, 3, 7, 5, 5]$,

$$A = \begin{bmatrix} 4.0 & 1.0 & 8.0 & 6.0 & 3.0 & 2.0 & 5.0 & 8.0 & 6.0 & 7.0 \\ 4.0 & 1.0 & 8.0 & 6.0 & 7.0 & 9.0 & 3.0 & 1.0 & 2.0 & 3.0 \\ 4.0 & 1.0 & 8.0 & 6.0 & 3.0 & 2.0 & 5.0 & 8.0 & 6.0 & 7.0 \\ 4.0 & 1.0 & 8.0 & 6.0 & 7.0 & 9.0 & 3.0 & 1.0 & 2.0 & 3.0 \end{bmatrix}$$

Range of starting points $0 < x_i < 10, i = 1, \ldots, n$.
Number of local minima: m local minima.
Global minima:

$$f(x^*)_{n,m} = \begin{cases} -10.1532, & \text{when } m = 5, \\ -10.4029, & \text{when } m = 7, \\ -10.5364, & \text{when } m = 10. \end{cases}$$

Global minima for three functions at $x^* = (4, 4, 4, 4)$.

13 GP*: Goldstein and Price function [57,80,87,88,92,94]

$u(x) = 1 + (x_1 + x_2 + 1)^2(19 - 14x_1 + 3x_1^2 - 14x_2 + 6x_1x_2 + 3x_2^2)$
$v(x) = 30 + (2x_1 - 3x_2)^2(18 - 32x_1 + 12x_1^2 + 48x_2 - 36x_1x_2 + 27x_2^2)$.

$$\min_x \left\{ v(x)u(x) \right\}.$$

Range of starting points $-2 < x_i < 2, i = 1, 2$.
Number of local minima: 4 local minima.
Global minima: $f(x^*) = 3$ at $x^* = (0, -1)$.

14 Ras*: Rastrigin function [93]

$$\min_x \Big\{ x_1^2 + x_2^2 - cos(18x_1) - cos(18x_2) \Big\}.$$

Range of starting points $-1 < x_i < 1, i = 1, 2$.
Number of local minima: many local minima.
Global minima: $f(x^*) = -2$ at $x^* = (0, 0)$.

15 Bh1*: Bohachevsky function [80]

$$\min_x \Big\{ x_1^2 + 2x_2^2 - 0.3cos(3\pi x_1) - 0.4cos(4\pi x_2) + 0.7 \Big\}.$$

Range of starting points $-100 < x_i < 100, i = 1, 2$.
Number of local minima: many local minima.
Global minima: $f(x^*) = 0$ at $x^* = (0, 0)$.

16 SH*: Shubert function in [57,80,87,88,92]

$$\min_x \Big\{ \Big(\sum_{i=1}^{5} icos\Big((i+1)x_1 + i\Big)\Big)\Big(\sum_{i=1}^{5} icos\Big((i+1)x_2 + i\Big)\Big) \Big\}.$$

Range of starting points $-5.12 < x_i < 5.12, i = 1, 2$.
Number of local minima: 760 local minima.
Global minima: $f(x^*) = -186.7309$ at 18 point different of x^*.

17 P8* Ref. [92]

$$\min_x \Big\{ \frac{\pi}{n}\Big(k_1 sin(\pi y_1)^2 + \sum_{i=1}^{n-1} (y_i - k_2)^2 \Big[1 + k_1 sin(\pi y_{i+1})^2\Big] + (y_n - k_2)^2\Big) \Big\},$$

with $y_i = 1 + \frac{1}{4}\Big(x_i + 1\Big)$, $k_1 = 10$ and $k_2 = 1$.
Range of starting points $-10 \le x_i \le 10, i = 1, 2, 3$.
Number of local minima: 5^3 local minima.
Global minima: $f(x^*) = 0$ at $x^* = (-1, -1, -1)$.

18 P16* Ref. [92]

$$\min_x k_3 \Big\{ sin^2(\pi k_4 x_1) + \sum_{i=1}^{n-1} (x_i - k_5)^2 \Big[1 + k_6 sin^2(\pi k_4 x_{i+1})\Big] + (x_n - k_5)^2 \Big[1 + k_6 sin^2(\pi k_7 x_n)\Big] \Big\},$$

where $k_3 = 0.1, k_4 = 3, k_5 = 1, k_6 = 1, k_7 = 2$.
Range of starting points $-5 \le x_i \le 5, i = 1, .., n$.
Number of local minima: 15^5 local minima.
Global minima: $f(x^*) = 0$ at $x^* = (1, 1, 1, 1, 1)$.

19 CB*: Camel back in [80] and camel function in [93]

$$\min_x \Big\{ 4x_1^2 - 2.1x_1^4 + \frac{1}{3}x_1^6 + x_1 x_2 - 4x_2^2 + 4x_2^4 \Big\}.$$

Range of starting points $-5 < x_i < 5, i = 1, 2$.
Number of local minima: many local minima.
Global minima: $f(x^*) = -1.0316285$ at $x^* = \{(0.089842, -0.71266), (-0.089842, 0.71266)\}$.

20 H3*: Hartmann function [57,80,87,88,92–94]

$$\min_x \Big\{ -\sum_{i=1}^{4} c_i exp\Big(-\sum_{j=1}^{3} a_{ij}(x_j - p_{ij})^2\Big) \Big\}.$$

Range of starting points $-1 < x_j < 1, j = 1,2,3$.
Number of local minima: 4 local minima.
Global minima: $f(x^*) = -3.86278$ at $x^* = \big(0.114614, 0.555649, 0.852547\big)$.

21 H6*: Hartmann function [57,80,87,88,92–94]

$$\min_x \left\{ -\sum_{i=1}^{4} c_i exp\Big(-\sum_{j=1}^{6} a_{ij}\big(x_j - p_{ij}\big)^2\Big)\right\}.$$

Range of starting points $-1 < x_j < 1, j = 1,2,\ldots,n$.
Number of local minima: 4 local minima.
Global minima: $f(x^*) = -3.32237$ at $x^* = \big(0.201690, 0.150011, 0.476874, 0.275332, 0.311652, 0.657300\big)$.

22 HM*: hump Function [57]

$$\min_x \left\{1.0316285 + 4x_1^2 - 2.1x_1^4 + \frac{1}{3}x_1^6 + x_1x_2 - 4x_2^2 + 4x_2^4\right\}.$$

Range of starting points $-5 < x_i < 5, i = 1,2$.
Number of local minima: 3 local minima.
Global minima: $f(x^*) = 0$ at $x^* = \{(0.0898, -0.7126), (-0.0898, 0.7126)\}$.

23 Le*: Levy function [95]

$$\min_x \left\{ sin^2(\pi w_1) + \sum_{i=1}^{n-1}(w_i - 1)^2\Big(1 + 10sin^2(\pi w_i + 1)\Big) + (w_n - 1)^2\Big[1 + sin^2(2\pi w_n)\Big]\right\},$$

where $w_i = 1 + \frac{x_i - 1}{4}$, for $i = 1,\ldots,n$.
Range of starting points $-10 < x_i < 10, i = 1,2,\ldots,n$.
Number of local minima: many local minima.
Global minima: $f(x^*) = 0$ at $x^* = \big(1, 1, \ldots, 1\big)$.

References

1. Abdel-Baset, M.; Hezam, I. A Hybrid Flower Pollination Algorithm for Engineering Optimization Problems. *Int. J. Comput. Appl.* **2016**, *140*, 10–23. [CrossRef]
2. Agrawal, P.; Ganesh, T.; Mohamed, A.W. A novel binary gaining–sharing knowledge-based optimization algorithm for feature selection. *Neural Comput. Appl.* **2021**, *33*, 5989–6008. [CrossRef]
3. Ayumi, V.; Rere, L.; Fanany, M.I.; Arymurthy, A.M. Optimization of Convolutional Neural Network using Microcanonical Annealing Algorithm. *arXiv* **2016**, arXiv:1610.02306.
4. Lobato, F.S.; Steffen, V., Jr. Fish swarm optimization algorithm applied to engineering system design. *Lat. Am. J. Solids Struct.* **2014**, *11*, 143–156. [CrossRef]
5. Mazhoud, I.; Hadj-Hamou, K.; Bigeon, J.; Joyeux, P. Particle swarm optimization for solving engineering problems: A new constraint-handling mechanism. *Eng. Appl. Artif. Intell.* **2013**, *26*, 1263–1273. [CrossRef]
6. Mohamed, A.W.; Sabry, H.Z. Constrained optimization based on modified differential evolution algorithm. *Inf. Sci.* **2012**, *194*, 171–208. [CrossRef]
7. Mohamed, A.W.; Hadi, A.A.; Mohamed, A.K. Gaining-sharing knowledge based algorithm for solving optimization problems: A novel nature-inspired algorithm. *Int. J. Mach. Learn. Cybern.* **2020**, *11*, 1501–1529. [CrossRef]
8. Rere, L.; Fanany, M.I.; Arymurthy, A.M. Metaheuristic Algorithms for Convolution Neural Network. *Comput. Intell. Neurosci.* **2016**, *2016*, 1537325. [CrossRef] [PubMed]
9. Samora, I.; Franca, M.J.; Schleiss, A.J.; Ramos, H.M. Simulated annealing in optimization of energy production in a water supply network. *Water Resour. Manag.* **2016**, *30*, 1533–1547. [CrossRef]
10. Shao, Y. Dynamics of an Impulsive Stochastic Predator–Prey System with the Beddington–DeAngelis Functional Response. *Axioms* **2021**, *10*, 323. [CrossRef]
11. Vallepuga-Espinosa, J.; Cifuentes-Rodríguez, J.; Gutiérrez-Posada, V.; Ubero-Martínez, I. Thermomechanical Optimization of Three-Dimensional Low Heat Generation Microelectronic Packaging Using the Boundary Element Method. *Mathematics* **2022**, *10*, 1913. [CrossRef]
12. Blum, C.; Roli, A. Metaheuristics in combinatorial optimization: Overview and conceptual comparison. *ACM Comput. Surv. (CSUR)* **2003**, *35*, 268–308. [CrossRef]
13. Aarts, E.; Korst, J. *Simulated Annealing and Boltzmann Machines: A Stochastic Approach to Combinatorial Optimization and Neural Computing*; John Wiley & Sons, Inc.: New York, NY, USA, 1989.

14. Hillier, F.S.; Price, C.C. *International Series in Operations Research & Management Science*; Springer: Berlin/Heidelberg, Germany, 2001.
15. Laarhoven, P.J.V.; Aarts, E.H. *Simulated Annealing: Theory and Applications*; Springer: Berlin/Heidelberg, Germany, 1987.
16. Farid, M.; Leong, W.J.; Hassan, M.A. A new two-step gradient-type method for large-scale unconstrained optimization. *Comput. Math. Appl.* **2010**, *59*, 3301–3307. [CrossRef]
17. Gilbert, J.C.; Nocedal, J. Global convergence properties of conjugate gradient methods for optimization. *SIAM J. Optim.* **1992**, *2*, 21–42. [CrossRef]
18. Hager, W.W.; Zhang, H. Algorithm 851: CG_DESCENT, a conjugate gradient method with guaranteed descent. *ACM Trans. Math. Softw. (TOMS)* **2006**, *32*, 113–137. [CrossRef]
19. Ruder, S. An overview of gradient descent optimization algorithms. *arXiv* **2016**, arXiv:1609.04747.
20. Shi, Z. A new memory gradient method under exact line search. *Asia-Pac. J. Oper. Res* **2003**, *20*, 275–284.
21. Zhang, L.; Zhou, W.; Li, D.H. A descent modified Polak–Ribière–Polyak conjugate gradient method and its global convergence. *IMA J. Numer. Anal.* **2006**, *26*, 629–640. [CrossRef]
22. Abubakar, A.B.; Malik, M.; Kumam, P.; Mohammad, H.; Sun, M.; Ibrahim, A.H.; Kiri, A.I. A Liu-Storey-type conjugate gradient method for unconstrained minimization problem with application in motion control. *J. King Saud Univ.-Sci.* **2022**, *34*, 101923. [CrossRef]
23. Dai, Y.; Yuan, Y. An efficient hybrid conjugate gradient method for unconstrained optimization. *Ann. Oper. Res.* **2001**, *103*, 33–47. [CrossRef]
24. Deng, S.; Wan, Z. A three-term conjugate gradient algorithm for large-scale unconstrained optimization problems. *Appl. Numer. Math.* **2015**, *92*, 70–81. [CrossRef]
25. Ma, G.; Lin, H.; Jin, W.; Han, D. Two modified conjugate gradient methods for unconstrained optimization with applications in image restoration problems. *J. Appl. Mathemat. Comput.* **2022**, 1–26. [CrossRef]
26. Mtagulwa, P.; Kaelo, P. An efficient modified PRP-FR hybrid conjugate gradient method for solving unconstrained optimization problems. *Appl. Numer. Math.* **2019**, *145*, 111–120. [CrossRef]
27. Waziri, M.Y.; Kiri, A.I.; Kiri, A.A.; Halilu, A.S.; Ahmed, K. A modified conjugate gradient parameter via hybridization approach for solving large-scale systems of nonlinear equations. *SeMA J.* **2022**, 1–23. [CrossRef]
28. Zhang, L.; Zhou, W.; Li, D. Global convergence of a modified Fletcher–Reeves conjugate gradient method with Armijo-type line search. *Numer. Math.* **2006**, *104*, 561–572. [CrossRef]
29. Alshamrani, A.M.; Alrasheedi, A.F.; Alnowibet, K.A.; Mahdi, S.; Mohamed, A.W. A Hybrid Stochastic Deterministic Algorithm for Solving Unconstrained Optimization Problems. *Mathematics* **2022**, *10*, 3032. [CrossRef]
30. Hager, W.W.; Zhang, H. A new conjugate gradient method with guaranteed descent and an efficient line search. *SIAM J. Optim.* **2005**, *16*, 170–192. [CrossRef]
31. Golub, G.H.; O'Leary, D.P. Some history of the conjugate gradient and Lanczos algorithms: 1948–1976. *SIAM Rev.* **1989**, *31*, 50–102. [CrossRef]
32. Hager, W.W.; Zhang, H. A survey of nonlinear conjugate gradient methods. *Pac. J. Optim.* **2006**, *2*, 35–58.
33. Fletcher, R.; Reeves, C.M. Function minimization by conjugate gradients. *Comput. J.* **1964**, *7*, 149–154. [CrossRef]
34. Powell, M.J. Nonconvex minimization calculations and the conjugate gradient method. In *Numerical Analysis*; Springer: Berlin/Heidelberg, Germany, 1984; pp. 122–141.
35. Al-Baali, M. Descent property and global convergence of the Fletcher Reeves method with inexact line search. *IMA J. Numer. Anal.* **1985**, *5*, 121–124. [CrossRef]
36. Powell, M.J.D. Restart procedures for the conjugate gradient method. *Math. Program.* **1977**, *12*, 241–254. [CrossRef]
37. Polak, E.; Ribiere, G. Note sur la convergence de méthodes de directions conjuguées. *ESAIM Math. Model. Numer. Anal.* **1969**, *3*, 35–43. [CrossRef]
38. Polyak, B.T. The conjugate gradient method in extremal problems. *USSR Comput. Math. Math. Phys.* **1969**, *9*, 94–112. [CrossRef]
39. Hestenes, M.R.; Stiefel, E. Methods of Conjugate Gradients for Solving. *J. Res. Natl. Bur. Stand.* **1952**, *49*, 409. [CrossRef]
40. Liu, Y.; Storey, C. Efficient generalized conjugate gradient algorithms, part 1: Theory. *J. Optim. Theory Appl.* **1991**, *69*, 129–137. [CrossRef]
41. Dai, Y.H.; Yuan, Y. A nonlinear conjugate gradient method with a strong global convergence property. *SIAM J. Optim.* **1999**, *10*, 177–182. [CrossRef]
42. Abubakar, A.B.; Kumam, P. A descent Dai-Liao conjugate gradient method for nonlinear equations. *Numer. Algorithms* **2019**, *81*, 197–210. [CrossRef]
43. Abubakar, A.B.; Muangchoo, K.; Ibrahim, A.H.; Muhammad, A.B.; Jolaoso, L.O.; Aremu, K.O. A new three-term Hestenes-Stiefel type method for nonlinear monotone operator equations and image restoration. *IEEE Access* **2021**, *9*, 18262–18277. [CrossRef]
44. Babaie-Kafaki, S.; Ghanbari, R. A descent family of Dai–Liao conjugate gradient methods. *Optim. Methods Softw.* **2014**, *29*, 583–591. [CrossRef]
45. Dai, Y.; Liao, L. New conjugacy conditions and related nonlinear conjugate gradient methods. *Appl. Math. Optim.* **2001**, *43*, 87–101. [CrossRef]
46. Ibrahim, A.H.; Kumam, P.; Kumam, W. A family of derivative-free conjugate gradient methods for constrained nonlinear equations and image restoration. *IEEE Access* **2020**, *8*, 162714–162729. [CrossRef]

47. Su, Z.; Li, M. A Derivative-Free Liu–Storey Method for Solving Large-Scale Nonlinear Systems of Equations. *Math. Probl. Eng.* **2020**, *2020*, 6854501. [CrossRef]
48. Yuan, G.; Zhang, M. A three-terms Polak–Ribière–Polyak conjugate gradient algorithm for large-scale nonlinear equations. *J. Comput. Appl. Math.* **2015**, *286*, 186–195. [CrossRef]
49. Yuan, G.; Jian, A.; Zhang, M.; Yu, J. A modified HZ conjugate gradient algorithm without gradient Lipschitz continuous condition for non convex functions. *J. Appl. Mathemat. Comput.* **2022**, 1–22. [CrossRef]
50. Zhou, Y.; Wu, Y.; Li, X. A new hybrid prpfr conjugate gradient method for solving nonlinear monotone equations and image restoration problems. *Math. Probl. Eng.* **2020**, *2020*, 6391321. [CrossRef]
51. Yuan, G.; Meng, Z.; Li, Y. A modified Hestenes and Stiefel conjugate gradient algorithm for large-scale nonsmooth minimizations and nonlinear equations. *J. Optim. Theory Appl.* **2016**, *168*, 129–152. [CrossRef]
52. Yuan, G.; Wei, Z.; Yang, Y. The global convergence of the Polak–Ribière–Polyak conjugate gradient algorithm under inexact line search for nonconvex functions. *J. Comput. Appl. Math.* **2019**, *362*, 262–275. [CrossRef]
53. Kan, A.R.; Timmer, G. Stochastic methods for global optimization. *Am. J. Math. Manag. Sci.* **1984**, *4*, 7–40. [CrossRef]
54. Alnowibet, K.A.; Alshamrani, A.M.; Alrasheedi, A.F.; Mahdi, S.; El-Alem, M.; Aboutahoun, A; Mohamed, A.W. A Efficient Modified Meta-Heuristic Technique for Unconstrained Optimization Problems. *Axioms* **2022**, *11*, 483. [CrossRef]
55. Alnowibet, K.A.; Mahdi, S.; El-Alem, M.; Abdelawwad, M.; Mohamed, A.W. Guided Hybrid Modified Simulated Annealing Algorithm for Solving Constrained Global Optimization Problems. *Mathematics* **2022**, *10*, 1312. [CrossRef]
56. EL-Alem, M.; Aboutahoun, A.; Mahdi, S. Hybrid gradient simulated annealing algorithm for finding the global optimal of a nonlinear unconstrained optimization problem. *Soft Comput.* **2021**, *25*, 2325–2350. [CrossRef]
57. Hedar, A.R.; Fukushima, M. Hybrid simulated annealing and direct search method for nonlinear unconstrained global optimization. *Optim. Methods Softw.* **2002**, *17*, 891–912. [CrossRef]
58. Pedamallu, C.S.; Ozdamar, L. Investigating a hybrid simulated annealing and local search algorithm for constrained optimization. *Eur. J. Oper. Res.* **2008**, *185*, 1230–1245. [CrossRef]
59. Yiu, K.F.C.; Liu, Y.; Teo, K.L. A hybrid descent method for global optimization. *J. Glob. Optim.* **2004**, *28*, 229–238. [CrossRef]
60. Zoutendijk, G. Nonlinear programming, computational methods. In *Integer and Nonlinear Programming*; Abadie, J., Ed.; North-Holland: Amsterdam, The Netherlands, 1970; pp. 37–86.
61. Wolfe, P. Convergence conditions for ascent methods. *SIAM Rev.* **1969**, *11*, 226–235. [CrossRef]
62. Wolfe, P. Convergence conditions for ascent methods. II: Some corrections. *SIAM Rev.* **1971**, *13*, 185–188. [CrossRef]
63. Conn, A.R.; Scheinberg, K.; Vicente, L.N. *Introduction to Derivative-Free Optimization*; SIAM: Philadelphia, PA, USA, 2009.
64. Kramer, O.; Ciaurri, D.E.; Koziel, S. Derivative-free optimization. In *Computational Optimization, Methods and Algorithms*; Springer: Berlin/Heidelberg, Germany, 2011; pp. 61–83.
65. Larson, J.; Menickelly, M.; Wild, S.M. Derivative-free optimization methods. *Acta Numer.* **2019**, *28*, 287–404. [CrossRef]
66. Shi, H.J.M.; Xie, Y.; Xuan, M.Q.; Nocedal, J. Adaptive Finite-Difference Interval Estimation for Noisy Derivative-Free Optimization. *arXiv* **2021**, arXiv:2110.06380.
67. Shi, H.J.M.; Xuan, M.Q.; Oztoprak, F.; Nocedal, J. On the numerical performance of derivative-free optimization methods based on finite-difference approximations. *arXiv* **2021**, arXiv:2102.09762.
68. Berahas, A.S.; Cao, L.; Choromanski, K.; Scheinberg, K. A theoretical and empirical comparison of gradient approximations in derivative-free optimization. *Found. Comput. Math.* **2022**, *22*, 507–560. [CrossRef]
69. Curtis, A.; Reid, J. The choice of step lengths when using differences to approximate Jacobian matrices. *IMA J. Appl. Math.* **1974**, *13*, 121–126. [CrossRef]
70. Calio, F.; Frontini, M.; Milovanović, G.V. Numerical differentiation of analytic functions using quadratures on the semicircle. *Comput. Math. Appl.* **1991**, *22*, 99–106. [CrossRef]
71. Gill, P.E.; Murray, W.; Saunders, M.A.; Wright, M.H. Computing forward-difference intervals for numerical optimization. *SIAM J. Sci. Stat. Comput.* **1983**, *4*, 310–321. [CrossRef]
72. Xie, Y. Methods for Nonlinear and Noisy Optimization. Ph.D. Thesis, Northwestern University, Evanston, IL, USA, 2021.
73. De Levie, R. An improved numerical approximation for the first derivative. *J. Chem. Sci.* **2009**, *121*, 935–950. [CrossRef]
74. Carter, R.G. On the global convergence of trust region algorithms using inexact gradient information. *SIAM J. Numer. Anal.* **1991**, *28*, 251–265. [CrossRef]
75. Rivet, A.; Souloumiac, A. *Introduction to Optimization*; Optimization Software, Publications Division: New Delhi, India, 1987.
76. Byrd, R.H.; Chin, G.M.; Nocedal, J.; Wu, Y. Sample size selection in optimization methods for machine learning. *Math. Program.* **2012**, *134*, 127–155. [CrossRef]
77. Cartis, C.; Scheinberg, K. Global convergence rate analysis of unconstrained optimization methods based on probabilistic models. *Math. Program.* **2018**, *169*, 337–375. [CrossRef]
78. Grapiglia, G.N. Quadratic regularization methods with finite-difference gradient approximations. *Comput. Optim. Appl.* **2022**, 1–21. [CrossRef]
79. Paquette, C.; Scheinberg, K. A stochastic line search method with expected complexity analysis. *SIAM J. Optim.* **2020**, *30*, 349–376. [CrossRef]
80. Ali, M.M.; Khompatraporn, C.; Zabinsky, Z.B. A numerical evaluation of several stochastic algorithms on selected continuous global optimization test problems. *J. Glob. Optim.* **2005**, *31*, 635–672. [CrossRef]

81. Barbosa, H.J.; Bernardino, H.S.; Barreto, A.M. Using performance profiles to analyze the results of the 2006 CEC constrained optimization competition. In Proceedings of the IEEE Congress on Evolutionary Computation, Barcelona, Spain, 18–23 July 2010; pp. 1–8.
82. Dolan, E.D.; Moré, J.J. Benchmarking optimization software with performance profiles. *Math. Program.* **2002**, *91*, 201–213. [CrossRef]
83. Moré, J.J.; Wild, S.M. Benchmarking derivative-free optimization algorithms. *SIAM J. Optim.* **2009**, *20*, 172–191. [CrossRef]
84. Vaz, A.I.F.; Vicente, L.N. A particle swarm pattern search method for bound constrained global optimization. *J. Glob. Optim.* **2007**, *39*, 197–219. [CrossRef]
85. Liang, J.; Runarsson, T.P.; Mezura-Montes, E.; Clerc, M.; Suganthan, P.N.; Coello, C.C.; Deb, K. Problem definitions and evaluation criteria for the CEC 2006 special session on constrained real-parameter optimization. *J. Appl. Mech.* **2006**, *41*, 8–31.
86. Mohamed, A.W.; Hadi, A.A.; Mohamed, A.K.; Awad, N.H. Evaluating the performance of adaptive gainingsharing knowledge based algorithm on cec 2020 benchmark problems. In Proceedings of the 2020 IEEE Congress on Evolutionary Computation (CEC), Glasgow, UK, 19–24 July 2020; pp. 1–8.
87. Bessaou, M.; Siarry, P. A genetic algorithm with real-value coding to optimize multimodal continuous functions. *Struct. Multidisc. Optim.* **2001**, *23*, 63–74. [CrossRef]
88. Chelouah, R.; Siarry, P. Tabu search applied to global optimization. *Eur. J. Oper. Res.* **2000**, *123*, 256–270. [CrossRef]
89. Fan, S.K.S.; Zahara, E. A hybrid simplex search and particle swarm optimization for unconstrained optimization. *Eur. J. Oper. Res.* **2007**, *181*, 527–548. [CrossRef]
90. Paulavičius, R.; Chiter, L.; Žilinskas, J. Global optimization based on bisection of rectangles, function values at diagonals, and a set of Lipschitz constants. *J. Glob. Optim.* **2018**, *71*, 5–20. [CrossRef]
91. Cardoso, M.F.; Salcedo, R.L.; De Azevedo, S.F. The simplex-simulated annealing approach to continuous non-linear optimization. *Comput. Chem. Eng.* **1996**, *20*, 1065–1080. [CrossRef]
92. Dekkers, A.; Aarts, E. Global optimization and simulated annealing. *Math. Program.* **1991**, *50*, 367–393. [CrossRef]
93. Tsoulos, I.G.; Stavrakoudis, A. Enhancing PSO methods for global optimization. *Appl. Math. Comput.* **2010**, *216*, 2988–3001. [CrossRef]
94. Siarry, P.; Berthiau, G.; Durbin, F.; Haussy, J. Enhanced Simulated-Annealing Algorithm for Globally Minimizing Functions of Many Continuous Variables. *ACM Trans. Math. Softw.* **1997**, *23*, 209–228. [CrossRef]
95. Laguna, M.; Martí, R. Experimental testing of advanced scatter search designs for global optimization of multimodal functions. *J. Glob. Optim.* **2005**, *33*, 235–255. [CrossRef]

mathematics

Article

An Optimized Discrete Dragonfly Algorithm Tackling the Low Exploitation Problem for Solving TSP

Bibi Aamirah Shafaa Emambocus [1], Muhammed Basheer Jasser [1,*], Angela Amphawan [1] and Ali Wagdy Mohamed [2]

[1] Department of Computing and Information Systems, School of Engineering and Technology, Sunway University, Petaling Jaya 47500, Selangor, Malaysia
[2] Operations Research Department, Faculty of Graduate Studies for Statistical Research, Cairo University, Giza 12613, Egypt
* Correspondence: basheerj@sunway.edu.my

Abstract: Optimization problems are prevalent in almost all areas and hence optimization algorithms are crucial for a myriad of real-world applications. Deterministic optimization algorithms tend to be computationally costly and time-consuming. Hence, heuristic and metaheuristic algorithms are more favoured as they provide near-optimal solutions in an acceptable amount of time. Swarm intelligence algorithms are being increasingly used for optimization problems owing to their simplicity and good performance. The Dragonfly Algorithm (DA) is one which is inspired by the swarming behaviours of dragonflies, and it has been proven to have a superior performance than other algorithms in multiple applications. Hence, it is worth considering its application to the traveling salesman problem which is a predominant discrete optimization problem. The original DA is only suitable for solving continuous optimization problems and, although there is a binary version of the algorithm, it is not easily adapted for solving discrete optimization problems like TSP. We have previously proposed a discrete adapted DA algorithm suitable for TSP. However, it has low effectiveness, and it has not been used for large TSP problems. In this paper, we propose an optimized discrete adapted DA by using the steepest ascent hill climbing algorithm as a local search. The algorithm is applied to a TSP problem modelling a package delivery system in the Kuala Lumpur area and to benchmark TSP problems, and it is found to have a higher effectiveness than the discrete adapted DA and some other swarm intelligence algorithms. It also has a higher efficiency than the discrete adapted DA.

Keywords: discrete optimization; dragonfly algorithm; metaheuristics; optimization; swarm intelligence algorithms; traveling salesman problem

MSC: 49M37

Citation: Emambocus, B.A.S.; Jasser, M.B.; Amphawan, A.; Mohamed, A.W. An Optimized Discrete Dragonfly Algorithm Tackling the Low Exploitation Problem for Solving TSP. *Mathematics* **2022**, *10*, 3647. https://doi.org/10.3390/math10193647

Academic Editors: Humberto Rocha and Ana Maria Rocha

Received: 21 July 2022
Accepted: 7 September 2022
Published: 5 October 2022

1. Introduction

Optimization is crucial in almost every domain where certain processes, designs, or systems are adjusted so as to be the most effective possible. Its aim is to find the best solution among a set of available solutions which either maximizes or minimizes an objective function. Hence, optimization algorithms are usually iterative where the objective function is evaluated repeatedly and the best solution found is chosen. Optimization algorithms can be classified into two types; deterministic algorithms, and heuristic algorithms [1]. Deterministic algorithms consist of exact methods which find the best solution to a problem. However, they are computationally costly and time-consuming, especially for large-scale real-world applications. Conversely, heuristic algorithms try to find a near-optimal solution in a feasible amount of time. Heuristic algorithms can be further developed to produce metaheuristic algorithms which use a combination of diversification and intensification techniques to perform better than simple heuristic algorithms. They can be classified as either trajectory-based or population-based algorithms [1].

Swarm intelligence algorithms are population-based metaheuristic algorithms which are inspired by various biological organisms. In nature, the simple and self-organizing interactions that individuals from a specific biological population have with each other and with their environment cause a functional global pattern to emerge [2]. Swarm intelligence algorithms are inspired by these simple interactions of different groups of animals or swarms of insects. They make use of a population of search agents which replicate the interactions of a biological population for solving complex optimization problems.

The Dragonfly Algorithm (DA) [3] is a swarm intelligence algorithm which is inspired by dragonfly swarms, in particular, their hunting and migrating behaviours. Dragonflies swarm statically and dynamically while hunting and migrating, respectively, and these two types of swarming aptly represent the two requisite phases of optimization algorithms; exploration, and exploitation. DA has been found to have a better performance than multiple other swarm intelligence algorithms in various applications [4]. In [3], three versions of the dragonfly algorithm are proposed, namely, the original continuous DA for solving continuous optimization problems, the binary DA (BDA) to cater for discrete or binary optimization problems, and the multi-objective DA (MODA) to cater for multi-objective optimization problems.

The Traveling Salesman Problem (TSP) is a combinatorial optimization problem with a discrete search space. A myriad of real-world problems can be represented as TSP and hence it has numerous real-world applications. Large TSP problems are difficult to be solved using exact algorithms and the solvable instances consume a significant amount of computational time and resources [5]. Hence, heuristic algorithms are often used to solve TSP by providing near-optimal solutions. A number of swarm intelligence algorithms such as the Particle Swarm Optimization (PSO) [6], and the Ant Colony Optimization (ACO) [7], and their variants [8–11] have been successfully employed for solving TSP.

Considering the good performance of DA and its superior performance over other swarm intelligence algorithms in multiple applications, it is worth applying it for solving TSP. The original DA was proposed to solve continuous optimization problems and although a binary version of the original DA is proposed in [3], it is difficult to be adapted for discrete problems like TSP. Hence, in [12], we proposed a discrete adapted dragonfly algorithm suitable for TSP. This is to propose a new discretized variant of the dragonfly algorithm which is suitable for solving discrete optimization problems like TSP and which can be further improved. However, this algorithm has not been applied to large TSP problems and it has low effectiveness.

In this paper, we propose to improve the low effectiveness of the adapted discrete dragonfly algorithm in [12] by employing the steepest ascent Hill Climbing (HC) algorithm as a local search to improve the exploitation phase. The algorithm is tested using a TSP problem consisting of 50 locations in the area of Kuala Lumpur. The TSP problem models a package delivery system where the shortest route to deliver parcels at specific locations and return to the initial location needs to be found. Moreover, the performance of the algorithm is compared to other swarm intelligence algorithms in solving benchmark TSP problems from TSPLIB. From the experiments conducted, the proposed algorithm is found to have higher effectiveness, that is, it produces solutions with a lower cost as compared to the adapted discrete DA [12], and the enhanced Swap Sequence based PSO (SSPSO) [8] algorithms, and it also has higher effectiveness as compared to some other swarm intelligence algorithms as it provides the optimal solution or close to the optimal solutions for benchmark TSP problems. Moreover, it has a higher efficiency than the adapted discrete DA as it converges to the optimal solution in a shorter amount of time.

The contributions of this paper include an optimized discrete dragonfly algorithm suitable for solving discrete optimization problems such as TSP, which provides optimal or near-optimal solutions for benchmark TSP problems, an application of the proposed algorithm to a TSP problem which models a package delivery system in the area of Kuala Lumpur, and a comparison of the performance of the proposed algorithm to that of other swarm intelligence algorithms in solving benchmark TSP problems.

The remaining of the paper is structured as follows: In Section 2, a background on swarm intelligence algorithms, the original dragonfly algorithm, and the hill climbing algorithm is provided. In Section 3, an explanation on the traveling salesman problem is given. In Section 4, some previous works using swarm intelligence algorithms for solving TSP, and the adapted discrete DA algorithm are presented. In Section 5, a description of the proposed algorithm is given. In Section 6, the results and discussions are presented and, finally, in Section 7, the conclusions and some future work are presented.

2. Background on Swarm Intelligence, Dragonfly Algorithm, and Hill Climbing Algorithm

2.1. Swarm Intelligence Algorithms

Swarm intelligence algorithms are metaheuristic optimization algorithms that are inspired by the simple and self-organizing interactions of biological organisms that give rise to a functional global pattern [2]. They make use of a number of search agents which replicate the actions of individuals in a specific biological population as they interact among themselves and their environment. Each search agent, which represents a solution, moves in the state space by considering a fitness function. This allows the algorithm to solve complex optimization problems. Some well-known swarm intelligence algorithms include the particle swarm optimization that is inspired by the swarming of bird flocks or fish schools, and the ant colony optimization that is based on the food searching behaviour of ants.

Swarm intelligence algorithms have a plethora of applications in various domains as they can be used for solving different types of optimization problems including continuous optimization problems, discrete optimization problems, and multi-objective optimization problems. A continuous optimization problem is one in which the solution can be any real value within a certain range of values whereas a discrete optimization problem is one in which the solution can be a specific one from a set of possible solutions. A multi-objective optimization problem is one which has more than one objective function.

Some recent applications of swarm intelligence algorithms include in agricultural technology drones used for improving the productivity of farming areas [13], in fog computing systems for task scheduling [14], in gene selection profile for the classification of microarray data [15], in feature selection [16], and in artificial neural networks for optimizing the parameters of the network [17,18]. Furthermore, they have various applications in data science [19], in Internet of Things (IoT) systems [20], in surveillance systems [21], in water resources engineering [22], and in supply chain management [23].

2.2. Dragonfly Algorithm

The dragonfly algorithm [3] is a metaheuristic optimization algorithm classified under swarm intelligence algorithms. It is inspired by the swarming behaviours of dragonflies during hunting and migrating. During hunting, the dragonflies swarm statically, that is they form small groups and fly over a small area by abruptly changing their flying path. This type of swarming is congruent with the exploration phase of optimization algorithms where the algorithm tries to find a good region of the search space. Conversely, during migration, the dragonflies fly together in a sole group and along one direction over long distances. This type of swarming is congruent with the exploitation phase of optimization algorithms where the algorithm tries to converge to the optimal solution. Figure 1 shows a static and a dynamic swarm of dragonflies.

Five factors are used to control the movement of the dragonflies in the search space during both the exploration and exploitation phases; separation, alignment, cohesion, attraction to food, and distraction from enemy. Each factor has a corresponding weight which is used to tune the factor to enable the algorithm to transition between the exploration and exploitation phases. The factors, along with the weights, also ensure the survival of the swarm by causing it to attract towards food sources and distract away from enemies.

The best solution which has been obtained by the population of search agents is taken as the food source and the worst solution is taken as the enemy.

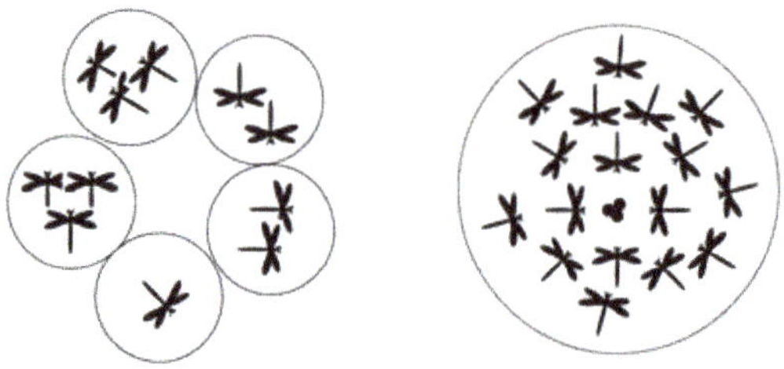

Figure 1. Static and dynamic dragonfly swarms.

The separation factor of a dragonfly is used to prevent its static collision with its neighbours, and is calculated using (1):

$$S_i = -\sum_{j=1}^{N} X_i - X_j \tag{1}$$

where X_i, and X_j are the current dragonfly's position and the j-th neighbour's position respectively, and N is the number of dragonflies in the neighbourhood.

The alignment factor of a dragonfly matches its velocity to that of its neighbours, and is calculated using (2):

$$A_i = \frac{\sum_{j=1}^{N} V_j}{N} \tag{2}$$

where V_j is the j-th neighbour's velocity and N is the number of dragonflies in the neighbourhood.

The cohesion factor of a dragonfly is its tendency towards the centre of mass of the neighbourhood, and is calculated using (3):

$$C_i = \frac{\sum_{j=1}^{N} X_j}{N} - X_i \tag{3}$$

where X_j is the j-th neighbour's position and N is the number of dragonflies in the neighbourhood.

The attraction to food factor of a dragonfly is its attraction towards a food source, and is calculated using (4):

$$F_i = X^+ - X_i \tag{4}$$

where X^+ is the food source's position.

The distraction from enemy factor of a dragonfly is its repulsion from an enemy, and is calculated using (5):

$$E_i = X^- + X_i \tag{5}$$

where X^- is the enemy's position.

To allow the dragonflies to move in the search space by considering these factors, two vectors are used: the step vector (ΔX) and the position vector (X).

The step vector determines the direction of the movement, and it is calculated using (6):

$$\Delta X_i^{t+1} = (sS_i + aA_i + cC_i + fF_i + eE_i) + w\Delta X_i^t \tag{6}$$

where S_i, A_i, C_i, F_i, and E_i are the separation, alignment, cohesion, food factor, and enemy factors of the i-th dragonfly respectively, s, a, c, f, and e are the separation, alignment, cohesion, food factor, and enemy factor's weights respectively, w is the inertia weight, and t is the iteration counter.

The position vector allows the movement in the search space, and it is calculated using (7):

$$X_i^{t+1} = X_i^t + \Delta X_i^{t+1} \tag{7}$$

In DA, in order to replicate the static and dynamic swarming behaviours of dragonflies, it is important to consider the neighbourhood of each search agent. This is achieved by considering a radius around each artificial dragonfly. The radius is increased proportionally to the iteration number so as to change the static swarms to dynamic ones until the whole population form one dynamic swarm and converges to the global optimum during the final iterations. This is another way by which the algorithm transitions from the exploration phase to the exploitation phase.

If a dragonfly has no neighbouring dragonfly, its position is updated using the Levy flight mechanism, which is a random walk employed to increase the stochasticity of the algorithm. The position vector of the dragonfly is calculated using (8):

$$X_i^{t+1} = X_i^t + Levy(d) \times X_i^t \tag{8}$$

where t is the current iteration number and d is the dimension of the position vectors.

The pseudocode of DA is given in Algorithm 1.

Algorithm 1: Dragonfly Algorithm

```
Initialize the population's positions randomly;
Initialize the step vectors;
while end condition do
    Calculate the objective values of all dragonflies;
    Update the food source and enemy;
    Update the weights;
    Calculate the factors using (1)–(5);
    Update radius of neighbourhoods;
    if dragonfly has one or more neighbours then
        Update step vector using (6);
        Update position vector using (7);
    else
        Update position vector using (8);
    end
    Check and correct new positions based on upper
     and lower bounds;
end
```

2.3. Hill Climbing Algorithm

The hill climbing algorithm is a heuristic local search algorithm which is used for optimization mainly in the field of artificial intelligence. Generally, starting from one position, it considers all the possible neighbouring solutions in a search region and selects the one which either maximizes or minimizes an objective function the most.

There are three main types of hill climbing, namely, the simple hill climbing algorithm, the steepest ascent hill climbing algorithm, and the stochastic hill climbing algorithm.

The simple hill climbing algorithm checks only one neighbouring solution at a time. If it is better than the current solution, the neighbouring solution is selected as the current solution. This type of hill climbing algorithm requires less computing power. However, a good solution is not guaranteed.

The steepest ascent hill climbing algorithm checks all the neighbouring solutions and then selects the best one. This type of hill climbing algorithm requires more computing power, however, it is more likely to find the most optimal solution.

The stochastic hill climbing algorithm checks a random neighbouring solution and if it is better than the current solution, the neighbouring solution is selected as the current solution, otherwise, another random neighbouring solution is selected.

3. Problem Formulation

The traveling salesman problem, a combinatorial optimization problem, is classified as a Non-Deterministic Polynomial-hard (NP-hard) problem as large TSP problems are difficult to solve [5]. It consists of a discrete state space with a finite set of possible solutions and the aim is to find the best solution from the set. TSP can be defined as follows: given a number of inputs called cities and the cost of travel between each possible pair, the aim is to find the route with the least cost to visit each city exactly once and to return to the starting city.

TSP has numerous real-world applications since a large number of real-world problems can be represented as TSP. For example, it can be used in X-Ray crystallography, computer wiring, vehicle routing, and order picking in warehouses [24]. Some more recent applications of TSP include route planning for Unmanned Aerial Vehicles (UAVs) [25], in delivery services using UAVs [26], in emergency air logistics [27], in robotic automated storage and retrieval system [28] and robot path planning [29].

4. Related Works

4.1. Existing Swarm Intelligence Algorithms Applied to TSP

In this section, some previous works in which swarm intelligence algorithms have been adapted and enhanced for solving TSP are presented.

In [30], the firefly algorithm is adapted and enhanced for solving TSP by using the method of swap sequences and Genetic Algorithm (GA). GA is first used to initialise the population of search agents. The serial number coding method is used for representing the discrete state space of TSP and the method of swap sequences is used to redefine the equations of the firefly algorithm in order to adapt the algorithm to TSP. Three neighbourhood structures are also used to redefine the disturbance mechanism of the firefly algorithm. The algorithm is tested using five TSP problems from TSPLIB and it is found to provide solutions which are close to the known optimal solutions.

In [31], a discrete sparrow search algorithm is proposed for solving TSP. For initialization of the population's positions, the roulette wheel selection is used. The path representation method is used for representing a TSP path. The position of the search agents is updated using the same equation as the original sparrow search algorithm and a decoding method called the order-based decoding is used to decode the solution obtained. To increase the diversity of the population, the Gaussian mutation perturbation is used together with swap operators. Furthermore, the 2-opt algorithm is used as a local search to improve the quality of the solutions and to increase the convergence rate. The algorithm is tested using 34 TSP instances and is found to be robust and have good convergence characteristics.

In [32], the Grey Wolf Optimization (GWO) is adapted for solving TSP by the use of swap sequences and swap operators. It makes use of the general steps of the Grey Wolf optimization algorithm. In the initialization stage, the population is initialized with a random TSP path and the agents with the three best solutions are chosen as the Alpha, Beta, and Delta wolves. The position of each search agent is then updated by considering the Alpha, Beta, and Delta wolves by making use of the method of swap sequences. The algorithm is used for solving benchmark TSP problems and is found to provide better results than ACO and GA for several TSP problems.

In [33], the chicken swarm optimization algorithm is adapted for solving TSP by using the methods of swap operators, order crossover, and reverse order mutation. The integer coding method is used for the solution representation, and the method of swap operators is used for updating the position of the search agents. The order crossover, and reverse order mutation are also used for updating the position of the search agents in the state space by increasing the population diversity. Five instances from TSPLIB are used for testing the algorithm which is found to be effective in solving TSP.

In [34], a discrete spider monkey optimization algorithm is proposed for solving TSP. It makes use of the method of swap sequences and swap operators for updating the position of the search agents. The position of each search agent is updated based on the position of a subgroup leader, known as the local leader, and the position of a main group leader, known as the global leader, and also their self-experience. The algorithm is tested using TSP instances from TSPLIB and its performance is compared with other swarm intelligence algorithms. It is found to have a good performance in solving TSP.

In [8], an enhanced swap sequence based PSO algorithm is proposed for solving TSP. It makes use of path representation for representing solutions in the state space, and the method of swap operators for updating the positions. It incorporates the three strategies of the continuous XPSO algorithm and uses the method of swap operators in order to be suitable for solving TSP. The three strategies include a forgetting ability for each particle, the adjustment of the acceleration coefficients by making use of the population's experience, and the use of both the global and local exemplars for learning. The algorithm is found to provide better results than the swap sequence-based PSO.

4.2. Discrete Adapted Dragonfly Algorithm

In this section, a description of the adapted discrete dragonfly algorithm that we proposed in [12] is given. The algorithm makes use of the path representation method to represent a potential solution, that is a TSP path, and it makes use of the method of swap sequence [35] which was originally used for adapting PSO to TSP. The adapted discrete DA adapts the original DA algorithm to be suitable for solving TSP by the following: firstly, the method of path representation is used to represent a potential solution, that is, a TSP path which is taken as the position of a search agent. Secondly, the equations for calculating the five factors used in DA, the step vector, and the position vector are adapted to be suitable for the path representation. Thirdly, the five factors, the step vector, and the position vector are calculated using the method of swap sequence. The pseudocode of the adapted discrete DA is given in Algorithm 2.

Algorithm 2: Adapted Discrete DA Algorithm for TSP

```
Initialize the population's positions with random TSP paths;
Initialize the step vectors with random swap sequences;
while end condition do
    Calculate the objective values of all dragonflies;
    Update the food source and enemy;
    Update the weights;
    Calculate the factors using (9), (10), (11), (12), (13);
    Update step vector using (14);
    Update position vector using (15);
end
```

In the discrete adapted DA algorithm, the position and step vector of the search agents are first initialized with a random TSP path and a random swap sequence respectively. Then in each iteration, the objective value of each search agent's position, that is the cost of the TSP path represented by the search agent's position, is calculated. The position with the lowest objective value is taken as the food source and that with the highest objective value is taken as the enemy. The separation, alignment, cohesion, attraction to food, and distraction from enemy factors are calculated, and their corresponding weights are updated. Finally, the position of the search agents is updated using the step and the position vectors. Contrary to the original continuous DA, the radius of neighbourhood of the dragonflies is not considered in the discrete adapted DA. This is because the Euclidean distance between the dragonflies in a discrete state space cannot be easily calculated and hence all the search agents are considered to be in the same neighbourhood.

4.3. Initialization

In the initialization phase, the positions of the search agents are first initialized with a random solution, that is a TSP path. The TSP path is represented using path representation, that is, a permutation of numbers representing cities. The numbers indicates the order of visit of the cities. An example of a TSP path for a TSP problem consisting of 4 cities can be: 3 2 1 4 3. This indicates that starting from the city denoted by the number 3, city 2 will be visited next, followed by city 1 and city 4, before returning to the starting point, that is, city 3.

The step vector of the search agents is then initialized with a velocity, which in this case, is a random swap sequence. A swap sequence consists of a number of swap operators, and each swap operator contains a pair of indices which represents cities in a TSP path. It indicates that the position of the two cities will be swapped when the swap operator is applied to a TSP path. For example, a swap operator, SO, can be represented by $SO(2,4)$, and a swap sequence, SS, can be represented by $SS = (SO(2,4), SO(1,2))$.

4.4. Calculation of Factors

The separation, alignment, cohesion, attraction to food, and distraction from enemy factors are calculated using Equations (9)–(13). These equations have been produced by adapting the Equations (1)–(5) from the original DA. This is because the factors of a search agent are calculated using the positions and step vectors of its neighbours, and in the adapted discrete DA, the positions and step vectors are TSP paths and swap sequences respectively. Hence the equations used in the original DA are not suitable for the adapted discrete DA.

The separation factor is calculated as follows:

$$S_i = Inv(\bigoplus_{j=1}^{N} X \ominus X_j) \tag{9}$$

where X, X_j, and N are the the current search agent's position, the j-th neighbour's position, and the total number of neighbours respectively.

The '$\ominus$' operator indicates the subtraction of two positions, that is two TSP paths, to produce a swap sequence. For example, the subtraction of two paths, $X = 1\ 3\ 4\ 2$ and $X_j = 2\ 3\ 1\ 4$, is $X \ominus X_j = SO(1,3), SO(3,4)$.

The '$\bigoplus$' operator indicates the merging of the swap sequences into only one swap sequence which contains all the swap operators from the different swap sequences in sequential order. For example, for the swap sequences $SS_1 = SO(2,3), SO(4,1)$ and $SS_2 = SO(1,3)$, $SS_1 \bigoplus SS_2 = SO(2,3), SO(4,1), SO(1,3)$.

'Inv' indicates that the swap sequence is inversed. For example, for $SS = SO(2,3)$, $SO(4,1)$, $Inv(SS) = SO(1,4), SO(3,2)$.

The alignment factor is calculated as follows:

$$A_i = V_{avg} \tag{10}$$

where V_{avg} is the step vector of the neighbour having the fitness closest to the average fitness in the neighbourhood.

The cohesion factor is calculated as follows:

$$C_i = X_{avg} \ominus X \tag{11}$$

where X_{avg} is the position of the neighbour having the fitness closest to the average fitness in the neighbourhood and X is the current search agent's position.

The attraction to the food source factor is calculated as follows:

$$F_i = X_f \ominus X \quad (12)$$

where X_f is the food source's position and X is the current search agent's position.

The distraction from the enemy factor is calculated using the procedure '$CalculateE_i()$' which provides a swap sequence.

$$E_i = CalculateE_i(X_e, X) \quad (13)$$

X_e is the enemy's position and X is the current search agent's position. The procedure compares each city in X_e and X and if a city is similar in both positions, it generates a swap operator which consists of that city and another different random city. This is to decrease the similarity between the two TSP paths.

4.5. Update of Positions

For updating the position of the search agents, the step vector is first calculated as follows:

$$\Delta X_{t+1} = (sS_i \oplus aA_i \oplus cC_i \oplus fF_i \oplus eE_i) \oplus w\Delta X_t \quad (14)$$

where S_i, A_i, C_i, F_i, and E_i are the separation, alignment, cohesion, food factor, and enemy factors of the i-th dragonfly respectively, s, a, c, f, and e are the separation, alignment, cohesion, food factor, and enemy factor's weights respectively, w is the inertia weight, and t is the iteration counter.

In this equation, '$\oplus$' indicates the merging of two swap sequences, resulting in one swap sequence with all the swap operators in the first swap sequence followed by all the swap operators in the second swap sequence.

The position vector of the search agent is then calculated as follows:

$$X_{t+1} = X_t \otimes \Delta X_{t+1} \quad (15)$$

In this equation, '$\otimes$' indicates that the swap sequence 'ΔX_{t+1}' will be applied to the path 'X_t'. The application of a swap sequence to a path means that each swap operator in the swap sequence will be applied to the path sequentially to produce a new path. An example of applying a swap operator to a path is given below.

Considering a path $X = 2\ 4\ 1\ 3$ and a swap operator $SO(2,3)$, the cities in the second and third positions of X will be swapped. Hence, the resultant path will be $X = 2\ 1\ 4\ 3$.

5. Proposed Enhanced Adapted Discrete DA

In this section, a description of the proposed enhanced discrete adapted DA algorithm is provided. The proposed algorithm improves the exploitation phase of the adapted discrete DA by using the steepest ascent hill climbing algorithm as a local search. After the position of the search agents is updated using Equation (15), the hill climbing algorithm is employed to further exploit the region obtained and to update the position of the search agent to a better one. The steepest ascent hill climbing algorithm starts at the position obtained by Equation (15) and then looks for every possible position in that area of the search space. It then selects the one with the lowest cost. Hence, the steepest ascent hill climbing algorithm is able to locate the optimum solution in the area initially obtained by the search agent. Moreover, to prevent the algorithm from getting stuck in a local optimum, the position of a search agent is changed to a random solution when it cannot be further improved. This is done by keeping track of the personal best solution that is found by a search agent. If the personal best solution does not change over a certain number of iterations, the search agent's position is changed to another random position so as to allow it to get out of the local optimum and to search other regions of the state space. The pseudocode of the proposed algorithm is given in Algorithm 3.

Algorithm 3: Enhanced Adapted Discrete DA Algorithm for TSP

```
Initialize the population's positions with random TSP paths;
Initialize the step vectors with random swap sequences;
PBest_Stagnancy = 0;
while end condition do
    Calculate the objective values of all dragonflies;
    Update the food source and enemy;
    Update the weights;
    Calculate the factors using (9), (10), (11), (12), (13);
    Update step vector using (14);
    Update position vector using (15);
    Initialize position as current position for hill climbing;
    while local optima is not reached do
        Generate neighbours;
        for each neighbour do
            if cost of neighbour <cost of current position then
                current position = neighbour position;
            end
        end
    end
    if cost of position <cost of personal best position then
        best position = position;
    else
        PBest_Stagnancy = PBest_Stagnancy + 1;
    end
    if PBest_Stagnancy > 3 then
        Change position to random position;
    end
end
```

In the initialization phase, the position and step vectors are initialized with a random TSP path, and a random swap sequence respectively. This step is similar to the adapted discrete DA in Section 4.2. In addition, a personal best stagnancy variable is initialized to zero. This variable is used to keep track of the number of iterations in which the best solution found by a search agent has not improved.

In the main loop of iteration, the objective cost of each search agent is first calculated, and the food and enemy are updated with the best and worst positions respectively. Then the separation, alignment, cohesion, attraction to food and distraction from enemy factors are calculated. Contrary to the adapted discrete DA in Section 4.2, in the proposed algorithm, these factors are calculated based on the path obtained after applying the previous factor; that is the separation factor is first calculated and the swap operators obtained are immediately applied to the TSP path represented by the position of the search agent to update the path. Then the alignment factor is calculated based on the updated path after applying the separation factor. Similarly, the path is updated and then the cohesion factor is calculated. The food factor is then calculated based on the updated path after applying the swap operators of the cohesion factor and the enemy factor is calculated based on the updated path after applying the food factor. The path with the lowest cost is then chosen as the next position of the search agent. This is done so as to increase the efficiency of the algorithm so that it can provide good solutions in a short amount of time.

After the position of a search agent is updated using Equation (15), the hill climbing algorithm is employed to further update the position as follows: the position obtained by (15) is taken as the current position for the hill climbing algorithm. Then a set of neighbouring solutions is generated and the one with the lowest cost is selected as the current position. The steps of generating neighbours and selecting the one with the lowest cost as the current position are repeated until a solution with a lower cost cannot be found. The use of the hill climbing algorithm is to improve the exploitation of the dragonfly algorithm.

In order to prevent the algorithm from being stuck in local optima, the personal best solution of each search agent, which is the best solution found by the search agent, is recorded. In each iteration, after the position of the search agent has been updated, the personal best solution is checked and updated if it has changed. If the personal best solution remains unchanged over a certain amount of iterations, then the position of the search agent is changed to a random solution. This is to allow the search agent to get out of the local optimum and explore other regions of the search space.

5.1. Solution Representation

There are several ways to represent a TSP path as the position of a search agent in a discrete state space such as binary, path, adjacency, ordinal, and matrix representations [31]. In this paper, the path representation is used to encode the TSP solutions since this is the most natural representation of a path. This is the same representation used in [12] for the adapted discrete DA algorithm.

5.2. Objective Function

The objective function is the cost of the TSP path, which in this case is the total distance of the TSP path. This is obtained by calculating the distance between each pair of adjacent cities in the TSP path. It is considered that the distance to travel from city i to city j is the same as the distance to travel from city j to city i.

5.3. Update of Positions

Similar to the adapted discrete DA algorithm in Section 4.2, the method of swap sequences is used for updating a position, that is a TSP path. A sequence contains a number of swap operators which indicate that the two cities in the position denoted by the swap operator will be swapped to produce a new TSP path. The swap operators in the swap sequence are applied sequentially to the TSP path. For example, for a TSP path 2 4 1 3, and swap sequence $SO(1,2)$, $SO(3,2)$, the TSP path 4 1 2 3 will be obtained when the swap sequence is applied to the TSP path.

5.4. Experimental Parameters

Table 1 shows the parameters used for the optimized DA algorithm, their description, and their values.

Table 1. Experimental parameters.

Parameter	Description	Value
X_i	The position of search agent i	A TSP path, example: 1 3 4 2 1
V_i	The step vector of search agent i	A swap sequence, example: $SO(1,3)$ $SO(2,1)$
X_f	The food position	A TSP path, example: 1 4 2 3 1
X_e	The enemy position	A TSP path, example: 1 2 4 3 1
S_i	The separation factor of the ith search agent	A swap sequence, example: $SO(2,4)$ $SO(1,2)$
A_i	The alignment factor of the ith search agent	A swap sequence, example: $SO(1,4)$ $SO(3,2)$
C_i	The cohesion factor of the ith search agent	A swap sequence, example: $SO(1,3)$ $SO(3,4)$
F_i	The food factor of the ith search agent	A swap sequence, example: $SO(3,4)$ $SO(1,2)$

Table 1. *Cont.*

Parameter	Description	Value
E_i	The enemy factor of the ith search agent	A swap sequence, example: $SO(3,2)\ SO(4,1)$
s	The separation weight	A real value, example: 1.5
c	The cohesion weight	A real value, example: 1.2
a	The alignment weight	A real value, example: 2.3
f	The attraction to food weight	A real value, example: 1.2
e	The distraction from enemy weight	A real value, example: 0.5
w	The step vector weight	A real value, example: 0.1

6. Experimental Results and Analysis

In this section, a description of the experimental datasets used, the experimental setup, and the results with some discussions are provided.

6.1. Experimental Dataset

The test data consist of a TSP problem with 50 nodes, where each node represents a location in the area of Kuala Lumpur (KL). The cost between each pair of nodes is the distance needed to travel from one node to the other. It is considered that the distance to travel from city i to city j is the same as the distance to travel from city j to city i. The aim is to find the shortest route to visit each node once and to return to the initial node. The distance between two locations is taken as the shortest distance in kilometers (km) that can be taken by a vehicle based on Google Maps.

To test the proposed algorithm with TSP problems of different sizes, the TSP data is changed to 10, 20, and 40 nodes by taking the first 10, 20, and 40 locations respectively.

Moreover, several benchmark datasets from TSPLIB are used to test the performance of the proposed algorithm. Specifically, the burma14, ulysses16, ulysses22, bays29, eil51, berlin52, st70, eil76, rat99 and kroA100 datasets consisting of 14, 16, 22, 29, 51, 52, 70, 76, 99, and 100 nodes respectively are used. The datasets are in the form of coordinates and the distance matrix of each dataset is constructed by calculating the Euclidean distance between the nodes.

6.2. Experimental Setup

The proposed algorithm is used for solving the TSP problem with 50, 40, 20, and 10 locations in KL and the results are recorded in terms of the cost of the solution obtained, the time taken to converge to the optimal solution, and the total time taken by the algorithm.

To compare the performance of the proposed algorithm, the discrete adapted dragonfly algorithm in [12], and the enhanced SSPSO in [8] are used for solving the same TSP problems of 50, 40, 20, and 10 locations and the results are recorded in terms of the cost of the solution obtained, the time taken to converge to the optimal solution, and the total time taken by the algorithm. The results of the proposed algorithm are compared to that of the discrete adapted dragonfly algorithm and the enhanced SSPSO algorithm. The enhanced SSPSO is used for comparing the performance of the proposed algorithm since it is our own algorithm which had been used for the same dataset, and TSP problem, that is the delivery system in the area of Kuala Lumpur. In order to have a better algorithm for the delivery system, the new optimized DA algorithm is proposed in this paper.

The experiments are repeated for different numbers of maximum iteration and search agents. Specifically, the number of maximum iterations used are 20, 50, 100, 200, and 500, and the number of search agents used is 5, 10, 20, and 40.

Furthermore, in order to compare the performance of the proposed algorithm to that of other swarm intelligence algorithms, the algorithm is applied to benchmark TSP problems from TSPLIB and the best solution obtained by the proposed algorithm is compared to the best solution obtained by Ant Colony Optimization (ACO), Velocity Tentative PSO (VTPSO), Artificial Bee Colony with Swap Sequence (ABCSS), Discrete Spider Monkey Optimization (DSMO), Genetic Algorithm (GA), Producer Scrounger Method (PSM), and Grey Wolf Optimizer (GWO) algorithms. The results of ACO, VTPSO, ABCSS, and DSMO are taken from [34]. The results of ACO GA, PSM, and GWO are taken from [32]. The maximum number of iterations used for the proposed optimized DA is 500 and the number of search agents used is between 20 and 100.

Both the proposed optimized discrete adapted DA and the discrete adapted DA were implemented in MATLAB and all experiments were conducted on a MacOs Monterey operating system, Apple M2 chip CPU, and 8 GB RAM.

6.3. Results and Discussion

6.3.1. Greater Kuala Lumpur TSP Problem

In this section, we present a comparison and discussion on the performance of the proposed algorithm when applied to our own dataset consisting of locations in the area of Greater Kuala Lumpur which models a delivery system.

Tables 2–5 show the results obtained when the proposed optimized discrete DA, the discrete adapted DA, and the enhanced swap sequence based PSO are used for solving the TSP problem with 50, 40, 20, and 10 locations respectively. The experiments are conducted by using different number of maximum iterations and search agents and the results are recorded in terms of the cost of the solution, that is TSP path, provided by the algorithms, the time taken to converge to the global optimal solution, and the total time taken by the algorithms.

Figures 2–5 show the convergence curve of the proposed optimized discrete adapted DA and the discrete adapted DA in solving a TSP of 50, 40, 20, and 10 locations respectively. The figures show the convergence of the algorithms for 5, 10, 20, and 40 search agents and for a maximum iteration of 200. The figures show the convergence rate of the two algorithms and also the cost of the solution, that is the TSP path provided.

An example of a TSP path for 20 locations in Kuala Lumpur area provided by the optimized discrete adapted DA algorithm is given in Figure 6. The cost of the TSP path is 105.4 and the path is as follows: 11 15 18 16 10 9 4 13 2 3 12 1 20 19 17 7 6 5 8 14 11. The numbers represent the cities and their order represent the order in which they will be visited.

From Tables 2–5 it can be deduced that the proposed optimized discrete adapted DA algorithm has a higher effectiveness than the discrete adapted DA algorithm as the cost of the TSP path provided by the proposed algorithm is significantly lower in all of the experiments conducted. For example, for a maximum iteration of 500 and 40 search agents, the costs of the TSP paths obtained by the discrete adapted DA for 50, 40, 20, and 10 locations are 507.8, 409.3, 161.9, and 69.6 respectively while those obtained by the proposed optimized adapted discrete DA are 200.0, 178.7, 105.4, and 65.9 respectively. This means that the optimized adapted discrete DA improves the solution obtained by the adapted discrete DA by 60.6%, 56.3%, 34.9%, and 5.3% for 50, 40, 20, and 10 locations respectively.

Even for smaller number of iterations and search agents, the proposed algorithm converges to solutions with lower costs than the discrete adapted DA. For example, for a maximum of 20 iterations and only five search agents, the costs of the TSP paths obtained by the discrete adapted DA for 50, 40, 20, and 10 locations are 556.2, 478.1, 191.9, and 82.9 respectively while those obtained by the optimized discrete adapted DA are 229.4, 217.8, 117.5, and 65.9 respectively. This indicates that the optimized adapted discrete DA provides solutions which are 58.8%, 54.4%, 38.8%, and 20.5% better than those provided by the adapted discrete DA for 50, 40, 20, and 10 locations respectively.

Table 2. Performance comparison of proposed optimized discrete DA and discrete adapted DA in solving TSP of 50 cities.

Maximum No. of Iterations	No. of Search Agents	Enhanced SSPSO			Discrete Adapted DA			Proposed Optimized Discrete Adapted DA		
		Cost of Solution (km)	Time Taken to Converge to Optimum (s)	Total Time Taken (s)	Cost of Solution (km)	Time Taken to Converge to Optimum (s)	Total Time Taken (s)	Cost of Solution (km)	Time Taken to Converge to Optimum (s)	Total Time Taken (s)
20	5	532.0	0.0359	0.0481	556.2	3.519	7.91	229.4	3.5325	6.6253
20	10	550.0	0.04071	0.1139	571.0	18.0617	27.6912	223.3	6.9313	9.0644
20	20	542.0	0.2214	0.2959	570.7	15.7653	130.8685	223.5	25.5084	26.5358
20	40	547.0	0.1365	0.5762	546.0	2.1136	770.9831	232.3	57.8731	58.7077
50	5	525.0	0.0892	0.3160	561.6	6.2466	54.9875	217.2	10.734	12.7241
50	10	525.0	0.0021	0.6858	539.0	3.3454	190.3065	225.4	19.015	24.0916
50	20	541.0	0.5838	1.2465	524.0	53.0974	823.7265	211.8	64.8796	67.1365
50	40	509.0	3.3133	6.02159	542.3	587.3994	4322.1249	206.0	147.4982	150.3943
100	5	525.0	1.0883	1.3751	537.5	6.8421	200.1581	223.5	7.1802	18.6268
100	10	510.0	0.1848	3.7681	538.0	178.4626	788.3791	213.7	46.6898	48.3536
100	20	518.0	3.1911	9.6636	539.2	1429.93	3302.5555	209.4	29.4048	122.456
100	40	504.0	7.3404	17.9399	534.8	426.2547	15,899.7002	210.6	229.6718	240.2979
200	5	530.0	4.8935	6.3373	537.0	180.8259	429.2883	217.9	48.1498	48.6066
200	10	501.0	0.7947	14.5220	529.2	798.1125	1676.7151	206.6	59.3682	109.5615
200	20	497.0	17.5760	38.0765	527.7	4210.0023	7259.5666	199.4	126.6875	195.9365
200	40	494.0	20.0064	82.9766	519.5	1934.8816	34,189.6828	206.7	324.0921	554.3716
500	5	478.0	26.4978	42.7225	519.6	504.9482	6311.1325	213.1	96.5744	100.8358
500	10	494.0	45.6093	101.8348	487.2	17,838.5761	26,586.0832	196.8	121.3713	245.6557
500	20	488.0	21.9401	309.6251	517.3	16,875.0334	113,766.0094	193.6	262.4892	500.1441
500	40	494.0	30.9125	588.0762	507.8	283,915.1445	546,679.3105	200.0	534.5106	1198.7924

Table 3. Performance comparison of proposed optimized discrete adapted DA and discrete adapted DA in solving TSP of 40 cities.

Maximum No. of Iterations	No. of Search Agents	Enhanced SSPSO			Discrete Adapted DA			Proposed Optimized Discrete Adapted DA		
		Cost of Solution (km)	Time Taken to Converge to Optimum (s)	Total Time Taken (s)	Cost of Solution (km)	Time Taken to Converge to Optimum (s)	Total Time Taken (s)	Cost of Solution (km)	Time Taken to Converge to Optimum (s)	Total Time Taken (s)
20	5	431.0	0.02126	0.03446	478.1	0.13406	11.4666	217.8	2.7567	4.3227
20	10	461.0	0.02927	0.0843	463.9	1.9294	29.2706	222.8	5.2488	5.4381
20	20	442.0	0.0140	0.1880	452.3	107.0152	131.4918	197.2	10.0264	12.3698
20	40	435.0	0.01673	0.4016	430.6	20.2805	686.2293	188.9	29.322	36.2265
50	5	448.0	0.06111	0.1847	428.9	32.3926	65.8366	207.3	3.9897	6.7687
50	10	415.0	0.4449	0.8789	436.8	95.0279	256.7351	195.5	12.7688	13.4064
50	20	427.0	0.1988	1.0586	449.8	87.115	1004.6713	200.4	14.9048	28.4775
50	40	418.0	1.8880	3.1482	444.3	2194.2967	4545.5709	194.3	65.8466	71.1544
100	5	422.0	0.0027	1.2992	422.6	114.8649	270.0771	190.7	7.247	11.6643
100	10	419.0	1.3789	3.1608	458.0	609.6101	950.9694	193.3	20.9147	35.4175
100	20	433.0	1.6942	4.0857	444.6	2547.628	3991.4558	185.0	40.165	49.4274
100	40	407.0	1.6516	15.0465	441.4	1860.1602	25,521.9187	194.0	103.962	119.8753
200	5	402.0	1.4485	5.3046	422.8	1.31	361.4874	186.8	13.2504	20.6362
200	10	421.0	2.9958	11.0945	444.0	862.4717	1301.4488	199.8	44.9657	57.7674
200	20	418.0	1.7099	22.8645	418.6	2661.0911	5393.6283	181.6	44.927	110.284
200	40	420.0	45.7771	56.4051	436.0	6848.2599	25,271.4722	178.7	212.0086	257.4372
500	5	421.0	30.1391	30.2641	435.5	2122.968	7159.2775	190.0	43.844	63.6809
500	10	417.0	14.5123	93.8408	429.8	4701.7931	23,959.7151	185.0	85.5298	137.7637
500	20	389.0	22.4112	194.3102	423.1	100,073.4386	124,577.1437	184.8	182.1964	292.4497
500	40	404.0	7.1911	498.5430	409.3	72,913.9465	357,630.9028	179.4	501.9859	681.6409

Table 4. Performance comparison of proposed optimized discrete adapted DA and discrete adapted DA in solving TSP of 20 cities.

Maximum No. of Iterations	No. of Search Agents	Enhanced SSPSO			Discrete Adapted DA			Proposed Optimized Discrete Adapted DA		
		Cost of Solution (km)	Time Taken to Converge to Optimum (s)	Total Time Taken (s)	Cost of Solution (km)	Time Taken to Converge to Optimum (s)	Total Time Taken (s)	Cost of Solution (km)	Time Taken to Converge to Optimum (s)	Total Time Taken (s)
20	5	195.0	0.004488	0.01227	191.9	2.0448	3.8275	117.5	0.32687	1.0266
20	10	173.0	0.02451	0.0298	187.6	5.9708	14.7066	109.1	1.7194	1.8762
20	20	182.0	0.0096	0.0752	192.7	7.3479	64.1203	107.1	2.2349	3.5711
20	40	168.0	0.09121	0.1348	186.1	18.9694	396.1467	106.1	3.7612	6.3904
50	5	173.0	0.0035	0.0603	175.0	6.4056	24.9236	105.4	0.63344	1.9217
50	10	176.0	0.1615	0.2060	160.1	0.0092196	97.1998	106.1	3.8454	3.9797
50	20	162.0	0.0033	0.4717	168.9	192.3332	501.511	105.4	3.4697	6.4025
50	40	151.0	1.0569	1.4717	168.8	341.0088	2616.6784	105.4	1.6023	16.6091
100	5	171.0	0.0048	0.2566	186.1	68.1113	125.241	108.8	1.0233	3.0716
100	10	173.0	0.2345	0.6208	171.6	269.7031	475.5543	106.8	3.3758	5.4659
100	20	167.0	0.3571	1.9250	171.1	7.419	2047.3412	105.7	11.401	12.6007
100	40	170.0	2.0494	4.0331	174.9	5646.5838	8242.855	105.4	1.6434	32.2757
200	5	163.0	0.9734	1.1322	180.1	89.7254	164.0154	105.7	0.8572	6.74
200	10	166.0	0.1077	3.3940	175.4	11.5837	624.1664	105.4	3.9401	10.1722
200	20	163.0	0.0767	6.4863	162.7	202.5637	2352.474	105.4	9.3677	21.9359
200	40	155.0	0.0475	19.5566	170.7	4165.0315	11,489.6363	105.4	23.7822	60.105
500	5	164.0	0.0311	10.7217	163.4	875.6302	3632.2101	105.4	9.4902	13.2498
500	10	164.0	3.2880	25.2614	175.8	1215.4295	15,191.4602	105.4	9.2816	30.2117
500	20	153.0	0.0044	56.4343	155.7	37,104.5216	61,427.5274	105.4	2.6679	64.409
500	40	155.0	1.7389	111.9266	161.9	83,639.0656	247,067.0486	105.4	79.663	157.4536

Table 5. Performance comparison of proposed optimized discrete adapted DA and discrete adapted DA in solving TSP of 10 cities.

Maximum No. of Iterations	No. of Search Agents	Enhanced SSPSO			Discrete Adapted DA			Proposed Optimized Discrete Adapted DA		
		Cost of Solution (km)	Time Taken to Converge to Optimum (s)	Total Time Taken (s)	Cost of Solution (km)	Time Taken to Converge to Optimum (s)	Total Time Taken (s)	Cost of Solution (km)	Time Taken to Converge to Optimum (s)	Total Time Taken (s)
20	5	80.0	0.0022	0.0028	82.9	0.50424	3.4169	65.9	0.26697	0.56227
20	10	78.0	0.0072	0.0074	71.2	4.8409	9.6044	65.9	0.25308	0.91832
20	20	75.0	0.0055	0.0171	78.0	21.0104	35.849	65.9	0.22486	1.2627
20	40	69.0	0.0248	0.0369	79.0	0.72344	146.1099	65.9	0.41628	2.3977
50	5	75.0	0.0124	0.01961	81.7	13.2891	15.7673	65.9	0.12756	1.0816
50	10	71.0	0.0055	0.0440	80.2	25.3691	48.5602	65.9	0.61573	1.4417
50	20	72.0	0.0996	0.1174	73.4	13.4782	222.6126	65.9	0.22748	2.5893
50	40	67.0	0.1659	0.2397	76.9	661.0827	915.7478	65.9	0.37537	5.4787
100	5	73.0	0.0715	0.0733	81.3	42.137	71.0761	65.9	0.21733	1.6297
100	10	72.0	0.2668	0.2721	77.3	27.525	229.6083	65.9	0.35535	2.3059
100	20	66.0	0.4829	0.4832	76.7	102.0761	959.875	65.9	0.22507	4.4552
100	40	68.0	0.3644	0.8834	72.1	607.6719	5944.7225	65.9	1.1377	11.2092
200	5	75.0	0.3177	0.3179	75.8	1.117	70.7217	65.9	0.47184	3.1885
200	10	65.0	0.0234	1.080	77.5	6.5001	253.8453	65.9	0.17053	4.7651
200	20	71.0	1.7071	1.7748	72.0	372.1562	991.3132	65.9	0.24187	8.242
200	40	67.0	3.4541	4.0137	68.8	416.1922	4307.4956	65.9	0.56337	19.9458
500	5	69.0	0.0880	2.4140	74.7	0.59851	1637.606	65.9	0.15524	5.3221
500	10	71.0	5.7717	5.9135	69.8	502.382	6421.5464	65.9	0.90636	9.7569
500	20	68.0	12.0173	12.7403	71.4	8413.495	28,095.8444	65.9	0.25608	19.0239
500	40	67.0	25.6158	25.8277	69.6	8756.6288	117,232.4453	65.9	0.37032	47.7772

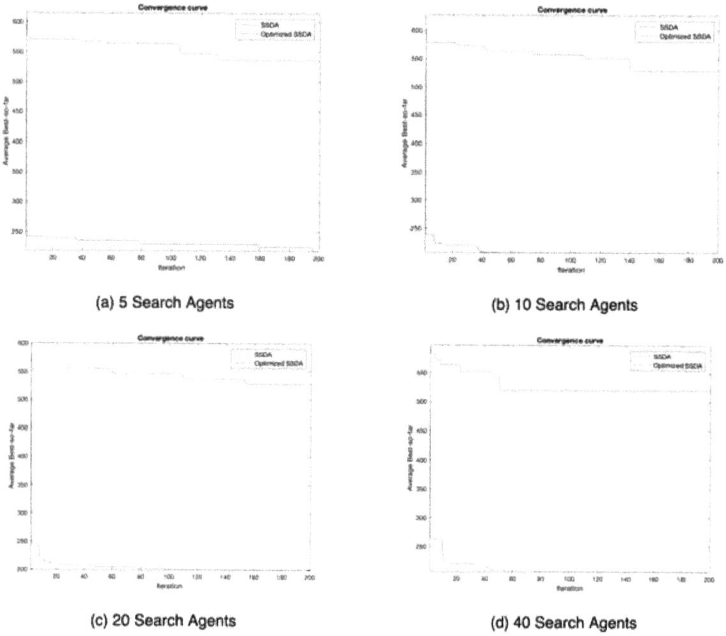

Figure 2. Convergence curve of optimized discrete adapted DA and discrete adapted DA in solving TSP of 50 cities.

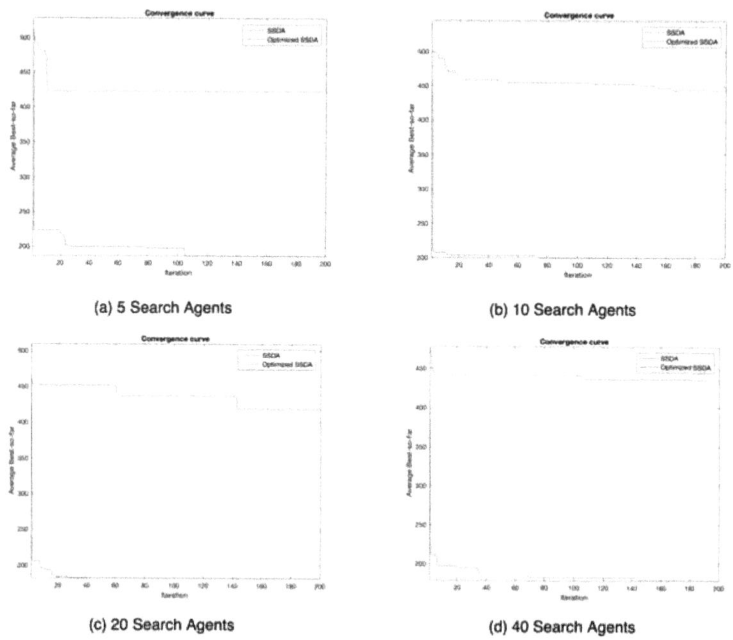

Figure 3. Convergence curve of optimized discrete adapted DA and discrete adapted DA in solving TSP of 40 cities.

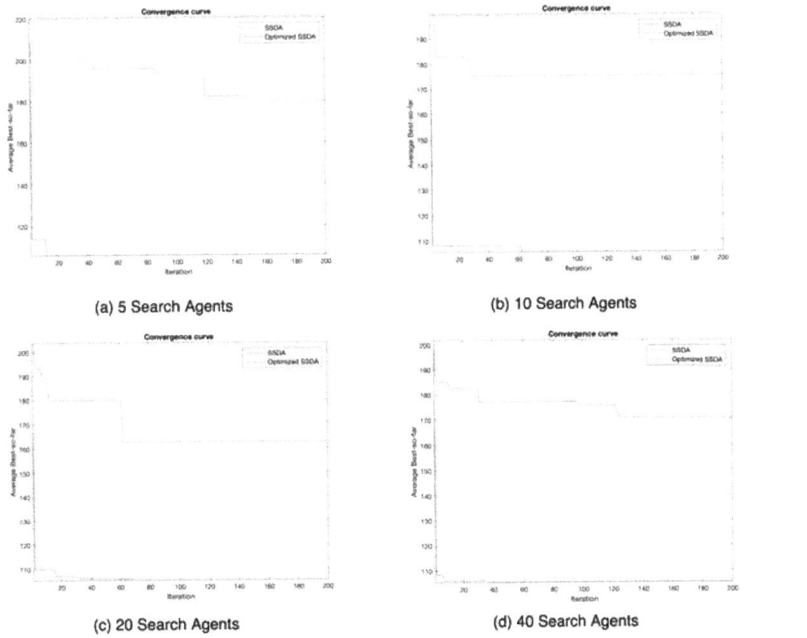

Figure 4. Convergence curve of optimized discrete adapted DA and discrete adapted DA in solving TSP of 20 cities.

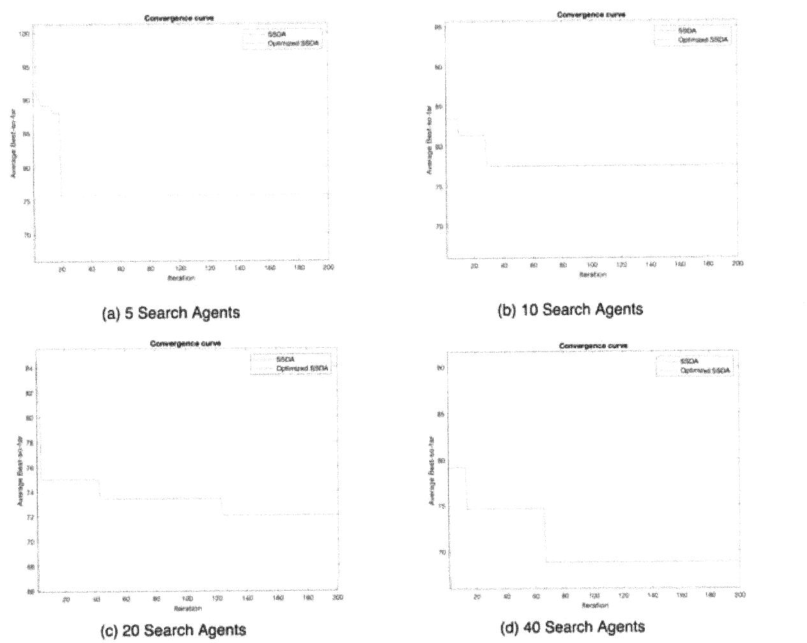

Figure 5. Convergence curve of optimized discrete adapted DA and discrete adapted DA in solving TSP of 10 cities.

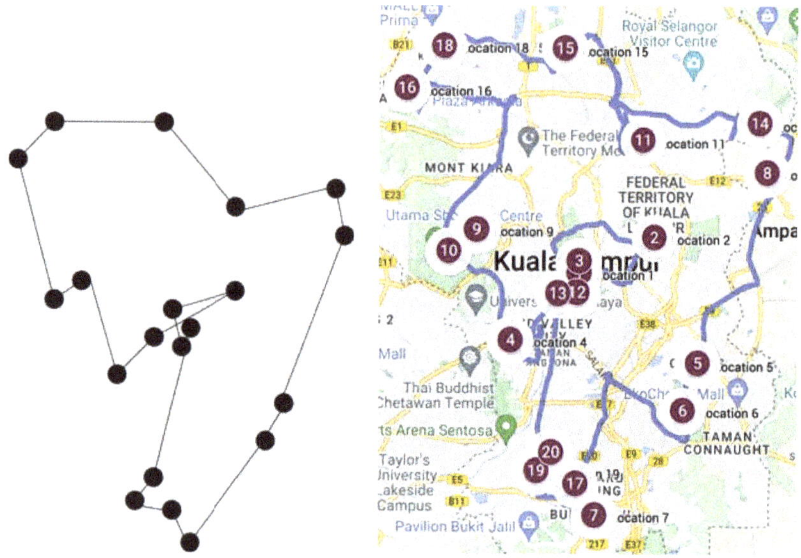

Figure 6. TSP path provided by optimized discrete adapted DA for 20 locations.

Moreover, the costs of the best TSP paths that can be provided by the adapted discrete DA for 50, 40, 20, and 10 locations are 487.2, 409.3, 155.7, and 69.6 respectively while the costs of the best TSP paths that can be provided by the proposed optimized adapted discrete DA for 50, 40, 20, and 10 locations are 193.6, 179.4, 105.4, and 65.9 respectively.

In comparison to the enhanced SSPSO algorithm in [8], the proposed optimized discrete DA algorithm has a higher effectiveness as it provides solutions with a lower cost in all the experiments conducted.

In terms of efficiency, it can be seen from Tables 2–5 that the proposed optimized adapted discrete DA algorithm takes less time than the adapted discrete DA algorithm for execution until the maximum number of iterations. Moreover, in terms of the time taken to converge to the global optimal solution, it can be seen that the proposed algorithm is better than the discrete adapted DA as it takes less time to converge. Hence, it can be deduced that the proposed algorithm has a higher convergence rate as compared to the adapted discrete DA algorithm.

From Figures 2–5, it can be seen that the proposed optimized adapted discrete DA algorithm provides better solution than the adapted discrete DA as the proposed algorithm converges to solutions with lower costs in all cases. Moreover, it can be seen that in multiple cases the proposed algorithm has a higher convergence rate than the adapted discrete DA as the proposed algorithm converges at earlier iterations.

6.3.2. Benchmark TSP Problems

In this section, we present a comparison and discussion on the performance of the proposed algorithm in solving benchmark TSP problems.

Table 6 shows a comparison of the performance of the proposed optimized DA algorithm and other swarm intelligence algorithms, namely ACO, GA, PSM, and GWO in solving benchmark TSP problems. The results of ACO, GA, PSM, and GWO are taken from [32]. The results are compared in terms of the cost of the best solution that can be obtained by the algorithm. The maximum number of iterations used for the proposed optimized DA, and the other swarm intelligence algorithms is 500, and the maximum number of search agents used is 100.

Table 7 shows a comparison of the performance of the proposed optimized DA algorithm and other swarm intelligence algorithms, namely ACO, VTPSO, ABCSS and DSMO in solving benchmark TSP problems. The results for ACO, VTPSO, ABCSS and DSMO are taken from [34]. The results are compared in terms of the cost of the best solution that can be obtained by the algorithm. The maximum number of iterations used for ACO, VTPSO, and DSMO is 500, while that for ABCSS is 1000. The number of search agents used for ACO, VTPSO, ABCSS and DSMO is between 100 and 300. For the proposed optimized DA, the maximum number of iterations used is 500, and the maximum number of search agents used is 100. Although the number of search agents and maximum iteration for the proposed algorithm and the other swarm intelligence algorithms are not the same, the results are shown for comparison purposes.

Table 6. Performance comparison of proposed optimized discrete adapted DA and other swarm intelligence algorithms in solving benchmark TSP using a maximum of 100 search agents.

TSP Instance	Cost of Best Solution Obtained				
	Proposed Optimized DA	**ACO**	**GA**	**PSM**	**GWO**
burma14	30.8785	31.21	30.87	30.87	30.87
ulysses16	73.9876	77.13	74.0	73.99	73.99
ulysses22	75.3097	86.74	76.09	75.51	75.51
bays29	9074.148	9964.78	9336.82	9076.98	9076.98
eil51	430.244	499.92	524.18	438.7	455.24
berlin52	7544.3659	8046.06	9184.19	8109.91	8048.91
st70	687.0724	734.19	1015.0	767.65	752.84
eil76	566.5564	595.58	805.78	591.89	604.32
kroA100	24,205.4508	24,504.9	51446.8	26,419.8	25,983.8

Table 7. Performance comparison of proposed optimized discrete adapted DA and other swarm intelligence algorithms in solving benchmark TSP using different number of search agents.

TSP Instance	Cost of Best Solution Obtained				
	Proposed Optimized DA	**ACO**	**VTPSO**	**ABCSS**	**DSMO**
burma14	30.8785	31.21	30.87	30.87	30.87
ulysses16	73.9876	77.13	73.99	73.99	73.99
ulysses22	75.3097	84.78	75.31	75.31	75.31
bays29	9074.148	9964.78	9074.15	9074.15	9074.15
eil51	430.244	499.92	429.51	428.98	428.86
berlin52	7544.3659	7870.45	7544.37	7544.37	7544.37
st70	687.0724	734.19	682.57	682.57	677.11
eil76	566.5564	581.42	559.25	550.24	558.68
rat99	1298.888	1366.3	1256.25	1242.32	1225.56
kroA100	24,205.4508	24,504.9	21,307.44	21,299.0	21,298.21

From Table 6, it can be seen that the proposed algorithm achieves the same optimal solution for burma14 as GA, PSM, and GWO. For the ulysses16 dataset, the proposed algorithm provides the same optimal solution as PSM, and GWO. As for all the other datasets, that is, ulysses22, bays29, eil51, berlin52, st70, eil76, and kroA100, the proposed optimized DA provides better solutions as compared to ACO, GA, PSM, and GWO when the same number of search agents and maximum iterations is used.

From Table 7, it can be seen that the proposed algorithm can achieve the optimal solution for five of the datasets, namely for burma14, ulysses16, ulysses22, bays29, and berlin52, even though a smaller number of search agents and maximum iteration is used as compared to ACO, VTPSO, ABCSS, and DSMO. Although in some cases, VTPSO, ABCSS, and DMSO can provide a solution with lower cost as compared to our proposed algorithm, it is expected that our proposed algorithm will be able to provide similar or even better solutions if the same number of search agents and maximum iteration is used.

7. Conclusions and Future Work

Swarm intelligence algorithms are popular metaheuristic algorithms for solving complex optimization problems owing to the simple interactions of the search agents which give rise to a global intelligent behaviour. The dragonfly algorithm is a swarm intelligence algorithm inspired by the swarming behaviours of dragonflies while hunting and migrating. It has been found to have a higher performance than multiple other swarm intelligence algorithms in various applications.

Since DA has been found to have a better performance than multiple swarm intelligence algorithms in various applications, it is worth considering its application to the traveling salesman problem which is a popular discrete optimization problem having a plethora of real-world applications. In [3], a binary version of the dragonfly algorithm is proposed. However, this algorithm is difficult to be adapted for discrete problems like TSP. Hence, in [12], we proposed a discrete adapted dragonfly algorithm suitable for TSP. However, this algorithm has not been applied to large TSP problems and it has low effectiveness.

In this paper, we propose an optimized adapted discrete dragonfly algorithm which improves the low effectiveness of the adapted discrete DA in [12]. The proposed algorithm improves the exploitation phase of the adapted discrete DA by using the steepest ascent hill climbing algorithm as a local search. The proposed optimized adapted discrete DA has been tested using TSP instances consisting of 50, 40, 30, 20, and 10 cities. It has been found to provide better solutions, that is TSP paths with a lower cost, as compared to the adapted discrete DA [12], and the enhanced swap sequence based PSO [8] algorithms. It also has a higher convergence rate than the adapted discrete DA. Moreover, it has been tested using several benchmark TSP problems and it has been found to provide optimal solutions or solutions close to the optimal solutions.

For future work, the proposed algorithm can be applied to other real-world applications such as channel routing, planning, scheduling, and logistics.

Author Contributions: Conceptualization, B.A.S.E. and M.B.J.; writing—original draft preparation, B.A.S.E. and M.B.J.; writing—review and editing, M.B.J., A.A. and A.W.M.; supervision, M.B.J.; project administration, M.B.J.; funding acquisition, M.B.J. All authors have read and agreed to the published version of the manuscript.

Funding: This research was funded by Sunway University Internal Grant Scheme 2022 grant number GRTIN-IGS-DCIS[S]-11-2022.

Institutional Review Board Statement: Not applicable.

Informed Consent Statement: Not applicable.

Data Availability Statement: Not applicable.

Conflicts of Interest: The authors declare no conflict of interest.

Abbreviations

The following abbreviations are used in this manuscript:

DA	Dragonfly Algorithm
TSP	Traveling Salesman Problem
BDA	Binary Dragonfly Algorithm
MODA	Multi-Objective Dragonfly Algorithm
PSO	Particle Swarm Optimization
ACO	Ant Colony Optimization
HC	Hill Climbing

GA	Genetic Algorithm
GWO	Grey Wolf Optimizer
XPSO	Expanded PSO
SO	Swap Operator
SS	Swap Sequence
VTPSO	Velocity Tentative PSO
ABCSS	Artificial Bee Colony with Swap Sequence
DSMO	Discrete Spider Monkey Optimization
PSM	Producer Scrounger Method

References

1. Yang, X.S. *Nature-Inspired Optimization Algorithms*; Academic Press: Cambridge, MA, USA, 2020.
2. Slowik, A.; Kwasnicka, H. Nature inspired methods and their industry applications—Swarm intelligence algorithms. *IEEE Trans. Ind. Inform.* **2017**, *14*, 1004–1015. [CrossRef]
3. Mirjalili, S. Dragonfly algorithm: A new meta-heuristic optimization technique for solving single-objective, discrete, and multi-objective problems. *Neural Comput. Appl.* **2015**, *27*, 1053–1073. [CrossRef]
4. Emambocus, B.A.S.; Jasser, M.B.; Mustapha, A.; Amphawan, A. Dragonfly algorithm and its hybrids: A survey on performance, objectives and applications. *Sensors* **2021**, *21*, 7542. [CrossRef] [PubMed]
5. Gutin, G.; Punnen, A.P. *The Traveling Salesman Problem and Its Variations*; Springer Science & Business Media: Berlin/Heidelberg, Germany, 2006; Volume 12.
6. Zhi, X.H.; Xing, X.; Wang, Q.; Zhang, L.; Yang, X.; Zhou, C.; Liang, Y. A discrete PSO method for generalized TSP problem. In Proceedings of the 2004 International Conference on Machine Learning and Cybernetics (IEEE Cat. No. 04EX826), Shanghai, China, 26–29 August 2004; IEEE: Piscataway, NJ, USA, 2004; Volume 4, pp. 2378–2383.
7. Yang, J.; Shi, X.; Marchese, M.; Liang, Y. An ant colony optimization method for generalized TSP problem. *Prog. Nat. Sci.* **2008**, *18*, 1417–1422. [CrossRef]
8. Emambocus, B.A.S.; Jasser, M.B.; Hamzah, M.; Mustapha, A.; Amphawan, A. An enhanced swap sequence-based particle swarm optimization algorithm to solve TSP. *IEEE Access* **2021**, *9*, 164820–164836. [CrossRef]
9. Yao, X.S.; Ou, Y.; Zhou, K.Q. TSP solving utilizing improved ant colony algorithm. *J. Phys. Conf. Ser.* **2021**, *2129*, 012026. [CrossRef]
10. Wei, B.; Xing, Y.; Xia, X.; Gui, L. A novel particle swarm optimization with genetic operator and its application to tsp. *Int. J. Cogn. Inform. Nat. Intell.* **2021**, *15*, 1–17. [CrossRef]
11. Rokbani, N.; Kumar, R.; Abraham, A.; Alimi, A.M.; Long, H.V.; Priyadarshini, I.; Son, L.H. Bi-heuristic ant colony optimization-based approaches for traveling salesman problem. *Soft Comput.* **2021**, *25*, 3775–3794. [CrossRef]
12. Emambocus, B.A.S.; Jasser, M.B.; Amphawan, A. A discrete adapted dragonfly algorithm for solving the traveling salesman problem. In Proceedings of the 2021 Fifth International Conference on Intelligent Computing in Data Sciences (ICDS), Fez, Morocco, 20–22 October 2021; IEEE: Piscataway, NJ, USA, 2021; pp. 1–6.
13. Spanaki, K.; Karafili, E.; Sivarajah, U.; Despoudi, S.; Irani, Z. Artificial intelligence and food security: Swarm intelligence of AgriTech drones for smart AgriFood operations. *Prod. Plan. Control.* **2021**, *32*, 1–19. [CrossRef]
14. Boveiri, H.R.; Khayami, R.; Elhoseny, M.; Gunasekaran, M. An efficient swarm-intelligence approach for task scheduling in cloud-based internet of things applications. *J. Ambient. Intell. Humaniz. Comput.* **2019**, *10*, 3469–3479. [CrossRef]
15. Jahwar, A.; Ahmed, N. Swarm intelligence algorithms in gene selection profile based on classification of microarray data: a review. *J. Appl. Sci. Technol. Trends* **2021**, *2*, 1–9. [CrossRef]
16. Brezočnik, L.; Fister, I.; Podgorelec, V. Swarm intelligence algorithms for feature selection: A review. *Appl. Sci.* **2018**, *8*, 1521. [CrossRef]
17. Bacanin, N.; Bezdan, T.; Tuba, E.; Strumberger, I.; Tuba, M. Optimizing convolutional neural network hyperparameters by enhanced swarm intelligence metaheuristics. *Algorithms* **2020**, *13*, 67. [CrossRef]
18. Emambocus, B.A.S.; Jasser, M.B. Towards an optimized dragonfly algorithm using hill climbing local search to tackle the low exploitation problem. In Proceedings of the 2021 International Conference on Software Engineering & Computer Systems and 4th International Conference on Computational Science and Information Management (ICSECS-ICOCSIM), Pekan, Malaysia, 24–26 August 2021; IEEE: Piscataway, NJ, USA, 2021; pp. 306–311.
19. Yang, J.; Qu, L.; Shen, Y.; Shi, Y.; Cheng, S.; Zhao, J.; Shen, X. Swarm intelligence in data science: Applications, opportunities and challenges. In *International Conference on Swarm Intelligence*; Springer: Berlin/Heidelberg, Germany, 2020; pp. 3–14.
20. Sun, W.; Tang, M.; Zhang, L.; Huo, Z.; Shu, L. A survey of using swarm intelligence algorithms in IoT. *Sensors* **2020**, *20*, 1420. [CrossRef]
21. Paramanandham, N.; Rajendiran, K. Infrared and visible image fusion using discrete cosine transform and swarm intelligence for surveillance applications. *Infrared Phys. Technol.* **2018**, *88*, 13–22. [CrossRef]
22. Janga Reddy, M.; Nagesh Kumar, D. Evolutionary algorithms, swarm intelligence methods, and their applications in water resources engineering: A state-of-the-art review. *H2Open J.* **2020**, *3*, 135–188. [CrossRef]

23. Soni, G.; Jain, V.; Chan, F.T.; Niu, B.; Prakash, S. Swarm intelligence approaches in supply chain management: Potentials, challenges and future research directions. *Supply Chain. Manag. Int. J.* **2018**, *24*, 107–123. [CrossRef]
24. Davendra, D. *Traveling Salesman Problem: Theory and Applications*; BoD–Books on Demand: Rijeka, Croatia, 2010.
25. Xu, Y.; Che, C. A brief review of the intelligent algorithm for traveling salesman problem in UAV route planning. In Proceedings of the 2019 IEEE 9th International Conference on Electronics Information and Emergency Communication (ICEIEC), Beijing, China, 12–14 July 2019; IEEE: Piscataway, NJ, USA, 2019; pp. 1–7.
26. Huang, S.H.; Huang, Y.H.; Blazquez, C.A.; Chen, C.Y. Solving the vehicle routing problem with drone for delivery services using an ant colony optimization algorithm. *Adv. Eng. Inform.* **2022**, *51*, 101536. [CrossRef]
27. Muren; Wu, J.; Zhou, L.; Du, Z.; Lv, Y. Mixed steepest descent algorithm for the traveling salesman problem and application in air logistics. *Transp. Res. Part Logist. Transp. Rev.* **2019**, *126*, 87–102. [CrossRef]
28. Foumani, M.; Moeini, A.; Haythorpe, M.; Smith-Miles, K. A cross-entropy method for optimising robotic automated storage and retrieval systems. *Int. J. Prod. Res.* **2018**, *56*, 6450–6472. [CrossRef]
29. Nedjatia, A.; Vizvárib, B. Robot path planning by traveling salesman problem with circle neighborhood: Modeling, algorithm, and applications. *arXiv* **2020**, arXiv:2003.06712.
30. Teng, L.; Li, H. Modified discrete firefly algorithm combining genetic algorithm for traveling salesman problem. *TELKOMNIKA (Telecommun. Comput. Electron. Control.)* **2018**, *16*, 424–431. [CrossRef]
31. Zhang, Z.; Han, Y. Discrete sparrow search algorithm for symmetric traveling salesman problem. *Appl. Soft Comput.* **2022**, *118*, 108469. [CrossRef]
32. Sopto, D.S.; Ayon, S.I.; Akhand, M.; Siddique, N. Modified grey wolf optimization to solve traveling salesman problem. In Proceedings of the 2018 International Conference on Innovation in Engineering and Technology (ICIET), Dhaka, Bangladesh, 27–28 December 2018; IEEE: Piscataway, NJ, USA, 2018; pp. 1–4.
33. Liu, Y.; Liu, Q.; Tang, Z. A discrete chicken swarm optimization for traveling salesman problem. *J. Phys. Conf. Ser.* **2021**, *1978*, 012034. [CrossRef]
34. Akhand, M.; Ayon, S.I.; Shahriyar, S.; Siddique, N.; Adeli, H. Discrete spider monkey optimization for travelling salesman problem. *Appl. Soft Comput.* **2020**, *86*, 105887. [CrossRef]
35. Wang, K.P.; Huang, L.; Zhou, C.G.; Pang, W. Particle swarm optimization for traveling salesman problem. In Proceedings of the 2003 International Conference on Machine Learning and Cybernetics (IEEE Cat. No. 03EX693), Xi'an, China, 5 November 2003; Volume 3, pp. 1583–1585. [CrossRef]

MDPI
St. Alban-Anlage 66
4052 Basel
Switzerland
Tel. +41 61 683 77 34
Fax +41 61 302 89 18
www.mdpi.com

Mathematics Editorial Office
E-mail: mathematics@mdpi.com
www.mdpi.com/journal/mathematics

www.ingramcontent.com/pod-product-compliance
Lightning Source LLC
Chambersburg PA
CBHW061848210926
51850CB00006B/998
* 9 7 8 3 0 3 6 5 6 8 9 0 4 *